THEORY OF MAGNETISM
Application to Surface Physics

THEORY OF MAGNETISM
Application to Surface Physics

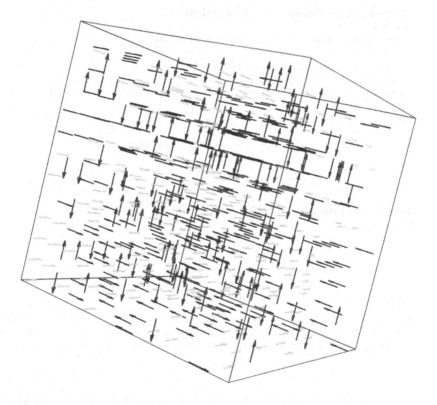

Hung T. Diep
University of Cergy-Pontoise, France

NEW JERSEY · LONDON · SINGAPORE · BEIJING · SHANGHAI · HONG KONG · TAIPEI · CHENNAI

Published by

World Scientific Publishing Co. Pte. Ltd.

5 Toh Tuck Link, Singapore 596224

USA office: 27 Warren Street, Suite 401-402, Hackensack, NJ 07601

UK office: 57 Shelton Street, Covent Garden, London WC2H 9HE

Library of Congress Cataloging-in-Publication Data
Diep, H. T., author.
    Theory of magnetism : application to surface physics / Hung-The Diep.
        pages cm
    Includes bibliographical references and index.
    ISBN 978-9814569941 (hardcover : alk. paper)
    1. Magnetism--Textbooks. 2. Statistical physics--Textbooks. 3. Surfaces (Physics)--Textbooks.
I. Title.
    QC753.2.D54 2013
    538.01--dc23
                                          2013036797

British Library Cataloguing-in-Publication Data
A catalogue record for this book is available from the British Library.

Printed in Singapore

To my wife and our children, Samuel, Tuan, Kim, Sarah,

and my mother

Foreword

The book is intended for graduate students and researchers who wish to learn the main properties of magnetic materials in the bulk state and at the nanometric scale such as thin films and multilayers. The book provides fundamental theories and methods of simulation to study and to understand these properties in an explicit manner. Exercises and problems are given for each chapter to help the reader apply the methods to discover new related phenomena and applications which are complementary to the lecture. Detailed solutions are provided for self-learning.

In the first part of the book, fundamental methods in magnetism are presented. The magnetism of systems of independent electrons and atoms is studied in chapter 1. The system of interacting electrons is studied in chapter 2 by the Hartree-Fock approximation which leads to the exchange interaction dependent on spin. This explains the origin of magnetic exchange interaction in magnetic materials. In chapter 3, we introduce the Heisenberg Hamiltonian which was the starting point of the modern theory of magnetism. The demonstration is made by using the method of second quantization. For readers who are not at ease with operator handling, this chapter can be omitted in the first reading of the book. The mean-field theory of systems of interacting spins is developed in chapter 4 where basic notions on the phase transition are given. The spin-wave theory, or theory of magnons, is studied in chapter 5 where detailed calculations of the magnon dispersion relation and low-temperature properties are shown. The Green's function method adapted for the study of magnetic systems is presented in chapter 6. This technique which can be applied in the whole range of temperature is complementary to the spin-wave theory. The phase transition theory is described in chapter 7 with the introduction of important concepts such as the Landau-Ginzburg theory, the renormalization group and the finite-size scaling. Monte Carlo simulation methods for the phase transition are described in chapter 8. Numerical methods constitute nowadays the third approach, next to theory and experiment, to study complicated and complex systems. Simulations are in particular necessary for testing theories and for quantitative comparisons with experiments.

The second part of the book is devoted to the application of the theory of magnetism to surface physics. In chapter 9 the magnetism at surfaces is shown by methods from the spin-wave theory, the Green's function technique and Monte Carlo simulations. Numerous examples covering typical cases in ferromagnets, antiferromagnets, ferrimagnets, helimagnets and frustrated spin systems are illustrated. Fundamental surface effects are shown and discussed. These simple models allow us to understand qualitatively experimental results observed in often more complicated real systems. The spin transport is described in chapter 10 where basic formulation of the Boltzmann's equation is recalled and recent methods of Monte Carlo simulation to deal with the spin resistivity are explained.

In the third part of the book, we present detailed solutions of problems given in each chapter. Many problems are important topics in magnetism.

An appendix on elements of statistical physics is also included to make the book self-contained. Finally, a simple Monte Carlo program is provided to facilitate the first step in the writing of a simulation program.

The material of this book can be used for one semester lectures of three hours weekly in a graduate program of physics. An equivalent amount of time is needed for students to solve problems with the help of a teaching assistant.

H. T. Diep
Professor of Physics
University of Cergy-Pontoise, France

Acknowledgments

I am thankful to my colleagues at the University of Cergy-Pontoise for sharing uncountable wonderful moments in my professional life and for their precious friendship over the years.

I would like to express here my deep affection to my former and current doctorate students whose enthusiasm in our search for the truth has given me energy and joy to carry out what I have done so far with them in our research activities. Many of their works are used in this book for illustration.

I am grateful to my former professors Ojiro Nagai and Isao Harada, to Jean-Claude S. Lévy and my numerous colleagues at the University Paris VII-Denis Diderot for their collaboration and friendship during these 40 years.

I would like to express my sincere thanks to my colleagues and friends Miron Kaufman and Tuong T. Truong for helpful discussions.

Contents

Application to Surface Physics 201

PART 1
Theory of Magnetism

Chapter 1

Magnetism of Free Electrons and Atoms

1.1 Introduction

Magnetism is a domain of physics where properties of matters due to the existence of spins of the particles are studied. The spin of a particle results from an intrinsic motion of the particle. It is often at the origin of principal behaviors of various systems from nuclear to condensed matter, passing by atoms and molecules.

The main properties of a system depend on the interactions between the constituent particles. Different kinds of interaction give rise to different kinds of behaviors. We take as an example the case of a crystalline solid. Elastic interactions between neighboring atoms give rise to vibrations of atoms which determine thermal properties of the crystal up to the melting. Exchange interactions, or magnetic interactions, between spins of neighboring atoms can give rise to a magnetic ordering which determines principal low-temperature properties of the system. We will consider various magnetic interactions and their consequences in the following chapters.

In this chapter, we study the behavior of several systems of independent spins. We show in particular the effect of the temperature and of an applied magnetic field. The spins considered here are those of conducting electrons in metals or those from an assembly of atoms.

Let us consider first an electron of charge $-e$ ($e > 0$), of mass m and of spin $\vec{S}$. The magnetic moment associated with the spin is written as

$$\vec{\mu}_s = -g\mu_B\vec{S} \tag{1.1}$$

where $g = 2.0023$ is the Landé factor and μ_B the Bohr magneton defined by

$$\mu_B = \frac{e\hbar}{2m} \tag{1.2}$$

The negative sign of (1.1) is due to the negative charge of the electron $(-e)$. For convenience, let $\vec{\sigma}$ be $2\vec{S}$. The components of $\vec{\sigma}$ are the well-known Pauli matrices

$$\sigma_x = \begin{pmatrix} 0 & 1 \\ 1 & 0 \end{pmatrix} \tag{1.3}$$

$$\sigma_y = \begin{pmatrix} 0 & -i \\ i & 0 \end{pmatrix} \tag{1.4}$$

$$\sigma_z = \begin{pmatrix} 1 & 0 \\ 0 & -1 \end{pmatrix} \tag{1.5}$$

Note that σ_z is diagonal. The eigenvalues of S_z are thus $-1/2$ and $1/2$. As can be verified using the above matrices, the spin operators obey the following commutation relations

$$\left[\sigma^+, \sigma^-\right] = 2\sigma_z \tag{1.6}$$
$$\left[\sigma_z, \sigma^\pm\right] = \pm 2\sigma^\pm \tag{1.7}$$
$$\left[\sigma_x, \sigma_y\right] = i\sigma_z + \text{relations by circular permutations } of \ x, y, z \tag{1.8}$$

where $\sigma^\pm = \sigma_x \pm i\sigma_y$.

In addition to the magnetic moment from its spin, the electron has also an orbital moment

$$\vec{\mu}_l = -\frac{e}{2}(\vec{r} \wedge \vec{v}) = -\frac{e}{2m}(\vec{r} \wedge \vec{p}) = -\mu_B \vec{l} \tag{1.9}$$

where $\vec{r}$ and $\vec{v}$ are the position and the velocity of the electron and $\vec{l}$ is the kinetic orbital moment $\vec{r} \wedge \vec{p}$ divided by $\hbar$. As will be seen below, this moment is the origin of the so-called diamagnetic phenomenon observed under the application of a magnetic field.

In the same manner, one can define a nuclear magnetic moment $\vec{\mu}_I$ associated to the nuclear spin I by a relation similar to (1.1) with μ_B replaced by $\mu_N = \frac{e\hbar}{2M}$ where M is the mass of the nucleus (proton or neutron) which is equal to $1836m$. Due to its heavy mass, the effect of $\vec{\mu}_I$ is very small with respect to that of $\vec{\mu}_s$. Therefore, when the two moments exist, one often neglects the effect of the nuclear moment.

We consider now a system of independent spins $\pm 1/2$. In the absence of an applied magnetic field, the random distribution of spin magnetic moments results in a zero total moment. Under an applied magnetic field $\vec{B}$, a number of spins will turn themselves into the field direction giving rise to a nonzero total moment $\vec{M}$. The susceptibility χ defined as $\chi = \frac{dM}{dB}$ is then positive. The system is called paramagnetic.

We consider another situation where atoms do not have initial nonzero magnetic moments. Under an applied field, the electrons of each atom may modify their states so as to create an induced moment to resist to the field effect. The induced moment $\vec{M}$ is in the opposite direction of the field giving rise to a negative susceptibility. The system is diamagnetic.

1.2 Paramagnetism of a free electron gas

When a magnetic field $\vec{H}$ is applied to a system of free electrons, the energy of an electron of spin parallel to $\vec{H}$ decreases by an amount $\mu_B H$, and that of an electron of spin antiparallel to $\vec{H}$ increases by the same amount. This effect is called "Zeeman effect". We write

$$E_\uparrow = E - \mu_B H \tag{1.10}$$
$$E_\downarrow = E + \mu_B H \tag{1.11}$$

The total magnetic moment induced by the field is

$$M = \mu_B(N_\uparrow - N_\downarrow) \tag{1.12}$$

where $N_\uparrow$ and $N_\downarrow$ are respectively the numbers of $\uparrow$ and $\downarrow$ spins.

Using the density of states for each type of spin given by Eq. (A.40) we have

$$M = \mu_B \int_0^\infty [\rho(E_\uparrow)f(E_\uparrow) - \rho(E_\downarrow)f(E_\downarrow)]dE \tag{1.13}$$

where $f(E_{\uparrow,\downarrow})$ is the Fermi-Dirac distribution function given by Eq. (A.37) [see Problems 3 and 4 in section 1.8]. In the case where H is small, we can replace $\rho(E_\uparrow) \simeq \rho(E_\downarrow) \simeq \rho(E)$ because $\rho(E)$ is a smooth function of E. In addition, we can replace $f(E_\uparrow)$ and $f(E_\downarrow)$ by their first-order Taylor expansion around E

$$f(E_\uparrow) \simeq f(E) + \frac{\partial f}{\partial E}(E_\uparrow - E) \tag{1.14}$$
$$f(E_\downarrow) \simeq f(E) + \frac{\partial f}{\partial E}(E_\downarrow - E) \tag{1.15}$$

Equation (1.13) becomes

$$M = \mu_B^2 H \int_0^\infty \rho(E)(-\frac{\partial f}{\partial E})dE \tag{1.16}$$

At low temperatures, $\frac{\partial f}{\partial E}$ is important only near E_F (see Problem 4 in section 1.8) so that

$$M \simeq 2\mu_B^2 H\rho(E_F) \int_0^\infty (-\frac{\partial f}{\partial E})dE = 2\mu_B^2 H\rho(E_F)[f(0) - f(\infty)]$$

$$= 2\mu_B^2 H\rho(E_F) \tag{1.17}$$

It is noted that $\rho(E_F)$ used here is for one kind of spin given by Eq. (A.40) without the spin degeneracy. We have $M == 2\mu_B^2 H\rho(E_F)$. The susceptibility is thus

$$\chi = \frac{dM}{dH} = 2\mu_B^2\rho(E_F) = \mu_B^2\rho_t(E_F) \tag{1.18}$$

where $\rho_t(E_F)$ is the "total" density of states with the spin degeneracy [Eq. (A.41)].

Equation (1.18) is called susceptibility of "Pauli paramagnetism" which is independent of T. To calculate M at higher orders in T, we can use a low-T expansion shown in Problem 6 in section 1.8. At high temperatures, χ is proportional to $1/T$ (Curie's law) as seen in that exercise. Experimental data for normal metals confirm Eq. (1.18) but strong variations of χ with temperature have been observed in some transition metals (Pd, Ti, ...) [79]. To explain these variations, it is necessary to take into account various interactions neglected in the free-electron gas used above.

1.3 Paramagnetism of a system of free atoms

We consider a system of N atoms each of which has a total moment $\vec{J} = \vec{S} + \vec{L}$. The modulus J of $\vec{J}$ is the sum of the amplitudes S and L: $J = S + L$. The magnetic moment of the i-th atom is

$$\vec{\mu}_i = -g\mu_B\vec{J}_i \tag{1.19}$$

where the Landé factor is given by

$$g = \frac{3}{2} + \frac{S(S+1) - L(L+1)}{2J(J+1)} \tag{1.20}$$

The effect of an applied magnetic field $\vec{B}$ is called Zeeman effect. If the field is applied along the z direction, the Zeeman energy is

$$\mathcal{H} = -\sum_{i=1}^N \vec{\mu}_i \cdot \vec{B} = \sum_{J_i^z} g\mu_B J_i^z B = \sum_{i=1}^N E_i \tag{1.21}$$

where $J_i^z = J, J - 1, ..., -J$ ($2J + 1$ values) and E_i is the energy of the i-th atom.

We consider the case where the atoms are independent. The partition function [see Eq. (A.9)] is written as the product of single-atom partition function z

$$Z = z^N \tag{1.22}$$

where

$$z = \sum_{J_i^z = -J}^{J} e^{-\beta E_i} = \sum_{J_i^z = -J}^{J} e^{-\beta g\mu_B B J_i^z} = \frac{\sinh[\beta g\mu_B B(J + 1/2)]}{\sinh(\beta g\mu_B B/2)} \tag{1.23}$$

The last equality was obtained by using the formula for the series of $2J + 1$ terms, of ratio $e^{-\beta g\mu_B B}$. For $J = 1/2$ one has

$$z = 2\cosh(\beta g\mu_B B/2) \tag{1.24}$$

The free energy F is given by (see Appendix A)

$$F = -k_B T \ln Z = -N k_B T \ln z \tag{1.25}$$

One considers the microscopic state l of energy

$$E_l = \sum_i E_i = -B \sum_i \mu_i^l = -B M_l \tag{1.26}$$

where μ_i^l is the magnetic moment of atom i in the state l and M_l the total magnetic moment the system in l. The average magnetic moment is written as

$$\begin{aligned}
\overline{M} &= \frac{1}{Z} \sum_l M_l e^{-\beta E_l} = \frac{1}{Z} \sum_l M_l e^{\beta B M_l} \\
&= \frac{1}{\beta} \frac{1}{Z} \frac{\partial}{\partial B} \sum_l e^{\beta B M_l} = \frac{1}{\beta} \frac{1}{Z} \frac{\partial Z}{\partial B} = \frac{1}{\beta} \frac{\partial \ln Z}{\partial B} \\
&= -\frac{\partial F}{\partial B}
\end{aligned} \tag{1.27}$$

Replacing Eq. (1.24) in Eq. (1.25) to obtain F, then replacing F in the last equality, one obtains for $J = 1/2$

$$\overline{M} = \frac{N g\mu_B}{2} \tanh(\frac{g\mu_B B}{2k_B T}) \tag{1.28}$$

The magnetization $\overline{m}$, defined as magnetic moment per unit volume, is

$$\overline{m} = \frac{N g\mu_B}{2V} \tanh(\frac{g\mu_B B}{2k_B T}) \tag{1.29}$$

At high T, one has $\tanh(\frac{g\mu_B B}{2k_B T}) \to \frac{g\mu_B B}{2k_B T}$, so that

$$\overline{m} = \frac{N}{V}(\frac{g\mu_B}{2})^2 \frac{B}{k_B T} \tag{1.30}$$

from which one obtains the susceptibility

$$\chi = \frac{\partial \overline{m}}{\partial B} = \frac{N}{V}(\frac{g\mu_B}{2})^2 \frac{1}{k_B T} \tag{1.31}$$

This is the Curie's law, similar to that of free electrons at high T (see Problem 6 in section 1.8).

At low T, one has $\tanh(\frac{g\mu_B B}{2k_B T}) \to 1$, namely the saturation of the magnetization $\overline{m} = \frac{N g\mu_B}{2V}$ in the case $J = 1/2$. This result is different from that of Pauli paramagnetism of free electrons at low T [see Eq. (1.18)].

The average energy of the system is written as (see Appendix A)

$$\begin{aligned}
\overline{E} &= -\frac{\partial \ln Z}{\partial \beta} \\
&= -\frac{N g\mu_B B}{2} \tanh(\frac{g\mu_B B}{2k_B T}) \quad \text{if } J = 1/2 \\
&= -N g\mu_B J B B_J(\frac{g\mu_B J B}{k_B T}) \quad \text{if } J \neq 1/2
\end{aligned} \tag{1.32}$$

One obtains the paramagnetic heat capacity for $J = 1/2$

$$C_V = N k_B (\frac{g\mu_B B}{2k_B T})^2 \frac{1}{\cosh^2(\frac{g\mu_B B}{2k_B T})} \tag{1.33}$$

Figure 1.1 shows C_V per atom versus $k_B T$ for $g\mu_B B = 1.5$. Note that for increasing B, the peak of C_V moves to the high-temperature side. Beyond the peak position, the system order due to the magnetic field effect is destroyed by the temperature.

1.4 Diamagnetism of many-electron atoms

In the case where the valence orbital of an atom is completely occupied, i.e. the atom has no permanent magnetic moment from its electrons, the applied magnetic field results in a diamagnetic effect as seen below. However, when the valence orbital is not completely occupied, the applied magnetic field gives rise to both paramagnetism and diamagnetism. The paramagnetism is studied in the previous section. In the following, we study the diamagnetism.

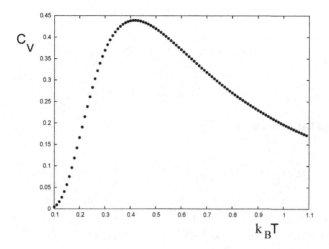

Fig. 1.1 C_V per atom versus $k_B T$ for $g\mu_B B = 1.5$.

We consider here an atom which has N_e electrons in its valence orbital. We suppose that the valence orbital is full: the orbital moment is $L = 0$ and the total spin is $S = 0$. The Hamiltonian of the valence electrons under the applied magnetic field $\vec{B}$ is written as

$$\mathcal{H} = \sum_{i=1}^{N_e} \{\frac{1}{2m}[\vec{p}_i + e\vec{A}(\vec{r}_i)]^2 + 2\mu_B \vec{B} \cdot \vec{S}_i + U(\vec{r}_i)\} \qquad (1.34)$$

where $\vec{A}(\vec{r}_i)$ is the vector potential associated with $\vec{B}$, namely $\text{rot}\vec{A}(\vec{r}_i) = \vec{B}$, and $U(\vec{r}_i)$ represents the interaction between electron i with the remaining electrons of the orbital. For $\vec{B} \parallel \vec{Oz}$, one can choose $\vec{A}(\vec{r}_i)$ as follows
$A_x = -yB/2$, $A_y = xB/2$, $A_z = 0$.
By expanding the square in Eq. (1.34), one obtains

$$\mathcal{H} = \mathcal{H}_0 + \mu_B(L_z + 2S_z)B + \frac{e^2 B^2}{8m} \sum_i (x_i^2 + y_i^2) \qquad (1.35)$$

where $\mathcal{H}_0$ is the Hamiltonian in zero field:

$$\mathcal{H}_0 = \sum_i [\frac{p_i^2}{2m} + U(\vec{r}_i)] \qquad (1.36)$$

L_z and S_z are the z components of the total orbital moment and the total spin:

$$\vec{L} = \frac{1}{\hbar} \sum_i \vec{r}_i \wedge \vec{p}_i \qquad (1.37)$$

$$\vec{S} = \sum_i \vec{S}_i \qquad (1.38)$$

The ground-state energy is

$$E_0(B) = E_0(B = 0) + \frac{e^2 B^2}{8m} < 0 | \sum_i (x_i^2 + y_i^2) | 0 > \qquad (1.39)$$

where $E_0(B = 0)$ is the energy in zero field. For a system of N free atoms, the total energy is equal to $NE_0(B)$. At $T = 0$, one has $F = E - TS = E$. The magnetization is

$$
\begin{aligned}
\overline{m} &= \frac{\overline{M}}{V} = -\frac{1}{V} \frac{\partial F}{\partial B} \\
&= -\frac{Ne^2 B}{4mV} < 0 | \sum_i (x_i^2 + y_i^2) | 0 > \qquad (1.40)
\end{aligned}
$$

The susceptibility is thus diamagnetic (< 0) and given by

$$\chi = -\frac{Ne^2}{4mV} < 0 | \sum_i (x_i^2 + y_i^2) | 0 > \qquad (1.41)$$

1.5 Free electron gas in a strong magnetic field: Landau diamagnetism

We show in this section that a system of free electrons under a strong magnetic field gives rise to a negative susceptibility. This phenomenon is called "Landau diamagnetism" which is different from the atom diamagnetism shown in the previous section. We have seen above that under a weak applied field, a perturbation treatment gives rise to the Pauli paramagnetism, Eq. (1.18), at low T. However, when the applied field is strong, we cannot use the perturbation theory. We have to incorporate the action of the field in the Hamiltonian via the vector potential $\vec{A}(\vec{r})$.

We suppose that the field $\vec{B}$ is applied along the z axis. We solve the problem in the following.

1.5.1 *Landau's levels*

The Schrödinger equation for an electron of effective mass m^* under the applied field $\vec{B}$ is written as

$$\frac{1}{2m^*}[\frac{\hbar}{i}\vec{\nabla} + e\vec{A}(\vec{r})]^2\Psi(\vec{r}_i) = E\Psi(\vec{r}_i) \tag{1.42}$$

where $\vec{B} = \overset{\rightarrow}{\mathrm{rot}}\vec{A}$. For simplicity, we choose $\vec{A} = (0, xB, 0)$. We have

$$\frac{\partial^2\Psi}{\partial x^2} + (\frac{\partial}{\partial y} - \frac{ieB}{\hbar}x)^2\Psi + \frac{\partial^2\Psi}{\partial z^2} + \frac{2m^*E}{\hbar^2}\Psi = 0 \tag{1.43}$$

The structure of this equation suggests a solution of the form

$$\Psi(x,y,z) = u(x)\exp[i(k_y y + k_z z)] \tag{1.44}$$

Equation (1.43) becomes

$$\frac{\partial^2 u(x)}{\partial x^2} + [\frac{2m^*E'}{\hbar^2} - (k_y - \frac{eB}{\hbar}x)^2]u(x) = 0 \tag{1.45}$$

where

$$E' = E - \frac{\hbar^2 k_z^2}{2m^*} \tag{1.46}$$

Thus, the motion of the electron in the z direction is that of a free electron. For the motion in the xy plane, we have to solve Eq. (1.45). We rewrite it as

$$-\frac{\hbar^2}{2m^*}\frac{\partial^2 u(x)}{\partial x^2} + \frac{1}{2}m^*(\frac{eB}{m^*}x - \frac{\hbar k_y}{m^*})^2 u(x) = E'u(x) \tag{1.47}$$

We recognize that the above equation is the Schrödinger equation of a harmonic oscillator of pulsation

$$\omega_c = \frac{eB}{m^*} \tag{1.48}$$

centered at

$$x_0 = \frac{1}{\omega_c}\frac{\hbar k_y}{m^*} \tag{1.49}$$

The energy of this oscillator is therefore

$$E' = (n + \frac{1}{2})\hbar\omega_c \tag{1.50}$$

where n is an integer ≥ 0. With Eq. (1.46), we obtain the total electron energy

$$E = E_n = (n + \frac{1}{2})\hbar\omega_c + \frac{\hbar^2 k_z^2}{2m^*} \tag{1.51}$$

where ω_c is called "cyclotron pulsation" and the different energy levels corresponding to different values of n are called "Landau's levels". These levels are shown in Fig. 1.2.

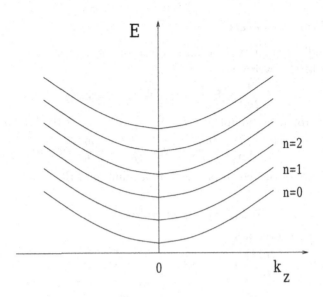

Fig. 1.2   Landau's levels given by Eq. (1.51).

1.5.2 *Degeneracy of Landau's levels*

The electron energy E is doubly quantized: the allowed values of k_z are given by the periodic condition in the z direction, and index n quantizes the energy of the harmonic motion in the xy plane. However, the quantization by n is subject to a double condition:

- the harmonic motion in the xy plane takes place only if the center of the oscillator x_0 lies inside the xy plane of the material, namely

$$0 < x_0 < L_x \tag{1.52}$$

where L_x is the length of the system in the x direction. Replacing x_0 by Eq. (1.49), the condition (1.52) becomes

$$0 < k_y < \frac{m^* \omega_c}{\hbar} L_x = \frac{eB}{\hbar} L_x \tag{1.53}$$

- the quantization of k_y by the periodic condition in the y direction, namely $k_y = 2\pi n_y / L_y$. The distance between two successive levels of k_y is $2\pi/L_y$. Therefore, the number of values of k_y inside the limits given by (1.53) is

$$d = \frac{\frac{eBL_x}{\hbar}}{\frac{2\pi}{L_y}} = \frac{eB}{2\pi\hbar} L_x L_y \tag{1.54}$$

d by definition is the degeneracy of the level E_n. We show now that this degeneracy of Landau's levels is the same as that in zero field. Without $\vec{B}$, the components of the wave vector $\vec{k}$ are quantized uniquely by the periodic conditions in three directions. The distances between successive values of k_x and k_y are $2\pi/L_x$ and $2\pi/L_y$. the number of states in a circular surface δA in the space (k_x, k_y) is

$$d' = \frac{\delta A}{\frac{(2\pi)^2}{L_x L_y}} = \frac{L_x L_y}{(2\pi)^2} 2\pi k_\parallel dk_\parallel \tag{1.55}$$

where $k_\parallel$ is the modulus of the wave vector in the (k_x, k_y) plane. Energy $E_\parallel$ of the state (k_x, k_y) is $E_\parallel = \frac{\hbar^2 k_\parallel^2}{2m^*}$. We have

$$d' = \frac{L_x L_y}{(2\pi)^2} \frac{2\pi m^*}{\hbar^2} dE_\parallel \tag{1.56}$$

To compare d' to d, we take $dE_\parallel = \hbar\omega_c$ which is the separation between two successive Landau's levels. We then have

$$d' = \frac{L_x L_y}{(2\pi)^2} \frac{2\pi m^*}{\hbar^2} \hbar\omega_c = \frac{eB}{2\pi\hbar} L_x L_y \tag{1.57}$$

We see that $d = d'$.

1.5.3 Quantization of electron orbit

Using the Bohr quantization relation

$$\oint \vec{p} \cdot d\vec{r} = (m + \gamma)2\pi\hbar \tag{1.58}$$

where m is an integer and $\gamma = 1/2$ a phase correction, we show that the projection of an electron trajectory in the k space on the (k_x, k_y) plane is given by

$$S_k = 2\pi\hbar^{-1}eB(m + \gamma) \tag{1.59}$$

Demonstration: We have $\vec{p} = \hbar\vec{k} + e\vec{A}$. We write

$$\oint \vec{p} \cdot d\vec{r} = \oint \hbar\vec{k} \cdot d\vec{r} + e \oint \vec{A} \cdot d\vec{r}$$

$$= e \oint \vec{r} \wedge \vec{B} \cdot d\vec{r} + e \int \vec{rot}\vec{A} \cdot d\vec{S}$$

$$= -e\vec{B} \cdot \oint \vec{r} \wedge d\vec{r} + e \int \vec{B} \cdot d\vec{S} \tag{1.60}$$

where we have transformed the circular integral on the trajectory into a surface integral by the Stokes's theorem and used a property of the mixed product between vectors. Since $\oint \vec{r} \wedge d\vec{r}$ is equal twice the surface S limited by the closed trajectory in the real space and $\int \vec{B} \cdot d\vec{S} = \vec{B} \cdot \int d\vec{S}$, we can write

$$\oint \vec{p} \cdot d\vec{r} = -2eBS + eBS = -e\phi \tag{1.61}$$

where $\phi = BS$ is the magnetic flux passing through the surface limited by the electron trajectory. Comparison of Eq. (1.61) to Eq. (1.58) gives

$$-e\phi = (m + \gamma)2\pi\hbar$$
$$-eBS = (m + \gamma)2\pi\hbar$$
$$S = -\frac{1}{eB}(m + \gamma)2\pi\hbar \tag{1.62}$$

Since the real space is connected to the reciprocal space by $\hbar\vec{k} = m^*\vec{v} = e\vec{B} \wedge \vec{r}$, we have

$$\hbar\Delta\vec{k} = e\vec{B} \wedge \Delta\vec{r}$$
$$\hbar^2(\Delta k)^2 = (eB)^2(\Delta r)^2$$
$$(\Delta r)^2 = \frac{\hbar^2}{(eB)^2}(\Delta k)^2$$
$$S = = \frac{\hbar^2}{(eB)^2}\mathcal{S}_k \tag{1.63}$$

The surface S of the trajectory in real space is connected to the surface $\mathcal{S}_k$ of the trajectory in reciprocal space by the last equality. Replacing this relation in Eq. (1.62), one obtains Eq. (1.59).

We have seen that each Landau's level is d-fold degenerate. Replacing $L_x L_y$ by the surface of the material L^2, we can express d of Eq. (1.54) as

$$d = \frac{BL^2}{\frac{2\pi\hbar}{e}} \tag{1.64}$$

The numerator is the magnetic flux passing through the surface of the sample and the denominator is the flux quantum. The Landau's degeneracy is thus the number of magnetic flux quanta crossing the surface of the material.

Putting $\zeta \equiv \frac{L^2}{\frac{2\pi\hbar}{e}}$, we write $d = \zeta B$.

1.5.4 *Diamagnetic susceptibility*

We show in the following that the susceptibility of electrons on Landau's levels is negative. The simplest way to do is to use the partition function for these levels

$$Z = \frac{z^N}{N!} \tag{1.65}$$

where z is the partition function for an electron. $N!$ expresses the indiscernibility of electrons. Taking into account the degeneracy of each level n, we have

$$
\begin{aligned}
z &= \sum_n d \exp(-\beta E_n) = \sum_n d \exp[-\beta(n + 1/2)\hbar\omega_c)] \\
&= \sum_{n=0}^{\infty} d \exp[-\beta(2n + 1)\mu_B B] = \frac{\exp(-\beta\mu_B B)}{1 - \exp(-2\beta\mu_B B)} \\
&= \frac{1}{2 \sinh(\beta\mu_B B)} \tag{1.66}
\end{aligned}
$$

where we have replaced ω_c by eB/m^* and $\mu_B = \frac{e\hbar}{2m^*}$. The formula of the geometric series has been used on the second line. The magnetic moment M of the system is written as [cf. Eq. (1.27)]

$$
\begin{aligned}
M &= -\frac{\partial F}{\partial B} = -k_B T \frac{\partial \ln Z}{\partial B} \\
&= -N k_B T \frac{\partial \ln z}{\partial B} \\
&= -N \mu_B \mathcal{L}(x) \tag{1.67}
\end{aligned}
$$

where

$$\mathcal{L}(x) = \coth(x) - \frac{1}{x} \tag{1.68}$$

with $x = \beta\mu_B B$.

At high T, one has $x << 1$, $\mathcal{L}(x) \simeq x/3$. This gives

$$M \simeq -\frac{N\mu_B^2 B}{3k_B T} \tag{1.69}$$

so that

$$\chi = -\frac{N\mu_B^2}{3k_B T} \tag{1.70}$$

The negative sign of χ shows the diamagnetic character of the electrons considered here. This diamagnetism is called "Landau's diamagnetism".

Remark: If we include the energy associated with k_z in E_n [cf. Eq. (1.51)], we obtain the following factor for the partition function z: $\int_{-\infty}^{\infty} \exp(-\beta \frac{\hbar^2 k_z^2}{2m^*}) dk_z = \sqrt{\frac{2\pi m^*}{\beta \hbar^2}}$. This factor does not depend on B.

1.6   De Haas-van Alphen effect

When B is very high, the degeneracy d is very large. The first Landau's level can contain all or almost all electrons. The higher levels are empty. The sum on all levels in Eq. (1.66) is not appropriate. We can calculate the number of electrons on the first few levels as follows.

Let l be the last fully occupied level. The level $l+1$ is partially occupied. As d is proportional to B, when B increases the degeneracy d increases: each level can receive more electrons. Electrons of higher levels come down to lower levels as B increases. There is a critical value of B which corresponds to the situation where all levels up to l are fully occupied and level $l+1$ is completely empty. There are thus $(l+1)$ full levels ($n = 0, 1, 2, ..., l$). Let B_l be the critical value. The total number of electrons N of the system verifies the relation

$$d\,(l+1) = \zeta B_l\,(l+1) = N \qquad (1.71)$$

from which

$$\frac{1}{B_l} = \frac{\zeta\,(l+1)}{N} \qquad (1.72)$$

Under this form, we see that the critical values $\frac{1}{B_l}$ are discrete ($l = 0, 1, ...$) with a distance $\frac{\zeta}{N}$ between two successive values.

For a given value of B, the number of electrons of $(l+1)$ full levels is $N_1 = \zeta B\,(l+1)$ and the partial level has $N_2 = N - N_1$ electrons. Omitting the energy associated to k_z, the system energy is written as

$$E = E_1 + E_2 \qquad (1.73)$$

where E_1 is the energy of full levels and E_2 that of the partial level. We have

$$E_1 = \sum_{n=0}^{l} \zeta B \hbar \omega_c (n + 1/2) = \frac{1}{2} \zeta B \hbar \omega_c (l+1)^2 \qquad (1.74)$$

where the sum is performed over the full levels (from 0 to l), and

$$E_2 = \hbar \omega_c (l + 1 + 1/2)[N - \zeta B\,(l+1)] \qquad (1.75)$$

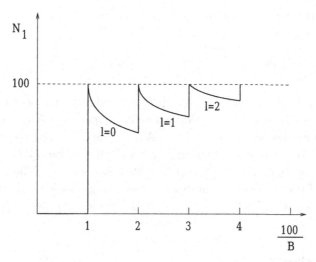

Fig. 1.3 N_1 versus $\frac{100}{B}$. We have taken $N = 100$, $\zeta = 1$. N_1 oscillates with $\frac{100}{B}$. At a given value of B, one has $N_2 = N - N_1$.

where the last factor $[N - \zeta B \, (l + 1)]$ is the number of electrons in the partially filled level $n = l + 1$.

We show in Fig. 1.3 N_1 versus $\frac{1}{B}$

N_1 oscillates with $\frac{1}{B}$, thus the energy oscillates with $\frac{1}{B}$. The magnetic moment of the system at $T = 0$ is calculated by (see previous section)

$$M = -\frac{\partial E}{\partial B} \tag{1.76}$$

We see that M oscillates with $\frac{1}{B}$.

We consider as an example the case where only the first level is occupied ($n = 0$). The system energy is

$$E = N\hbar\omega_c = N\hbar\frac{eB}{m^*} = N\mu_B B \tag{1.77}$$

The magnetic moment is

$$M = -\frac{\partial E}{\partial B} = -N\mu_B \tag{1.78}$$

Now if the first level is completely filled and the second one is partially filled, then

$$E_1 = \zeta B\frac{1}{2}\hbar\omega_c = \zeta B\frac{1}{2}\hbar\frac{eB}{m^*} = \zeta\mu_B B^2 \tag{1.79}$$

$$E_2 = (1 + 1/2)\hbar\omega_c[N - \zeta B] = \frac{3}{2}\hbar\frac{eB}{m^*}[N - \zeta B] = 3\mu_B[N - \zeta B] \tag{1.80}$$

The magnetic moment is thus

$$M = -\frac{\partial(E_1 + E_2)}{\partial B} = -3N\mu_B + 4\mu_B\zeta B \qquad (1.81)$$

This result means that when $\zeta B = N$, the magnetic moment is equal to $N\mu_B$, changing from the situation where it is equal to $-N\mu_B$ [see (1.78)] to the situation where it is equal to $+N\mu_B$ when $\zeta B = N$. The change of the sign of M takes place every time when B satisfies Eq. (1.71).

The susceptibility $\chi = \frac{dM}{dB}$ oscillates therefore with $\frac{1}{B}$. The oscillatory behavior of the magnetic moment of an electron gas at low T is termed as "de Haas-van Alphen effect". The oscillation period is $(\frac{1}{B_{l-1}} - \frac{1}{B_l}) = \frac{\zeta}{N}$.

1.7   Conclusion

We have studied in this chapter the behavior of a system of independent particles under the application of a magnetic field. We have examined four cases.

The first case concerns free electrons at low temperatures. We obtained the so-called Pauli paramagnetism where the susceptibility is a constant at the first-order approximation.

The second case is the system of free atoms where each atom has a permanent magnetic moment. Under the application of a magnetic field, the susceptibility is positive (paramagnetic) and proportional to $1/T$ (Curie's law).

The third case is a diamagnetic case: the reaction of the electrons in an atom to an applied magnetic field gives rise to a negative susceptibility. This phenomenon is called "atomic diamagnetism".

The fourth case concerns the effect of a strong applied magnetic field on a free electron gas of effective mass m^*. The electrons behave as quantum harmonic oscillators in the plane perpendicular to the field. Their energy is quantized in this plane: the energy levels are called "Landau's levels" with a degeneracy proportional to the field strength. The susceptibility is negative (Landau diamagnetism). The magnetic moment oscillates with the varying applied field. It changes the sign when the field coincides with a critical value. This phenomenon is called "de Haas-van Alphen effect".

In the following chapters we study different systems of interacting spins by various methods.

1.8 Problems

Problem 1. Orbital and spin moments of an electron:

Using the theory of angular momentum, calculate the orbital and spin moments of an electron. Determine the total magnetic moment.

Problem 2. Zeeman effect:

a) Calculate the magnetic moment per atom for Fe, provided the saturated magnetization under an applied magnetic field equal to $1.7 \times 10^6 \ A/m$, the mass density of Fe $\rho = 7970 \ kg/m^3$ and the atomic mass of Fe $M = 56$.

b) Calculate ΔE the separation of the energy levels due to the Zeeman effect on the atomic level corresponding to the wavelength $\lambda = 643.8 \ nm$ of a cadmium atom. Calculate the variation of frequency $\Delta \nu$ of the initial level.

Numerical application: Calculate ΔE and $\Delta \nu$ for the following fields $\mu_B H = 0.5$, 1, and 2 Tesla.

Problem 3. Density of states:

Calculate the density of states $\rho(E)$ of a free electron of energy E in three dimensions. Show that $\rho(E)$ is given by Eq. (A.41).

Problem 4. Fermi-Dirac distribution for free-electron gas:

Electrons are fermions which obey the Pauli's exclusion principle. Microscopic states follow the Fermi-Dirac statistics. The Fermi-Dirac distribution is given by (see Appendix A)

$$f(E, T, \mu) = \frac{1}{e^{\beta(E-\mu)} + 1} \tag{1.82}$$

where μ is the chemical potential, $\beta = \frac{1}{k_B T}$, k_B the Boltzmann constant and T the temperature. The function $f(E, T, \mu)$ is the number of electrons of the microscopic state of energy E at temperature T.

Give the properties of $f(E, T, \mu)$ at $T = 0$. Plot $f(E, T, \mu)$ as a function of E for an arbitrary $\mu(> 0)$, at $T = 0$ and at low T.

Problem 5. Sommerfeld's expansion:

Consider the function

$$I = \int_0^\infty h(E) f(E) dE \tag{1.83}$$

where $h(E)$ is a function differentiable at any order with respect to E.

Show that $h(E)$ can be expanded in powers of T at low T as follows:

$$I = \int_0^\mu h(E)dE + \frac{\pi^2}{6}(k_BT)^2 h^{(1)}(E)|_{E=\mu}$$
$$+\frac{7\pi^4}{360}(k_BT)^4 h^{(3)}(E)|_{E=\mu} + ... \tag{1.84}$$

where $h^{(n)}(E)|_{E=\mu}$ is the n-th derivative of $h(E)$ at $E = \mu$.

Problem 6. Pauli paramagnetism:

Calculate the susceptibility of a three-dimensional electron gas in an applied magnetic field $\vec{B}$, at low and high temperatures. One supposes that B is small.

Problem 7. Paramagnetism of free atoms for arbitrary $\vec{J}$:

Consider a gas of N atoms of moment $\vec{J}$ in a volume V. Show that the average of the total magnetic moment per volume unit of the gas is

$$\overline{m} = \frac{Ng\mu_BJ}{V}B_J(\frac{g\mu_BBJ}{k_BT}) \tag{1.85}$$

where $B_J(x)$ is the Brillouin function given by

$$B_J(x) = \frac{2J+1}{2J}\coth(\frac{2J+1}{2J}x) - \frac{1}{2J}\coth(\frac{x}{2J}) \tag{1.86}$$

Show that at high temperature one has

$$\chi = \frac{N}{V}(g\mu_B)^2\frac{J(J+1)}{3}\frac{1}{k_BT} \tag{1.87}$$

Find the limit of $\overline{m}$ at $T = 0$.

Problem 8. Langevin's theory of diamagnetism:

Consider an electron in an atom. In the theory of diamagnetism by Langevin, the motion of the electron around the nucleus is equivalent to the motion of a magnetic moment $\vec{m}$ generated by a current i which circulates in a closed loop of surface A.

a) Write a relation between i, $\vec{m}$ and A.

b) Show that the magnetic moment of the electron is written as $m = evr/2$ where e is the charge of the electron, v its velocity and r its orbital radius.

c) Show that an applied magnetic field $\vec{H}$, perpendicular to the orbital plane, gives rise to the following variation of its magnetic moment $\Delta m = -\frac{\mu_B e^2 r^2 \vec{H}}{4m_e}$ (m_e: electron mass). Comment on the negative sign.

d) What will be the result if $\vec{H}$ makes an angle θ with the surface normal ?

e) Calculate the susceptibility of a material of mass density ρ made of atoms of Z electrons, of mass M.

Numerical application: $\rho = 2220 \ kg/m^3$, $e = 1.6 \times 10^{-19} \ C$, $Z = 6$, $r = 0.7 \times 10^{-10} \ m$.

Problem 9. Langevin's theory of paramagnetism:

Consider an atom of permanent magnetic moment $\vec{m}$ (atom having an odd number of electrons). Using the Maxwell-Boltzmann statistics, show that the magnetic moment resulting from the application of a magnetic field $\vec{H}$ on a material of N atoms per volume unit, in an arbitrary direction is given by

$\vec{M} = N\vec{m}\mathcal{L}(\frac{\mu_B \vec{m}\cdot\vec{H}}{k_B T})$

where $\mathcal{L}(x) = \coth(x) - \frac{1}{x}$ (Langevin function). Calculate the susceptibility in the case of a weak field.

Problem 10. Calculate the variation of the forbidden energy gap in a semiconductor under an applied field $\vec{B}$, supposing the effective mass m^* equal to rest mass.

Problem 11. Paramagnetic resonance:

Consider a magnetic moment of an i-th electron $\vec{m}_i$ in an applied magnetic field $\vec{H}_0$.

a) Write the equation of motion of the spatially average value $\vec{M} = \sum_i \vec{m}_i/N$ (N:number of electrons)

b) Consider the case where $\vec{H} = \vec{H}_0 + \vec{H}_1(t)$, $\vec{H}_0$ being the z component and $\vec{H}_1(t)$ the time-dependent alternating field of frequency ω_1, of magnitude H_1 applied along x axis (or for convenience in the xy plane). Show that the paramagnetic resonance takes place when $H_1 = \hbar\omega_1/\gamma$ where γ (gyromagnetic factor)$= g\mu_B$ (g: Landé factor). Comment.

Problem 12. Nuclear Magnetic Resonance (NMR):

Consider a quantum nuclear spin $1/2$ in a magnetic field $\vec{H}_0$ applied along the z axis. The spin rotates around the z axis with the Larmor frequency ω as seen in the Electron Paramagnetic Resonance studied in Problem 11. The nuclear magnetic resonance has the same principle. Here we use a quantum treatment: the nuclear spin is given by the Pauli matrices σ_i ($i = x, y, z$). One applies a transverse rotating magnetic field $\vec{H}_1$ of radio-frequency ω_r, of magnitude H_1, in the xy plane. The motion of the spin is caused by the resulting field $\vec{H} = \vec{H}_0 + \vec{H}_1(t)$.

a) Write the Schrödinger equation in the case $H_1 = 0$. Determine the eigen-frequency of the spin precession around the z axis (Larmor frequency).

b) Write the Schrödinger equation in the case $H_1 \neq 0$. Solve the equation to obtain the eigen-frequency of the system. Determine the resonance frequency.

Chapter 2

Exchange Interaction in an Electron Gas

2.1 Introduction

The free-electron theory for the conducting electrons in metals leads to simple results. Let us recall some expressions here before considering the Coulomb interaction between electrons.

We consider a gas of N free electrons, of volume $\Omega = L^3$. We have

$$N = \int_0^\infty \rho(E) f(E, T, \mu) dE \tag{2.1}$$

where $\rho(E)$ is the density of states at the energy E and $f(E, T, \mu)$ the Fermi-Dirac distribution [see definitions (A.41) and (A.37)]. Note that the number of electrons at the energy level E at temperature T is $\rho(E) f(E, T, \mu)$. The above integral counts thus the electrons on all energy levels. The average energy is written as

$$\overline{E} = \int_0^\infty E \rho(E) f(E) dE \tag{2.2}$$

We have

$$\rho(E) = (2s + 1) \frac{\Omega}{4\pi^2} \left(\frac{2m}{\hbar^2} \right)^{3/2} E^{1/2} \equiv A E^{1/2} \tag{2.3}$$

where $(2s + 1)$ is the spin degeneracy. At $T = 0$ one has $f(E, T = 0, \mu) = 1$ for $E < \mu$ and $f(E, T = 0, \mu) = 0$ for $E > \mu$. We write

$$N = \int_0^{E_F} \rho(E) dE \tag{2.4}$$

where $E_F \equiv \mu$ at $T = 0$ is called Fermi energy. We have

$$N = \frac{2}{3} A E_F^{3/2} \tag{2.5}$$

23

Thus, using A given in (2.3), we have

$$E_F = \frac{\hbar^2}{2m}(3\pi^2 \frac{N}{\Omega})^{2/3} = \frac{\hbar^2 k_F^2}{2m} \qquad (2.6)$$

This result is important. It shows that the Fermi level is a function of the electron density $n = \frac{N}{\Omega}$. The total kinetic energy at $T = 0$ is

$$\overline{E}_0 = \int_0^{E_F} E\rho(E)dE$$

$$= A\int_0^{E_F} E^{3/2}dE = \frac{2}{5}AE_F^{5/2} \qquad (2.7)$$

Using (2.5) we obtain

$$\overline{E}_0 = \frac{3}{5}NE_F \qquad (2.8)$$

At low temperatures, using the Sommerfeld's expansion (1.84) we can show that

$$\mu = E_F[1 - \frac{\pi^2}{12}(\frac{k_B T}{E_F})^2 - O(T^4)] \qquad (2.9)$$

The chemical potential thus decreases with increasing T. With (1.84), we obtain from (2.2)

$$\overline{E} = \overline{E}_0[1 + \frac{5\pi^2}{12}(\frac{k_B T}{E_F})^2 + O(T^4)] \qquad (2.10)$$

where we have replaced μ by (2.9). We see that the electron energy increases with T. This yields the following heat capacity

$$C_V = \frac{1}{3}\pi^2\rho(E_F)k_B^2 T = \gamma T \qquad (2.11)$$

Note that the classical ideal gas has C_V independent of T.

The above results have been obtained for free electrons. We see in this chapter that these results are modified when we take into account the Coulomb interaction between electrons. The energy then has two more terms: a direct Coulomb term and an exchange term. The latter depends on the spins of the interacting electron pairs. The Hartree-Fock approximation is used in the following to calculate the direct and exchange electron energies.

2.2 Antisymmetric wave function

When the spin of a particle is to be taken into account, we have to use the symmetry properties of the wave function: i) the wave function does not change its sign with the permutation of two particles in their state if their spin is integer including zero (systems of bosons) ii) the wave function changes its sign with the permutation of two particles if their spin is half-integer (systems of fermions). The electron has the spin one-half, therefore we have to use an antisymmetric wave function.

We consider a system of N interacting electrons. The general Hamiltonian is written as

$$\mathcal{H} = \sum_i H(\vec{r}_i) + \frac{1}{2} \sum_{i,j} V(\vec{r}_i, \vec{r}_j) \qquad (2.12)$$

where the first sum is made over the one-particle terms $H(\vec{r}_i)$ such as kinetic energy and Zeemann's energy, while the second sum is performed on the two-particle interaction terms $V(\vec{r}_i, \vec{r}_j)$ between the particles at $\vec{r}_i$ and $\vec{r}_j$ (the case $i = j$ is excluded). In the case of electrons, $V(\vec{r}_i, \vec{r}_j)$ is the Coulomb interaction between the electron charges. We have

$$V(\vec{r}_i, \vec{r}_j) = \frac{e^2}{|\vec{r}_i - \vec{r}_j|} \qquad (2.13)$$

Note that even when the potential is spin-independent, the wave function depends on the particle spins as seen below.

The individual state of an electron is defined by both its orbital and its spin state. For a convenient notation, we write

$$\psi_{\vec{k},\sigma}(\vec{r}_i, \zeta_i) = \phi_{\vec{k}}(\vec{r}_i) S_\sigma(\zeta_i) \qquad (2.14)$$

where $\phi_{\vec{k}}(\vec{r}_i)$ is the orbital state and $S_\sigma(\zeta_i)$ denotes the spin part. We define the following notations

$\sigma = + \text{ or } -$

$\zeta = \uparrow \text{ or } \downarrow$

and

$S_+(\uparrow) = 1,\ S_+(\downarrow) = 0,\ S_-(\uparrow) = 0,\ S_-(\downarrow) = 1$

With these notations, it is clear that a particle with $\uparrow$ spin (respectively $\downarrow$ spin) cannot occupy the spin state $-$ (respectively $+$). For example, the wave function $\psi_{\vec{k},+}(\vec{r}_i, \downarrow)$ is zero. We say that $\psi_{\vec{k},\sigma}(\vec{r}_i, \zeta_i)$ is "the wave function of the electron at $\vec{r}_i$ of spin ζ occupying the state $(\vec{k}, \sigma)$".

The simplest form of the wave function for the system of N electrons is the product of individual wave functions $\psi_{\vec{k},\sigma}(\vec{r}_i, \zeta_i)$ written as

$$\psi_{f_1}(q_1)\psi_{f_2}(q_2)...\psi_{f_i}(q_i)\psi_{f_j}(q_j)...\psi_{f_N}(q_N) \qquad (2.15)$$

where, for simplicity, we used the notations $q_i = (\vec{r}_i, \zeta_i)$ and $f_i = (\vec{k}_i, \sigma_i)$.

Due to the electron indiscernibility, we should include all possible permutations of the electrons in the states. To satisfy the Pauli's exclusion principle, for each permutation we should change the sign of the product. The form which satisfies this requirement is a determinant. We write Ψ as

$$\Psi = A\bar{D} \qquad (2.16)$$

where A is a constant to be determined by the normalization of Ψ and $\bar{D}$ is the following determinant

$$\begin{vmatrix} \psi_{f_1}(q_1) & \psi_{f_1}(q_2) & ... & \psi_{f_1}(q_N) \\ \psi_{f_2}(q_1) & \psi_{f_2}(q_2) & ... & \psi_{f_2}(q_N) \\ ... & ... & ... & ... \\ \psi_{f_N}(q_1) & \psi_{f_N}(q_2) & ... & \psi_{f_N}(q_N) \end{vmatrix}$$

Under this form, we see that the permutation of two electrons $q_i \leftrightarrow q_j$ is equivalent to the permutation of two columns which results in a change of the sign of the determinant, namely of Ψ. In addition, we see that if two electrons occupy the same state, i. e. $f_i = f_j$, then the two corresponding lines of the determinant are identical, Ψ is zero. The Pauli's principle is thus satisfied with a determinant wave function. The determinant $\bar{D}$ is called "Slater determinant".

We can express Ψ by the expansion of $\bar{D}$:

$$\Psi = A\sum_{p}(-1)^p \mathcal{P}\psi_{f_1}(q_1)\psi_{f_2}(q_2)...\psi_{f_N}(q_N) \qquad (2.17)$$

where $\mathcal{P}$ is the "permutation operator" and p the parity associated with $\mathcal{P}$. Let us use the following convention to determine p: to permute two particles q_i and q_l we "count" the numbers of particles found on the left of each of them in $\psi_{f_1}(q_1)\psi_{f_2}(q_2)...\psi_{f_N}(q_N)$. Let n_i and n_l be these numbers. If $|n_l - n_i|$ is an even number then $p = 0$. Otherwise, $p = 1$. We have for the case of two particles $\Psi = A[\psi_{f_1}(q_1)\psi_{f_2}(q_2) - \psi_{f_1}(q_2)\psi_{f_2}(q_1)]$.

To get familiar with the notation, let us calculate the normalization constant A:

$$1 = \int |\Psi|^2 dq_1 dq_2 ... dq_N$$

$$= A^2 \int \left[\sum_{p'} (-1)^{p'} \mathcal{P}' \psi_{f_1}^*(q_1) \psi_{f_2}^*(q_2) ... \psi_{f_N}^*(q_N) \right]$$

$$\times \left[\sum_{p} (-1)^p \mathcal{P} \psi_{f_1}(q_1) \psi_{f_2}(q_2) ... \psi_{f_N}(q_N) \right] dq_1 dq_2 ... dq_N$$

$$= A^2 \int [N! \text{ terms}][N! \text{ terms}] dq_1 dq_2 ... dq_N \qquad (2.18)$$

If the individual wave functions $\psi_{f_i}(q_i)$ are orthonormal, then integrals of direct products are each equal to 1, integrals of cross products are zero. We get

$$1 = A^2 N!$$
$$\text{so that} \quad A = \sqrt{\frac{1}{N!}} \qquad (2.19)$$

We take the case of two particles. We calculate A explicitly as follows:

$$1 = A^2 \int [\psi_{f_1}^*(q_1) \psi_{f_2}^*(q_2) - \psi_{f_1}^*(q_2) \psi_{f_2}^*(q_1)]$$

$$\times [\psi_{f_1}(q_1) \psi_{f_2}(q_2) - \psi_{f_1}(q_2) \psi_{f_2}(q_1)] dq_1 dq_2$$

$$= A^2 [\int \psi_{f_1}^*(q_1) \psi_{f_1}(q_1) dq_1 \int \psi_{f_2}^*(q_2) \psi_{f_2}(q_2) dq_2$$

$$+ \int \psi_{f_1}^*(q_2) \psi_{f_1}(q_2) dq_2 \int \psi_{f_2}^*(q_1) \psi_{f_2}(q_1) dq_1$$

$$- \int \psi_{f_1}^*(q_1) \psi_{f_2}(q_1) dq_1 \int \psi_{f_2}^*(q_2) \psi_{f_1}(q_2) dq_2$$

$$- \int \psi_{f_1}^*(q_2) \psi_{f_2}(q_2) dq_2 \int \psi_{f_2}^*(q_1) \psi_{f_1}(q_1) dq_1]$$

$$= A^2 [1 + 1 - 0 - 0]$$

$$A = \sqrt{\frac{1}{2}} \qquad (2.20)$$

Remark: In the case of bosons, the wave function of the system is invariant with permutations. It can have the form of (2.17) but $p = 0$ for all permutations.

2.3   Hartree-Fock equation

In the Hartree-Fock approximation, we look for a wave function Ψ which minimizes the average energy of the system $E = < \Psi|\mathcal{H}|\Psi >$ under the normalization constraint (2.18). The form of Ψ is already given by (2.17) and $\mathcal{H}$ is given by (2.12). This is a variational problem with constraint. The Lagrange method can be used:

$$\delta_i \left[< \Psi|\mathcal{H}|\Psi > - \sum_j \lambda_j < \psi_{f_j}|\psi_{f_j} > \right] = 0 \qquad (2.21)$$

where we can apply the normalization constraint to each of the individual wave function $< \psi_{f_j}|\psi_{f_j} >= 1$. λ_j is the corresponding Lagrange multiplier. In (2.21), δ_i indicates that we make vary an individual wave function one by one. We have here a system of N coupled equations for $\psi_{f_i}, i = 1, N$.

Let us consider single-electron terms and interaction terms, separately.

2.3.1   *Single-electron terms*

The single-electron terms of $\mathcal{H}$ in (2.12) are $\sum_i H(q_i)$ where $\sum_i \dots = \sum_{\vec{k}_i} \sum_{\sigma_i} \dots$. We have

$$
\begin{aligned}
< \Psi| \sum_i H(q_i)|\Psi > &= \frac{1}{N!} \sum_{p,p'} (-1)^{p+p'} \int [\mathcal{P}'\psi^*_{f_1}(q_1)\psi^*_{f_2}(q_2)...\psi^*_{f_N}(q_N)] \\
&\quad \times [\sum_i H(q_i)] [\mathcal{P}\psi_{f_1}(q_1)\psi_{f_2}(q_2)...\psi_{f_N}(q_N)] \, dq_1 dq_2...dq_N \\
&= \frac{1}{N!} \sum_i \sum_{p,p'} (-1)^{p+p'} \int \mathcal{P}'\{\psi^*_{f_1}(q_1)\psi^*_{f_2}(q_2)...\psi^*_{f_N}(q_N) \\
&\quad \times H(q_i)(\mathcal{P}')^{-1}\mathcal{P}[\psi_{f_1}(q_1)\psi_{f_2}(q_2)...\psi_{f_N}(q_N)]\}dq_1...dq_N \\
&= \frac{1}{N!} \sum_i \sum_{p,p'} (-1)^{2p+r} \int \mathcal{P}'\{\psi^*_{f_1}(q_1)\psi^*_{f_2}(q_2)...\psi^*_{f_N}(q_N) \\
&\quad \times H(q_i)\mathcal{R}[\psi_{f_1}(q_1)\psi_{f_2}(q_2)...\psi_{f_N}(q_N)]\}dq_1 dq_2...dq_N
\end{aligned}
$$

where $\mathcal{R} = (\mathcal{P}')^{-1}\mathcal{P}$ is a permutation operator with parity $r = -p' + p$. It is clear in the last line that the integral of any product with a permutation $\mathcal{R}$ is zero:

$$\int \psi^*_{f_1}(q_1)\psi^*_{f_2}(q_2)...\psi^*_{f_N}(q_N) H(q_i)\mathcal{R}[\psi_{f_1}(q_1)\psi_{f_2}(q_2)...\psi_{f_N}(q_N)]dq_1 dq_2...dq_N$$

is not zero only when $\mathcal{R} = \mathcal{I}$, I being the identity operator (no permutation). We thus have $r = 0$ so that the parity is $(-1)^{2p+r} = 1$. All integrals on variables other than q_i give 1 (orthonormality of ψ_{f_j}). There remains $\int \psi_{f_i}^*(q_i)H(q_i)\psi_{f_i}(q_i)dq_i$. The permutations $\mathcal{P}'$ give $N!$ such integrals. Finally we have

$$< \Psi| \sum_i H(q_i)|\Psi > = \frac{1}{N!}N! \sum_{f_i} \int \psi_{f_i}^*(q_i)H(q_i)\psi_{f_i}(q_i)dq_i$$

$$= \sum_{f_i} < \psi_{f_i}|H(q_i)|\psi_{f_i} > \qquad (2.22)$$

2.3.2 Interaction terms

The interaction terms are

$$< \Psi|\frac{1}{2}\sum_{i,j} V(q_i, q_j)|\Psi >$$

$$= \frac{1}{N!}\sum_{p,p'}(-1)^{p+p'}\int [\mathcal{P}'\psi_{f_1}^*(q_1)\psi_{f_2}^*(q_2)...\psi_{f_N}^*(q_N)][\frac{1}{2}\sum_{i,j}V(q_i,q_j)]$$

$$\times [\mathcal{P}\psi_{f_1}(q_1)\psi_{f_2}(q_2)...\psi_{f_N}(q_N)]\,dq_1 dq_2...dq_N$$

$$= \frac{1}{N!}\frac{1}{2}\sum_{i,j}\sum_{p,p'}(-1)^{p+p'}\int \mathcal{P}'\{\psi_{f_1}^*(q_1)\psi_{f_2}^*(q_2)...\psi_{f_N}^*(q_N)V(q_i,q_j)$$

$$\times (\mathcal{P}')^{-1}\mathcal{P}[\psi_{f_1}(q_1)\psi_{f_2}(q_2)...\psi_{f_N}(q_N)]\}dq_1 dq_2...dq_N$$

$$= \frac{1}{N!}\frac{1}{2}\sum_{i,j}\sum_{p,p'}(-1)^{2p+r}\int \mathcal{P}'\{\psi_{f_1}^*(q_1)\psi_{f_2}^*(q_2)...\psi_{f_N}^*(q_N)V(q_i,q_j)$$

$$\times \mathcal{R}[\psi_{f_1}(q_1)\psi_{f_2}(q_2)...\psi_{f_N}(q_N)]\}dq_1 dq_2...dq_N \qquad (2.23)$$

We consider two particles q_i and q_j. Inside $\{...\}$ of (2.23), all permutations $\mathcal{R}$ of two particles other than q_i and q_j inside $[...]$ give zero due to the orthogonality. The same is true for a permutation between q_i or q_j with another particle q_l $(l \neq j, i)$ because $... \int \psi_{f_l}^*(q_l)\psi_{f_i}(q_l)dq_l... = 0$. The only two possibilities which yield a non zero integral are when $\mathcal{R} = \mathcal{I}$ (no permutation) and when $\mathcal{R}$ permutes q_i and q_j. In these cases, integrals on variables other than q_i and q_j are equal to 1, while integrals on q_i and q_j can be done if we know the expression for $V(q_i, q_j)$. Thus, we have

If $\mathcal{R} = \mathcal{I}$

$$\to K_{ij} = \int \psi_{f_i}^*(q_i)\psi_{f_j}^*(q_j)V(q_i,q_j)\psi_{f_i}(q_i)\psi_{f_j}(q_j)dq_idq_j \qquad (2.24)$$

If $\mathcal{R} = q_i \leftrightarrow q_j$

$$\to -J_{ij} = -\delta_{\sigma_i,\sigma_j}\int \psi_{f_i}^*(q_i)\psi_{f_j}^*(q_j)V(q_i,q_j)\psi_{f_i}(q_j)\psi_{f_j}(q_i)dq_idq_j \quad (2.25)$$

We see that

- the negative sign in the second case results from the permutation of q_i and q_j
- the permutation of q_i et q_j is possible only when q_i and q_j have the same spin state. This is indicated by the Kronecker symbol in the integral J_{ij}.

Now, when operating $\mathcal{P}'$ in (2.23) we obtain $N!$ integrals (2.24) and (2.25). We have

$$< \Psi|\frac{1}{2}\sum_{i,j}V(q_i,q_j)|\Psi >$$

$$= \frac{1}{2}\sum_{f_i,f_j}\{\int \psi_{f_i}^*(q_i)\psi_{f_j}^*(q_j)V(q_i,q_j)\psi_{f_i}(q_i)\psi_{f_j}(q_j)dq_idq_j$$

$$-\delta_{\sigma_i,\sigma_j}\int \psi_{f_i}^*(q_i)\psi_{f_j}^*(q_j)V(q_i,q_j)\psi_{f_i}(q_j)\psi_{f_j}(q_i)dq_idq_j\}$$

$$= \frac{1}{2}\sum_{f_i,f_j}[< \psi_{f_i}(q_i)\psi_{f_j}(q_j)|V(q_i,q_j)|\psi_{f_i}(q_i)\psi_{f_j}(q_j) >$$

$$-\delta_{\sigma_i,\sigma_j}< \psi_{f_i}(q_i)\psi_{f_j}(q_j)|V(q_i,q_j)|\psi_{f_i}(q_j)\psi_{f_j}(q_i) >]$$

$$= \frac{1}{2}\sum_{f_i,f_j}[K_{ij} - J_{ij}] \qquad (2.26)$$

We have in mind that $\sum_{f_i,f_j} \cdots = \sum_{\vec{k}_i,\vec{k}_j}\sum_{\sigma_i,\sigma_j}\cdots$ and that the variables q_i and q_j are running variables in the above integrals. Thus we can put $q_i = \vec{r}_1$ and $q_j = \vec{r}_2$ in order to avoid a confusion with the indices for the states. The sums on the particle spins are replaced by the sums on the spin states σ_i et σ_j that the particles occupy.

Let us come back to (2.21). Using (2.22) and (2.26) in (2.21), we obtain

$$0 = \delta_i \left[<\Psi|\mathcal{H}|\Psi> - \sum_i \lambda_j < \psi_{f_j}|\psi_{f_j} > \right]$$

$$= \int d\vec{r}_1 \delta\psi^*_{\vec{k}_i,\sigma_i}(\vec{r}_1)\{H(\vec{r}_1)\psi_{\vec{k}_i,\sigma_i}(\vec{r}_1)$$

$$+ \sum_{j\neq i} \left[\int d\vec{r}_2 \psi^*_{\vec{k}_j,\sigma_j}(\vec{r}_2)V(\vec{r}_1,\vec{r}_2)\psi_{\vec{k}_j,\sigma_j}(\vec{r}_2) \right] \psi_{\vec{k}_i,\sigma_i}(\vec{r}_1)$$

$$- \sum_{j\neq i} \delta_{\sigma_i,\sigma_j} \left[\int d\vec{r}_2 \psi^*_{\vec{k}_j,\sigma_j}(\vec{r}_2)V(\vec{r}_1,\vec{r}_2)\psi_{\vec{k}_i,\sigma_i}(\vec{r}_2) \right] \psi_{\vec{k}_j,\sigma_j}(\vec{r}_1)$$

$$-\lambda_i\psi_{\vec{k}_i,\sigma_i}(\vec{r}_1)\} + c.c. \qquad (2.27)$$

where *c.c.* denotes "complex conjugate". Since $\delta\psi^*_{\vec{k}_i,\sigma_i}(\vec{r})$ is arbitrary, the quantity in the parentheses {...} should be zero. We obtain the following Hartree-Fock equation

$$H(\vec{r}_1)\psi_{\vec{k}_i,\sigma_i}(\vec{r}_1) + \sum_{j\neq i} \left[\int d\vec{r}_2 \psi^*_{\vec{k}_j,\sigma_j}(\vec{r}_2)V(\vec{r}_1,\vec{r}_2)\psi_{\vec{k}_j,\sigma_j}(\vec{r}_2) \right] \psi_{\vec{k}_i,\sigma_i}(\vec{r}_1)$$

$$- \sum_{j\neq i} \delta_{\sigma_i,\sigma_j} \left[\int d\vec{r}_2 \psi^*_{\vec{k}_j,\sigma_j}(\vec{r}_2)V(\vec{r}_1,\vec{r}_2)\psi_{\vec{k}_i,\sigma_i}(\vec{r}_2) \right] \psi_{\vec{k}_j,\sigma_j}(\vec{r}_1)$$

$$= \lambda_i\psi_{\vec{k}_i,\sigma_i}(\vec{r}_1) \qquad (2.28)$$

or under a more compact form

$$H(\vec{r}_1)\psi_{\vec{k}_i,\sigma_i}(\vec{r}_1) + \sum_{j\neq i} < \psi_{\vec{k}_j,\sigma_j}(\vec{r}_2)|V(\vec{r}_1,\vec{r}_2))|\psi_{\vec{k}_j,\sigma_j}(\vec{r}_2) > \psi_{\vec{k}_i,\sigma_i}(\vec{r}_1)$$

$$- \sum_{j\neq i} \delta_{\sigma_i,\sigma_j} < \psi_{\vec{k}_j,\sigma_j}(\vec{r}_2)|V(\vec{r}_1,\vec{r}_2)|\psi_{\vec{k}_i,\sigma_i}(\vec{r}_2) > \psi_{\vec{k}_j,\sigma_j}(\vec{r}_1)$$

$$= \lambda_i\psi_{\vec{k}_i,\sigma_i}(\vec{r}_1) \qquad (2.29)$$

or, by multiplying on the left by $\psi^*_{\vec{k}_i,\sigma_i}$ and then integrating on $\vec{r}_1$,

$$< \psi_{\vec{k}_i,\sigma_i}|H(\vec{r}_1)|\psi_{\vec{k}_i,\sigma_i}(\vec{r}_1) > + \sum_{j\neq i}[K_{ij} - J_{ij}] = \lambda_i \qquad (2.30)$$

where K_{ij} and J_{ij} are defined by (2.24) and (2.25). K_{ij} is called "direct interaction" or "Coulomb interaction" and J_{ij} "exchange interaction".

2.3.2.1 *Remarks*

- Equation (2.29) does not have the form of the Schrödinger equation $\mathcal{H}\psi_i = \lambda_i \psi_i$ because of the exchange term.
- We can show that λ_i is equal the the energy necessary to remove the electron of the state $\vec{k}_i$ from the system (see Koopmann's theorem in Problem 2 in section 2.6).
- For a practical purpose, we can include in the sum the case $\vec{k}_i = \vec{k}_j$ because the direct term then cancels the exchange term.
- We see that $< \psi_{\vec{k}_j,\sigma_j}(\vec{r}_2)|V(\vec{r}_1,\vec{r}_2))|\psi_{\vec{k}_j,\sigma_j}(\vec{r}_2) >$ is nothing but the value of the interaction of the particle $\vec{k}_i$ with the particle $\vec{k}_j$ which is averaged over the space. The exchange interaction is also a spatially averaged value.
- Hartree-Fock equation (2.30) is written for a particle in the state $\vec{k}_i$. For N particles, we have to write N similar equations. These equations have to be self-consistently solved.
- If we suppose that the interaction $V(\vec{r}_1, \vec{r}_2)$ has the form (2.13), then for a given distance $|\vec{r}_1 - \vec{r}_2|$, the interaction energy between two parallel spins is lower than that between two antiparallel spins, due to the negative exchange term.
- For a small distance $|\vec{r}_1 - \vec{r}_2| \to 0$ only states of antiparallel spins are possible because of the Pauli's exclusion principle which forbids two parallel spins to occupy the same place. Thus, there is a Fermi hole in the proximity of each spin where a parallel spin cannot come in (see Problem 1 in section 2.6).
- When the space is available, the spins prefer to stay far from each other so that the particle energy is decreased by the exchange term. This explains in particular the first Hund's empirical law according to which the electrons in an atomic orbital arrange themselves so as to maximize the total spin. We will see in the following that this is also true for an interacting electron gas.

2.4   Results for an interacting electron gas

For a gas of N interacting electrons of volume Ω, the Hamiltonian is written as

$$\mathcal{H} = -\frac{\hbar^2}{2m}\sum_i \nabla_i^2 + \frac{1}{2}\sum_{i,j} V(\vec{r}_i, \vec{r}_j) \qquad (2.31)$$

where the first sum is taken over the electron kinetic energies and the second sum is performed over the Coulomb interactions $V(\vec{r}_i, \vec{r}_j)$ given by (2.13).

2.4.1 *Direct-interaction energy and exchange-interaction energy*

We use the Hartree-Fock equation (2.29) with the following individual wave function

$$\psi_{\vec{k}_i,\sigma_i}(\vec{r}, \zeta) = \frac{1}{\sqrt{\Omega}} e^{i\vec{k}_i \cdot \vec{r}} S_{\sigma_i}(\zeta) \tag{2.32}$$

Since ζ should be compatible with state σ_i, i. e. $S_{\sigma_i}(\zeta) = 1$, we can omit it in the following to simplify the writing. We just keep the index σ_i of $\psi_{\vec{k}_i,\sigma_i}(\vec{r})$. Note that when $V(\vec{r}_i, \vec{r}_j)$ depends only on the relative distance between the two electrons as it is the case here, the wave function of the form of a plane wave (2.32) is always a solution of the Hartree-Fock equation. Replacing (2.32) in (2.29) we obtain

$$(H_0 + H_d + H_e)\psi_{\vec{k}_i,\sigma_i}(\vec{r}) = \epsilon_i \psi_{\vec{k}_i,\sigma_i}(\vec{r}) \tag{2.33}$$

where H_0, H_d and H_e are respectively kinetic energy, direct-interaction energy and exchange-interaction energy, of electron i (we have used the conventional notation ϵ_i in lieu of λ_i). Explicitly,

$$H_0 = \frac{\hbar^2 k_i^2}{2m} \tag{2.34}$$

$$H_d = \sum_{\vec{k}_j} \sum_{\sigma_j = +,-} \frac{e^2}{\Omega} \int_\Omega \frac{d\vec{r}_1}{|\vec{r} - \vec{r}_1|} \tag{2.35}$$

where the sum over $\vec{k}_j$ is taken in the first Brillouin zone, and

$$\begin{aligned}
H_e \psi_{\vec{k}_i,\sigma_i}(\vec{r}) &= -\sum_{\vec{k}_j,\sigma_j} \delta_{\sigma_i,\sigma_j} \frac{e^2}{\Omega} \int_\Omega \frac{d\vec{r}_1 e^{-i\vec{k}_j \cdot \vec{r}_1 + i\vec{k}_i \cdot \vec{r}_1}}{|\vec{r} - \vec{r}_1|} \frac{1}{\sqrt{\Omega}} e^{i\vec{k}_j \cdot \vec{r}} \\
&= -\sum_{\vec{k}_j,\sigma_j} \delta_{\sigma_i,\sigma_j} \frac{e^2}{\Omega} \int_\Omega \frac{d\vec{r}_1 e^{i(\vec{k}_i - \vec{k}_j) \cdot (\vec{r}_1 - \vec{r})} e^{i(\vec{k}_i - \vec{k}_j) \cdot \vec{r}}}{|\vec{r} - \vec{r}_1|} \frac{1}{\sqrt{\Omega}} e^{i\vec{k}_i \cdot \vec{r}}
\end{aligned} \tag{2.36}$$

Putting

$$v(\vec{k}) = \int_\Omega d\vec{r}' e^{i\vec{k}\cdot\vec{r}_1} V(\vec{r}_1) = \int_\Omega \frac{d\vec{r}_1 e^{i\vec{k}\cdot\vec{r}_1}}{|\vec{r}_1|} \tag{2.37}$$

where $v(\vec{k})$ is the Fourier transform of $V(\vec{r}_1)$, we obtain

$$\left\{ \frac{\hbar^2 k_i^2}{2m} + \frac{e^2}{\Omega} \sum_{\vec{k}_j} \sum_{\sigma_j=+,-} \left[v(0) - \delta_{\sigma_i,\sigma_j} v(\vec{k}_i - \vec{k}_j) \right] \right\} \psi_{\vec{k}_i,\sigma_i}(\vec{r}) = \epsilon_i \psi_{\vec{k}_i,\sigma_i}(\vec{r}) \tag{2.38}$$

We see here that we have transformed the Hartree-Fock equation into the form of a Schrödinger equation $\{...\}\psi_{\vec{k}_i,\sigma_i}(\vec{r}) = \epsilon_i\psi_{\vec{k}_i,\sigma_i}(\vec{r})$. The energy of electron i is thus

$$\epsilon_i = \frac{\hbar^2 k_i^2}{2m} + \frac{e^2}{\Omega} \sum_{\vec{k}_j} \sum_{\sigma_j=+,-} \left[v(0) - \delta_{\sigma_i,\sigma_j} v(\vec{k}_i - \vec{k}_j) \right] \tag{2.39}$$

To complete our calculation, we have to calculate $v(0)$ and $v(\vec{k}_i - \vec{k}_j)$. If we suppose that the electron gas is superposed on a background of N positive charges of the ions as in the case of a metal model, and that these positive charges are uniformly distributed with a charge density

$$\rho^+ = \frac{Ne}{\Omega} \tag{2.40}$$

then the interaction between electron i with the positive-charge background is given by

$$-e\rho^+ \int_\Omega \frac{d\vec{r}_i}{|\vec{r} - \vec{r}_i|} = -\frac{Ne^2}{\Omega} v(0) \tag{2.41}$$

where the notation (2.37) has been used. We see that the above term cancels the direct-interaction energy given in (2.39):

$$\frac{e^2}{\Omega} \sum_{\vec{k}_j} \sum_{\sigma_j=+,-} v(0) = \frac{e^2}{\Omega} v(0) \sum_{\vec{k}_j} \sum_{\sigma_j=+,-} 1 = \frac{e^2}{\Omega} v(0) N \tag{2.42}$$

Therefore, the remaining energy of the electron in (2.39) is the kinetic energy and the exchange energy. We have

$$\epsilon_i = \frac{\hbar^2 k_i^2}{2m} - \frac{e^2}{\Omega} \sum_{\vec{k}_j} \sum_{\sigma_j=+,-} \delta_{\sigma_i,\sigma_j} v(\vec{k}_i - \vec{k}_j) \tag{2.43}$$

We calculate now $v(\vec{k}_i - \vec{k}_j)$. Putting $\vec{k}' = \vec{k}_i - \vec{k}_j$, we have

$$
\begin{aligned}
v(\vec{k}') &= \int_\Omega \frac{d\vec{r}'\, e^{i\vec{k}'\cdot\vec{r}'}}{r'} \\
&= 2\pi \int_0^\infty r'^2 dr' \int_0^\pi \sin\theta\, d\theta \frac{e^{ik'r'\cos\theta}}{r'} \\
&= 2\pi \lim_{\mu\to 0} \int_0^\infty e^{-\mu r'} r'\, dr' \int_1^{-1} d(-\cos\theta)\, e^{ik'r'\cos\theta} \\
&= \frac{2\pi}{ik'} \lim_{\mu\to 0} \int_0^\infty dr' [e^{(-\mu+ik')r'} - e^{(-\mu-ik')r'}] \\
&= \frac{2\pi}{ik'} \lim_{\mu\to 0} \left[\frac{-1}{-\mu+ik'} + \frac{-1}{\mu+ik'} \right] \\
&= \lim_{\mu\to 0} \frac{4\pi}{k'^2 + \mu^2} \qquad\qquad\qquad\qquad (2.44) \\
&= \frac{4\pi}{k'^2} \qquad\qquad\qquad\qquad\qquad\qquad (2.45)
\end{aligned}
$$

In the above calculation, we have taken the infinite upper limit for the volume while introducing the factor $e^{-\mu r'}$ to avoid the otherwise divergence at $r' = \infty$. However, at the end of the calculation we have taken the $\lim_{\mu\to 0}$ to recover the original integrand. We will see below that this mathematical trick has a physical meaning: the factor $e^{-\mu r'}$ represents a screening effect due to the presence of other charges in the system. This screened Coulomb interaction $e^{-\mu r}/r$ makes the interaction damped faster with increasing distance r'. μ, the inverse of the screening length, is shown to be proportional to $n^{2/3}$ (n: electron density), in the Thomas-Fermi approximation (see Problem 3 in section 2.6). Equation (2.43) becomes

$$
\epsilon_i = \frac{\hbar^2 k_i^2}{2m} - \frac{4\pi e^2}{\Omega} \sum_{\vec{k}_j} \sum_{\sigma_j=+,-} \delta_{\sigma_i,\sigma_j} \frac{1}{|\vec{k}_i - \vec{k}_j|^2} \qquad (2.46)
$$

Note that the sum on σ_j gives a factor 1 because of the condition $\sigma_j = \sigma_i$.

The sum on $\vec{k}_j$, transformed into an integral, gives

$$
\frac{4\pi e^2}{\Omega} \sum_{\vec{k}_j} \frac{1}{|\vec{k}_i - \vec{k}_j|^2} = \frac{4\pi e^2}{\Omega} \frac{\Omega}{(2\pi)^3} \int \frac{d\vec{k}_j}{(\vec{k}_i - \vec{k}_j)^2}
$$

$$
= \frac{e^2}{\pi} \int_0^{k_F} k_j^2 dk_j \int_1^{-1} \frac{d(-\cos\theta)}{k_i^2 + k_j^2 - 2k_i k_j \cos\theta}
$$

$$
= \frac{e^2}{\pi k_i} \int_0^{k_F} k_j dk_j \ln \left| \frac{k_j - k_i}{k_j + k_i} \right|
$$

$$
= \frac{e^2}{\pi k_i} \left[\frac{k_F^2 - k_i^2}{2} \ln \left| \frac{k_F + k_i}{k_F - k_i} \right| + k_i k_F \right] \qquad (2.47)
$$

where in the last line we have used an integration by parts by putting $u' = k_j dk_j$ and $v = \ln|...|$. This result is for $T = 0$ because we have limited the upper limit of the wave vector at the Fermi wave vector k_F.

Equation (2.43) becomes finally

$$
\epsilon_i = \frac{\hbar^2 k_i^2}{2m} - \frac{e^2}{2\pi} \left[\frac{k_F^2 - k_i^2}{k_i} \ln \left| \frac{k_F + k_i}{k_F - k_i} \right| + 2k_F \right] \qquad (2.48)
$$

This is an important result: the exchange interaction between electrons shown by the second term has been neglected in the free-electron theory.

There is however a paradox: we can verify that the density of states $\rho(\epsilon_i)$ (see definition in Appendix A) calculated by using (2.48) is equal to zero at $k_i = k_F$. But at $T = 0$, all energy levels up to the Fermi level are occupied by definition. This paradox can be solved by introducing the screening effect in the Coulomb interaction, as is seen in Problem 4 in section 2.6. Therefore, this paradox is not an artefact of the Hartree-Fock approximation. It is a problem of the unscreened Coulomb potential (2.13).

2.4.2   Effective mass

When an electron interacts weakly with its environment, we can consider it as a free electron but with a modified mass which takes into account the interaction. The modified mass is called "effective mass". It can be defined by the following tensor

$$
\frac{1}{m_{\alpha\beta}^*} = \frac{1}{\hbar^2} \frac{\partial^2 \epsilon_{\vec{k}_i}}{\partial k_\alpha \partial k_\beta} \qquad (2.49)
$$

where $\alpha, \beta = x, y, z$. The effect of the interaction with the environment is contained in $\epsilon_{\vec{k}_i}$. It can be an interaction with other electrons of the gas as in what described above or an interaction with a periodic potential as in the theory of almost-free electrons, or an interaction with electrons of neighboring atoms in the tight-binding theory. The energy of electron i can be written as

$$\epsilon_{\vec{k}_i} = \frac{\hbar^2 k_i^2}{2m^*}$$

In the case of an electron gas, using (2.48) we can calculate m^*. m^* contains the electron-electron interaction as seen in Problem 4 of section 2.6.

2.4.3 Total energy of an interacting electron gas

We consider an interacting electron gas at $T = 0$.

2.4.3.1 Total kinetic energy

We have

$$E_c = \sum_{\vec{k}_i} \sum_{\sigma_i} \frac{\hbar^2 k_i^2}{2m} = 2 \sum_{\vec{k}_i} \frac{\hbar^2 k_i^2}{2m}$$

$$= 2 \frac{\Omega}{(2\pi)^3} \int d\vec{k}_i \frac{\hbar^2 k_i^2}{2m}$$

$$= 2 \frac{\Omega}{(2\pi)^3} \frac{\hbar^2}{2m} \int_0^{k_F} 4\pi k_i^4 dk_i = \frac{\Omega}{2\pi^2} \frac{k_F^5}{5}$$

$$= \frac{3}{5} N \frac{\hbar^2 k_F^2}{2m} = \frac{3}{5} N \epsilon_F \tag{2.50}$$

where we have used [see (2.8), (2.5) and (2.6)]

$$N = \sum_{\vec{k}_i} \sum_{\sigma_i} 1 = 2 \frac{\Omega}{(2\pi)^3} \int d\vec{k}_i = 2 \frac{\Omega}{(2\pi)^3} \int_0^{k_F} 4\pi k_i^2 dk_i$$

$$= 2 \frac{\Omega}{(2\pi)^3} \frac{4\pi k_F^3}{3} \tag{2.51}$$

For a later comparison, let us express E_c in the atomic unity Rydberg: from (2.51) we have

$$k_F^3 = 3\pi^2 \frac{N}{\Omega} = 3\pi^2 \frac{1}{\frac{4\pi r_0^3}{3}} = \frac{9\pi}{4} \frac{1}{r_0^3}$$

$$k_F = \frac{1}{ar_0} \tag{2.52}$$

where we have defined r_0 by $\frac{\Omega}{N}$ = volume of an electron = $\frac{4\pi r_0^3}{3}$ and $a = (\frac{4}{9\pi})^{1/3} \simeq 0.52$. The average kinetic energy per electron is thus

$$\bar{\epsilon}_c = E_c/N = \frac{3}{5}\frac{\hbar^2}{2m}[\frac{1}{ar_0}]^2 = \frac{3}{5}\frac{me^4}{2\hbar^2}\frac{1}{(ar_s)^2}$$

$$= \frac{3}{5}\frac{1}{(ar_s)^2} \quad \text{Rydberg} \tag{2.53}$$

where $r_s = \frac{r_0}{a_H}$ with $a_H = \frac{\hbar^2}{me^2}$ (Bohr radius) and 1 Rydberg$=\frac{me^4}{2\hbar^2}$. We have then

$$\bar{\epsilon}_c = \frac{2.21}{r_s^2} \quad \text{Rydberg} \tag{2.54}$$

2.4.3.2   Total exchange energy

In principle, we can integrate the exchange term of (2.48) with respect to $\vec{k}_i$ to find the total exchange energy E_{ex}. There is however an alternative way which is simpler as seen in the following. From (2.46) we have

$$E_{ex} = -\frac{1}{2}\frac{4\pi e^2}{\Omega}\sum_{\vec{k}_i,\vec{k}_j}\sum_{\sigma_i,\sigma_j}\delta_{\sigma_i,\sigma_j}\frac{1}{|\vec{k}_i - \vec{k}_j|^2}$$

$$= -2\frac{2\pi e^2}{\Omega}\frac{\Omega^2}{(2\pi)^6}\int d\vec{k}_i \int d\vec{k}_j \frac{1}{|\vec{k}_i - \vec{k}_j|^2} \tag{2.55}$$

where we added a factor $1/2$ in the first line to remove the double counting, and a factor 2 in the second line for the sum on spin σ_i. We write

$$\frac{1}{|\vec{k}_i - \vec{k}_j|^2} = \frac{1}{k_i^2 + k_j^2 - 2k_ik_j\cos\theta} = \frac{1}{k_i^2}\frac{1}{1 + (\frac{k_j}{k_i})^2 - 2\frac{k_j}{k_i}\cos\theta} \tag{2.56}$$

If we suppose $k_j < k_i$, then we recognize that the last fraction is the square of the generating function of the Legendre polynomials $P_l(\cos\theta)$

$$\frac{1}{(1 - 2zx + x^2)^{1/2}} = \sum_{l=0}^{\infty}x^l P_l(z) \tag{2.57}$$

where $x < 1$, $-1 \leq z \leq 1$. We have

$$\frac{1}{|\vec{k}_i - \vec{k}_j|^2} = \frac{1}{k_i^2}[\sum_{l=0}^{\infty}(\frac{k_j}{k_i})^l P_l(\cos\theta)]^2$$

$$= \frac{1}{k_i^2}\sum_{l=0}^{\infty}\sum_{l'=0}^{\infty}(\frac{k_j}{k_i})^{l+l'} P_l(\cos\theta)P_{l'}(\cos\theta) \tag{2.58}$$

Using this relation, we write

$$
I = \int d\vec{k}_i \int d\vec{k}_j \frac{1}{|\vec{k}_i - \vec{k}_j|^2} = 2 \int_{k_i \leq k_F} d\vec{k}_i \int_{k_j < k_i} 2\pi (\frac{k_j}{k_i})^2 dk_j \int_1^{-1} d(-\cos\theta)
$$
$$
\times \sum_{l=0}^{\infty} \sum_{l'=0}^{\infty} (\frac{k_j}{k_i})^{l+l'} P_l(\cos\theta) P_{l'}(\cos\theta) \tag{2.59}
$$

where factor 2 in the first line is for taking into account the symmetric case $k_j > k_i$. Using the following orthogonality of Legendre polynomials

$$
\int_{-1}^{1} d(\cos\theta) P_l(\cos\theta) P_{l'}(\cos\theta) = \frac{2}{2l+1} \delta_{l,l'} \tag{2.60}
$$

we obtain

$$
\begin{aligned}
I &= 32\pi^2 \int_0^{k_F} k_i^2 dk_i \int_0^{k_i} dk_j \sum_{l=0}^{\infty} (\frac{k_j}{k_i})^{2l+2} \frac{1}{2l+1} \\
&= 32\pi^2 \int_0^{k_F} k_i^2 dk_i k_i \sum_{l=0}^{\infty} \frac{1}{(2l+1)(2l+3)} \\
&= 8\pi^2 k_F^4 \sum_{l=0}^{\infty} \frac{1}{(2l+1)(2l+3)} \\
&= 8\pi^2 k_F^4 \frac{1}{2} \sum_{l=0}^{\infty} [\frac{1}{2l+1} - \frac{1}{2l+3}] \\
&= 8\pi^2 k_F^4 \frac{1}{2} [1 - 1/3 + 1/3 - 1/5 + 1/5 - 1/7...] \\
&= 4\pi^2 k_F^4 \tag{2.61}
\end{aligned}
$$

Equation (2.55) becomes

$$
E_{ex} = -\frac{2e^2 \Omega}{(2\pi)^3} k_F^4 \tag{2.62}
$$

The exchange energy per electron is thus

$$
\bar{\epsilon}_{ex} = E_{ex}/N = -\frac{2\pi e^2}{(2\pi)^3} \frac{\Omega}{N} k_F^4 = -\frac{2e^2}{(2\pi)^3} \frac{4\pi r_0^3}{3} k_F^4 \tag{2.63}
$$

Replacing k_F by (2.52) and r_0 by $r_s a_H$ we get, in the same manner as for (2.54),

$$
\bar{\epsilon}_{ex} = -\frac{0.916}{r_s} \quad \text{Rydberg} \tag{2.64}
$$

The total energy per electron is

$$\bar{\epsilon} = \frac{2.21}{r_s^2} - \frac{0.916}{r_s} \qquad \text{Rydberg} \qquad (2.65)$$

We recall that r_s, in unity of Bohr radius, is the radius of the average volume occupied by an electron. When r_s is large, namely when the electron density is small, the exchange term dominates in the total energy.

2.4.4   *Paramagnetic-ferromagnetic transition*

We show in the following that the ferromagnetic phase is more stable than the paramagnetic when the electron density is smaller than a critical value.

We have calculated the total energy of an electron gas in a phase where all spin orientations are allowed: we have used k_F by supposing that each energy level is occupied by two antiparallel spins (spin degeneracy=2). Let us calculate now the total energy in the ferromagnetic phase where all spins are parallel. The same calculation as it has been done for (2.51) without spin degeneracy leads to

$$\tilde{k}_F = (6\pi^2 \frac{N}{\Omega})^{1/3} \qquad (2.66)$$

where factor 6 replaces factor 3 in the parentheses found above for the paramagnetic case. The relation between k_F and $\tilde{k}_F$ is $\tilde{k}_F = 2^{1/3}k_F$. Thus, the ferromagnetic kinetic energy is related to the paramagnetic one is

$$E_c' = \frac{3}{5}N\frac{\hbar^2\tilde{k}_F^2}{2m} = 2^{2/3}E_c \qquad (2.67)$$

The exchange energies of the two phases are connected by

$$E_{ex}' = -\frac{e^2\Omega}{(2\pi)^3}\tilde{k}_F^4 = 2^{1/3}E_{ex} \qquad (2.68)$$

where we have taken off the factor 2 due to the spin degeneracy in (2.55) and replaced k_F by $\tilde{k}_F$ in (2.63).

The difference of the total energy of the two phases is thus

$$\Delta E = E_{ex}' - E_{ex} + E_c' - E_c = (2^{1/3} - 1)E_{ex} + (2^{2/3} - 1)E_c$$
$$= -\frac{0.916}{r_s}(2^{1/3} - 1) + \frac{2.21}{r_s^2}(2^{2/3} - 1) \qquad (2.69)$$

The ferromagnetic phase is favorable if $\Delta E < 0$, namely if

$$r_s > 5.46 = r_s^c \tag{2.70}$$

We conclude that when the electron density is smaller than that corresponding to r_s^c the electron gas undergoes a transition to the ferromagnetic state.

2.5   Conclusion

We have considered in this chapter a gas of interacting electrons. The introduction of a determinant wave function to take into account the fermion character of the electrons has allowed us to calculate the electron interaction energy in the Hartree-Fock approximation. This interaction energy has two terms: the direct-interaction term (or Coulomb term) and the exchange-interaction term. While the first term does not depend on the electron spins, the exchange term is not zero only for pairs of parallel spins. It is at the origin of the ferromagnetic phase which exists for dilute gas of interacting electrons. The exchange interaction explains moreover the electron distribution in atomic orbitals.

The Hartree-Fock approximation gives an energy correction at the first order in term of interaction V between electrons. Other methods such as the second quantization and the Green's function allow us to calculate higher-order corrections.

2.6   Problems

Problem 1. System of two electrons — Fermi hole:

    a) We consider two electrons of a He atom. The Coulomb interaction between them at positions $\vec{r}$ and $\vec{r}'$ is of the form $V(\vec{r} - \vec{r}') = e^2/|\vec{r} - \vec{r}'|$ (e: electron charge). Write down the determinant wave function for this two-electron He atom in the ground state . Write down the integral allowing to calculate the total energy in terms of individual wave functions $\varphi_{1s}(\vec{r})$.

    b) We consider now two electrons in a box of volume $\Omega = L^3$ where L is the linear dimension of the box. Using the plane wave functions for the electrons, show that the probability to find two electrons of opposite spins does not depend on their relative distance $\vec{r} - \vec{r}'$, but the probability to find two electrons of parallel spins depends on their distance. The small probability at short distance is called the "Fermi hole".

Problem 2. Theorem of Koopmann:

Consider a system of $2N$ electrons with the following Hamiltonian

$$\mathcal{H}_{2N} = \sum_{i=1}^{2N} \frac{\vec{p}_i^2}{2m} + \frac{1}{2} \sum_{i,j=1}^{2N} \frac{e^2}{|\vec{r}_i - \vec{r}_j|} \qquad (2.71)$$

a) Using a determinant wave function for Ψ_{2N}, calculate the average energy $E_{2N} \equiv\, <\Psi_{2N}|\mathcal{H}_{2N}|\Psi_{2N}>$ of the ground state as functions of $T(i)$, $K(i,j)$ and $J(i,j)$ defined by

$$T(i) =\, < \varphi_i(\vec{r}_i)|\frac{\vec{p}_i^2}{2m}|\varphi_i(\vec{r}_i) >$$

$$K(i,j) = e^2 < \varphi_i(\vec{r}_i)\varphi_j(\vec{r}_j)|\frac{1}{|\vec{r}_i - \vec{r}_j|}|\varphi_i(\vec{r}_i)\varphi_j(\vec{r}_j) >$$

$$J(i,j) = e^2 < \varphi_i(\vec{r}_i)\varphi_j(\vec{r}_j)|\frac{1}{|\vec{r}_i - \vec{r}_j|}|\varphi_i(\vec{r}_j)\varphi_j(\vec{r}_i) >$$

where $\varphi_i(\vec{r})$ is the individual wave function of the state i.

b) Calculate the work W necessary to remove one electron from the system ($W = E_{2N-1} - E_{2N}$), supposing that the states of the remaining electrons do not change. Show that $W = -\lambda_i$ where λ_i is the energy of an electron given by the Hartree-Fock equation.

Problem 3. Screened Coulomb potential, Thomas-Fermi approximation:

Consider an extra charge q (trial charge) placed in an electron gas. The density of the gas around the charge reacts in a way so as to screen the effect of q. This phenomenon is called "screening effect".

a) Write down the Poisson equation for the potential $\varphi(\vec{r})$ at $\vec{r}$ from the trial charge.

b) In the perturbed region around the trial charge, the chemical potential is given by $\mu = \epsilon_F(\vec{r}) - e\varphi(\vec{r})$ ($\epsilon_F(\vec{r})$: local Fermi energy, $-e$: electron charge). One supposes that the perturbation is small and spherical, show that

$$\varphi(r) = \frac{q\exp(-\lambda r)}{r} \qquad (2.72)$$

where λ given by

$$\lambda^2 = \frac{4n_0^{1/3}}{a_H} \qquad (2.73)$$

with a_H (Bohr's radius)$=\hbar^2/me^2$ and n_0 the non perturbed electron density. Equation (2.72) is known as the screened Coulomb potential.

Problem 4. Paradox of the Hartree-Fock approximation:

a) Show that the density of states $\rho(\epsilon_i)$ of an electron calculated using the Hartree-Fock result (2.48) is equal to 0 at $k_i = k_F$. Calculate the electron effective mass m^*.

b) Introduce the screening factor $e^{-\mu r}$ to the Coulomb potential where μ is a screening constant [see Eq. (2.72)]. Using (2.44) without taking the limit $\mu \to 0$, calculate the electron energy using the method similar to that used to obtain (2.48). Show that the density of states does not diverge at $k_i = k_F$. Calculate the effective mass and the density of states at $k_i = k_F$.

Problem 5. Hydrogen molecule:

Consider a hydrogen molecule composed of two atoms A and B having respectively electrons a and b. The distance between the nucleus is R_{AB}, that between the electrons is R_{ab}, and the same notations are used for other distances.

a) Show that the separation of two energy levels of the molecular state is written as

$$\Delta E = -2(J' - \alpha^2 Q)/(1 - \alpha^4) \tag{2.74}$$

where

$$\alpha = \int \varphi_A^*(\vec{r})\varphi_B(\vec{r})d\vec{r} \tag{2.75}$$

$$Q = \int \varphi_A^*(\vec{r}_1)\varphi_B^*(\vec{r}_2)\left(\frac{e^2}{R_{ab}} - \frac{e^2}{R_{Ab}} - \frac{e^2}{R_{Ba}}\right)$$
$$\times \varphi_A(\vec{r}_1)\varphi_B(\vec{r}_2)d\vec{r}_1 d\vec{r}_2 \tag{2.76}$$

$$J' = \int \varphi_A^*(\vec{r}_1)\varphi_B^*(\vec{r}_2)\left(\frac{e^2}{R_{ab}} - \frac{e^2}{R_{Ab}} - \frac{e^2}{R_{Ba}}\right)$$
$$\times \varphi_A(\vec{r}_2)\varphi_B(\vec{r}_1)d\vec{r}_1 d\vec{r}_2 \tag{2.77}$$

$\varphi_A(\vec{r})$ and $\varphi_B(\vec{r})$ being the individual wave functions of the electrons of A and B in their atomic state.

b) Show that $\Delta E = -2J$ where J is the exchange interaction of the Heisenberg model $\mathcal{H} = -2J\vec{S}_a \cdot \vec{S}_b$, $\vec{S}_a$ and $\vec{S}_b$ being the spins of electrons a and b, respectively.

Chapter 3

Magnetic Exchange Interactions

We show in this chapter the origin of exchange interactions between spins in crystalline solids. One of the most popular models for spin-spin interactions is the Heisenberg model. This model along with other spin models such as the Ising, XY and Potts models have been largely used to study magnetic properties of materials.

For a demonstration commodity, we shall use the second quantization method. We briefly introduce it in this chapter for that purpose, but we also recommend it for the study of spin-wave excitations shown in chapter 5.

3.1 Introduction

The method of second quantization is very useful in the study of systems of weakly interacting particles. In particular, the second quantization is an efficient tool to describe collective elementary excitations such as phonons and magnons.

3.1.1 *Second Quantization*

To describe correctly a system of interacting particles, we have to include the interaction in the Schrödinger equation. In principle, the wave function contains all the information on the system. However, a solution of the Schrödinger equation is often impossible, either by a complicated form of the potential or by a large number of particles. The second quantization provides a technique to avoid the Schrödinger equation. Of course, we can show the equivalence of the two methods by going from the Schrödinger equation to the second quantization by just changing the way to describe the microscopic states of the system.

The second quantization method allows to transform the Hamiltonian in terms of creation and annihilation operators as seen below. It has the advantage that the Bose-Einstein or Fermi-Dirac statistics of the particles are implicitly incorporated in the Hamiltonian through the operators so that we do not need to use cumbersome products of individual wave functions to construct boson symmetric or fermion antisymmetric wave functions for the system (see chapter 2).

We recall that the first quantization is the quantization of the particle energy by the boundary conditions applied to the solution of the Schrödinger equation. The wave function is not an operator. In the second quantization, the wave function is replaced by a field operator which is in fact a quantization of the wave function.

We consider a system of N identical, indiscernible particles. The Hamiltonian is written as

$$\mathcal{H} = \sum_i H_i + \frac{1}{2} \sum_{i,j} V(\vec{r}_i, \vec{r}_j) \qquad (3.1)$$

where H_i is a single-particle term such as the kinetic energy of the particle i and $V(\vec{r}_i, \vec{r}_j)$ the interaction between two particles at $\vec{r}_i$ and $\vec{r}_j$. Since the particles are identical and indiscernible, we can imagine that they have the same "list" of individual states: each of them takes one state of the list. A state i is characterized by some physical parameters such as the wave vector and the spin state, $\vec{k}_i$ and σ_i ($i = 1, ..., N$). This state of the list is occupied by n_i particles. We can define a microscopic state of the system by the numbers of particles $\{n_i\}$ in the individual states $(\vec{k}_i, \sigma_i)$, $i = 1, ..., N$. All possible different particle distributions $\{n_i\}$ constitute the ensemble of microscopic states of the system. We say that the system is defined by a "state vector" given by

$$|\Psi> = |n_1, n_2, ..., n_k, ..., n_N> \qquad (3.2)$$

where n_k is the number of particles occupying the individual state k. This state vector replaces the wave function of the Schrödinger equation. As for wave functions, we impose that the state vectors form a complete set of orthogonal states. We have

$$< \Psi'|\Psi> = < n_1', ..., n_k', ...|n_1, ..., n_k, ... > = \delta_{n_1', n_1}...\delta_{n_k', n_k}... \qquad (3.3)$$

where δ_{n_k', n_k} is the Kronecker symbol.

In the first quantization, the postulate on the symmetry of the wave function allows us to distinguish bosons and fermions: for bosons the permutation of two particles in their states does not change the sign of the corresponding wave function, while for fermions the permutation does change its sign. One of the consequences of the symmetry postulate is that in the case of bosons an individual state can contain any number of particles while in the case of fermions, each individual state can have zero or one particle (see Appendix A). This is known as the Pauli's principle. In the method of second quantization, it is the symmetry of the operators which allows us to distinguish bosons and fermions as we will see in the following.

3.1.2 *Bosons*

We introduce the operators a_k and a_k^+ by the following relations

$$a_k^+|\Psi> = \sqrt{n_k + 1}|..., n_k + 1, ... > \tag{3.4}$$

$$a_k|\Psi> = \sqrt{n_k}|..., n_k - 1, ... > \tag{3.5}$$

As seen in the kets, operator a_k^+ creates a particle while operator a_k destroys a particle in the state k when they operate on $|\Psi>$. For this reason, they are called creation and annihilation operators, respectively. By the above definitions, we see that

$$a_k^+ a_k|\Psi> = n_k|..., n_k, ... > \tag{3.6}$$

This relation shows that operator $a_k^+ a_k$ has the eigenvalue n_k which is the number of particles in the state k. We call therefore $a_k^+ a_k$ operator "occupation number".

In addition, using (3.4) and (3.5), we get

$$a_k a_k^+|\Psi> = (n_k + 1)|..., n_k, ... > \tag{3.7}$$

Comparing this relation to (3.6), we obtain

$$a_k a_k^+ - a_k^+ a_k = 1 \tag{3.8}$$

Now, if $k \neq k'$, by using (3.3) we have

$$< \Psi|a_{k'} a_k^+|\Psi> = \sqrt{(n_k + 1)n_{k'}} < ..., n_k, n_{k'}, ...|..., n_k + 1, n_{k'} - 1, ... >$$

$$= 0 \tag{3.9}$$

$$< \Psi|a_k^+ a_{k'}|\Psi> = \sqrt{(n_k + 1)n_{k'}} < ..., n_k, n_{k'}, ...|..., n_k + 1, n_{k'} - 1, ... >$$

$$= 0 \tag{3.10}$$

Combining with (3.8) we can write

$$[a_{k'}, a_k^+] \equiv a_{k'}a_k^+ - a_k^+ a_{k'} = \delta_{k,k'} \tag{3.11}$$

We can show in the same manner that for arbitrary k and k', we have

$$[a_{k'}, a_k] = 0 \quad , \quad [a_{k'}^+, a_k^+] = 0 \tag{3.12}$$

Relations (3.11) and (3.12) are called "commutation relations".

We can show that [50] in the case of bosons Hamiltonian (3.1) can be written in the second quantization as

$$\hat{\mathcal{H}} = \sum_{p,r} \langle r|H|p \rangle a_r^+ a_p + \frac{1}{2} \sum_{pqrs} \langle rs|V|pq \rangle a_r^+ a_s^+ a_p a_q \tag{3.13}$$

where

$$\langle r|H|p \rangle = \int d\vec{r}\phi_r^*(\vec{r})H(\vec{r})\phi_p(\vec{r}) \tag{3.14}$$

$$\langle rs|V|pq \rangle = \int\int d\vec{r}d\vec{r}'\,\phi_r^*(\vec{r})\phi_s^*(\vec{r}')V(\vec{r},\vec{r}')\phi_p(\vec{r})\phi_q(\vec{r}') \tag{3.15}$$

The wave function $\phi_i(\vec{r})$ describes the individual state i of the particle at $\vec{r}$. For example, in the case of a plane wave we have $\phi_i(\vec{r}) = \exp(i\vec{k}_i \cdot \vec{r})/\sqrt{\Omega}$ where $\vec{k}_i$ is the wave vector and Ω the system volume.

3.1.3 *Fermions*

In the case of fermions, we define creation and annihilation operators b_f^+ and b_f by

$$b_f|\Psi \rangle = b_f|...n_f... \rangle = (-1)^{[f]}n_f|...n_f - 1... \rangle \tag{3.16}$$

$$b_f^+|\Psi \rangle = b_f^+|...n_f... \rangle = (-1)^{[f]}(1 - n_f)|...n_f + 1... \rangle \tag{3.17}$$

where $[f]$ is, by convention, the number of particles occupying the states on the left of the state f in the ket. It is noted that in some books the coefficients in front of the ket of (3.16)-(3.17) are given by $\sqrt{n_f}$ and $\sqrt{1 - n_f}$ instead of n_f and $(1 - n_f)$. However, we can verify that they are equivalent because n_f is 0 or 1. We have

$$b_f b_g|...n_f...n_g...) = (-1)^{[g]}n_g b_f|...n_f,...n_g - 1,... \rangle$$
$$= (-1)^{[g]+[f]}n_g n_f|...n_f - 1,...n_g - 1,... \rangle$$
$$b_g b_f|...n_f...n_g...) = (-1)^{[f]}n_f b_g|...n_f - 1...n_g,... \rangle$$
$$= (-1)^{[f]+[g]-1}n_f n_g|...n_f - 1,...n_g - 1,... \rangle$$

from which we have $b_f b_g = -b_g b_f$, or equivalently

$$[b_f, b_g]_+ \equiv b_f b_g + b_g b_f = 0 \tag{3.18}$$

In the same manner, we obtain for arbitrary f and g

$$\left[b_f^+, b_g^+\right]_+ = 0 \tag{3.19}$$

and

$$\left[b_f^+, b_g\right]_+ = 0 \quad \text{if} \quad f \neq g \tag{3.20}$$

Now if $f = g$, we have

$$\begin{aligned}
b_f^+ b_f |...n_f... > &= (-1)^{[f]} n_f b_f |...n_f - 1... > \\
&= (-1)^{2[f]} n_f (1 - n_f + 1) |...n_f... > \\
&= n_f (2 - n_f) |...n_f... > = n_f |...n_f... >
\end{aligned} \tag{3.21}$$

where in the last line, we have used $n_f(2 - n_f) = n_f$ because

$$n_f = \left\{ \begin{array}{l} 0 \Rightarrow n_f(2 - n_f) = 0 \\ 1 \Rightarrow n_f(2 - n_f) = 1 \end{array} \right\} \Rightarrow n_f(2 - n_f) = n_f \tag{3.22}$$

We call $b_f^+ b_f$ operator "occupation number" because its eigenvalue when operating on the ket is n_f. Besides,

$$\begin{aligned}
b_f b_f^+ |...n_f...) &= (-1)^{[-f]+[f]} (n_f + 1)(1 - n_f) |...n_f... > \\
&= (1 + n_f)(1 - n_f) |...n_f... >
\end{aligned} \tag{3.23}$$

$$\begin{array}{l} \text{If } n_f = 0 \Rightarrow (1 + n_f)(1 - n_f) = 1 \\ \text{If } n_f = 1 \Rightarrow (1 + n_f)(1 - n_f) = 0 \end{array} \right\} \Rightarrow (1 + n_f)(1 - n_f) = 1 - n_f$$

We deduce

$$b_f b_f^+ |...n_f... > = (1 - n_f) |..n_f... > \tag{3.24}$$

Comparing (3.24) to (3.21), we obtain $b_f b_f^+ = 1 - b_f^+ b_f$, namely $\left[b_f, b_f^+\right]_+ = 1$. Combining with (3.20), we can write

$$\left[b_g, b_f^+\right]_+ = \delta_{g,f} \tag{3.25}$$

Relations (3.18), (3.19) and (3.25) are called "fermion anticommutation relations".

Hamiltonian (3.1) in the case of fermions is written in the second quantization as [50]

$$\hat{\mathcal{H}} = \sum_{i,k} \langle k|H|i\rangle b_k^+ b_i - \frac{1}{2} \sum_{i,j,k,l} \langle kl|V|ij\rangle b_k^+ b_l^+ b_i b_j \qquad (3.26)$$

where

$$\langle k|H|i\rangle = \int d\vec{r}\, \phi_k^*(\vec{r}) H(\vec{r}) \phi_i(\vec{r}) \qquad (3.27)$$

$$\langle kl|V|ij\rangle = \int \int d\vec{r} d\vec{r}'\, \phi_k^*(\vec{r}) \phi_l^*(\vec{r}') V(\vec{r},\vec{r}') \phi_i(\vec{r}) \phi_j(\vec{r}') \qquad (3.28)$$

Due to the anticommutation of the operators, we should respect the order of the operators as well as that of the arguments $\vec{r}$ and $\vec{r}'$ of ϕ functions in $\langle kl|V|ij\rangle$. A permutation of the operators should obey the anticommutation relations (3.18), (3.19) and (3.25).

Hamiltonians (3.13) and (3.26) are very useful in the study of systems of weakly interacting particles [79].

Remarks:

(1) The state of the fermion system, written by a ket containing the occupation numbers of occupied individual states [see (3.2)], can be expressed by the operators b and b^+ as

$$|n_{k_1} n_{k_2} ... n_{k_N}\rangle \equiv b_{k_1}^+ b_{k_2}^+ ... b_{k_N}^+ |000...\rangle$$

where $|000...\rangle$ represents the vacuum state and $n_{k_i} = 1$ $(i = 1, ..., N)$.

(2) We have the following relations

$$b_k|...0_k...\rangle = 0$$
$$b_k^+|...0_k...\rangle = |...1_k...\rangle$$
$$b_k^+|...1_k...\rangle = 0$$
$$b_k|...1_k...\rangle = |...0_k...\rangle$$

(3) If the individual state of a fermion is characterized by its wave vector $\vec{k}$ and its spin σ, then the anticommutation relations (3.18), (3.19) and (3.25) are explicitly written as

$$[b_{k\sigma}, b_{k'\sigma'}^+]_+ = \delta_{k,k'}\delta_{\sigma,\sigma'}$$
$$[b_{k\sigma}, b_{k'\sigma'}]_+ = [b_{k\sigma}^+, b_{k'\sigma'}^+]_+ = 0, \quad \forall(k\sigma,\ k'\sigma')$$

(4) The ensemble of state vectors $|n_{k_1} n_{k_2} ... n_{k_N}\rangle$, with $n_{k_i}(i = 1, N) = 0$ or 1, constitute a complete set of orthonormal states :

$$\langle n'_{k_1} n'_{k_2} ... n'_{k_i} ... | n_{k_1} n_{k_2} ... n_{k_i} ... \rangle = \delta_{n'_{k_1}, n_{k_1}} \delta_{n'_{k_2}, n_{k_2}} ... \delta_{n'_{k_i}, n_{k_i}} ...$$

$$\sum_{n_{k_1} ...} |n_{k_1} n_{k_2} ... \rangle \langle n_{k_1} n_{k_2} ...| = 1 \quad (n_{k_i} = 0 \text{ or } 1)$$

We have similar results for bosons. Note however that the sum over $n_{k_1}, n_{k_2}, ...$ is performed with all possible values of each n_{k_i}: for instance, in a system of N bosons, one has $n_{k_i} = 0, 1, ..., N$.

(5) It is noted that the state vector $|n_{k_1} n_{k_2} ... n_{k_N}\rangle$ is not a wave function.

3.1.4 *Field operators*

Using $b_{k\sigma}^+$ and $b_{k\sigma}$, the destruction and the creation of a fermion in the state $(\vec{k}, \sigma)$ at $\vec{r}$ are given by the expressions

$$\frac{1}{\Omega^{1/2}} b_{k\sigma} e^{i\vec{k}.\vec{r}} \quad \text{and} \quad \frac{1}{\Omega^{1/2}} b_{k\sigma}^+ e^{-i\vec{k}.\vec{r}}$$

Note that in these notations we have combined operators $b_{k\sigma}^+$ and $b_{k\sigma}$ with the corresponding wave functions, namely $\frac{1}{\Omega^{1/2}} e^{i\vec{k}.\vec{r}}$ where Ω is the system volume. We introduce now the "field operators" defined by

$$\hat{\Psi}_\sigma(\vec{r}) = \frac{1}{\Omega^{1/2}} \sum_{\vec{k}} b_{k\sigma} e^{i\vec{k}.\vec{r}} \tag{3.29}$$

$$\hat{\Psi}_\sigma^+(\vec{r}) = \frac{1}{\Omega^{1/2}} \sum_{\vec{k}} b_{k\sigma}^+ e^{-i\vec{k}.\vec{r}} \tag{3.30}$$

$\hat{\Psi}_\sigma(\vec{r})$ and $\hat{\Psi}_\sigma^+(\vec{r})$ are called annihilation and creation field operators at $\vec{r}$. From these relations, we have

$$b_{k\sigma} = \frac{1}{\Omega^{1/2}} \int d\vec{r} \hat{\Psi}_\sigma(\vec{r}) e^{-i\vec{k}.\vec{r}} \tag{3.31}$$

$$b_{k\sigma}^+ = \frac{1}{\Omega^{1/2}} \int d\vec{r} \hat{\Psi}_\sigma^+(\vec{r}) e^{i\vec{k}.\vec{r}} \tag{3.32}$$

Using the anticommutation relations of b^+ and b, (3.18)-(3.19) and (3.25), we obtain

$$[\hat{\Psi}_\sigma(\vec{r}_1), \hat{\Psi}_{\sigma'}^+(\vec{r}_2)]_+ = \delta_{\vec{r}_1, \vec{r}_2} \delta_{\sigma, \sigma'} \tag{3.33}$$

$$[\hat{\Psi}_\sigma(\vec{r}_1), \hat{\Psi}_{\sigma'}(\vec{r}_2)]_+ = [\hat{\Psi}_\sigma^+(\vec{r}_1), \hat{\Psi}_{\sigma'}^+(\vec{r}_2)]_+ = 0 \tag{3.34}$$

The definitions for bosons are similar to (3.29)-(3.30) with $b_{k\sigma}^+$ and $b_{k\sigma}$ replaced by a_k^+ and a_k (without spin). In addition, in the boson case, we have the commutation relations between $\hat{\Psi}(\vec{r})$ and $\hat{\Psi}^+(\vec{r})$ instead of (3.33)-(3.34).

In what follows, we limit ourselves to the fermion case. The boson case can be obtained in the same manner. We can express various physical quantities in terms of field operators $\hat{\Psi}_\sigma(\vec{r})$ and $\hat{\Psi}_\sigma^+(\vec{r})$. We have seen above that the Hamiltonian of the first quantization is expressed in the second quantization by (3.26) with the help of operators b^+ and b. We can now express it in terms of $\hat{\Psi}$ and $\hat{\Psi}^+$ as follows, putting $\Omega = 1$,

$$
\begin{aligned}
\hat{\mathcal{H}} = &\sum_\sigma \int d\vec{r}\,\hat{\Psi}_\sigma^+(\vec{r})H(\vec{r})\hat{\Psi}_\sigma(\vec{r}) \\
&-\frac{1}{2}\sum_{\sigma,\sigma'} \int\int d\vec{r}_1 d\vec{r}_2\,\hat{\Psi}_\sigma^+(\vec{r}_1)\hat{\Psi}_{\sigma'}^+(\vec{r}_2)V(\vec{r}_1,\vec{r}_2)\hat{\Psi}_\sigma(\vec{r}_1)\hat{\Psi}_{\sigma'}(\vec{r}_2)
\end{aligned} \quad (3.35)
$$

The operator "total occupation number" is given by

$$
\hat{N} = \sum_{i,\sigma} b_{i\sigma}^+ b_{i\sigma} \quad (3.36)
$$

Using (3.31)-(3.32), $\hat{N}$ can be rewritten as

$$
\begin{aligned}
\hat{N} &= \sum_{i,\sigma} \int d\vec{r}_1 d\vec{r}_2\,\hat{\Psi}_\sigma^+(\vec{r}_2)e^{i\vec{k}_i\cdot\vec{r}_2}e^{-i\vec{k}_i\cdot\vec{r}_1}\hat{\Psi}_\sigma(\vec{r}_1) \\
&= \sum_{i,\sigma} \int d\vec{r}_1 d\vec{r}_2\,\hat{\Psi}_\sigma^+(\vec{r}_2)e^{i\vec{k}_i\cdot(\vec{r}_2-\vec{r}_1)}\hat{\Psi}_\sigma(\vec{r}_1) \\
&= \sum_\sigma \int d\vec{r}_1 d\vec{r}_2\,\hat{\Psi}_\sigma^+(\vec{r}_2)\delta_{\vec{r}_2,\vec{r}_1}\hat{\Psi}_\sigma(\vec{r}_1) = \sum_\sigma \int d\vec{r}_1\,\hat{\Psi}_\sigma^+(\vec{r}_1)\hat{\Psi}_\sigma(\vec{r}_1) \\
&= \sum_\sigma \int \hat{\rho}_\sigma(\vec{r}_1)d\vec{r}_1
\end{aligned} \quad (3.37)
$$

where

$$
\hat{\rho}_\sigma(\vec{r}) \equiv \hat{\Psi}_\sigma^+(\vec{r})\hat{\Psi}_\sigma(\vec{r}) = \quad \text{operator density of particles} \quad (3.38)
$$

Operator density of states $\hat{\rho}(\vec{k})$ is given by the Fourier transform

$$\hat{\rho}(\vec{k}) = \sum_\sigma \int d\vec{r} \hat{\rho}_\sigma(\vec{r}) e^{-i\vec{k}\cdot\vec{r}} = \sum_\sigma \sum_{\vec{k}_1,\vec{k}_2} \int d\vec{r} e^{i(\vec{k}_2-\vec{k}_1)\cdot\vec{r}} b^+_{\vec{k}_1\sigma} b_{\vec{k}_2\sigma} e^{-i\vec{k}\cdot\vec{r}}$$

$$= \sum_\sigma \sum_{\vec{k}_1,\vec{k}_2} \int d\vec{r} e^{i(\vec{k}_2-\vec{k}_1-\vec{k})\cdot\vec{r}} b^+_{\vec{k}_1\sigma} b_{\vec{k}_2\sigma}$$

$$= \sum_\sigma \sum_{\vec{k}_1,\vec{k}_2} \delta_{\vec{k}_2-\vec{k}_1-\vec{k},0} b^+_{\vec{k}_1\sigma} b_{\vec{k}_2\sigma} = \sum_\sigma \sum_{\vec{k}_1} b^+_{\vec{k}_1\sigma} b_{\vec{k}_1+\vec{k},\sigma} \tag{3.39}$$

where we have used $\vec{k}_1$ instead of k_1 etc. for subscripts of b and b^+ operators to avoid confusion when performing the sums. We apply in the following the second quantization method to show the origin of magnetic interactions in materials.

3.2 Origin of ferromagnetism

In this paragraph, we show the origin of the magnetic interaction which leads to the Heisenberg spin model given in (4.1). We suppose that the reader has read chapter 2 on the Hartree-Fock approximation.

3.2.1 *Heisenberg model*

We consider the Coulomb interaction between two electrons written in the second quantization [see (3.35)]

$$\hat{\mathcal{H}} = -\frac{1}{2} \sum_{\sigma;\sigma'} \int \int \hat{\psi}^+_\sigma(\vec{r}_1) \hat{\psi}^+_{\sigma'}(\vec{r}_2) \frac{e^2}{|\vec{r}_1-\vec{r}_2|} \hat{\psi}_\sigma(\vec{r}_1) \hat{\psi}_{\sigma'}(\vec{r}_2) d\vec{r}_1 d\vec{r}_2 \tag{3.40}$$

where $\hat{\psi}_\sigma$ and $\hat{\psi}^+_\sigma$ are field operators defined by

$$\hat{\psi}_\sigma(\vec{r}) = \sum_{m,n} b_{mn\sigma} \varphi_{nm}(\vec{r}) \tag{3.41}$$

$$\hat{\psi}^+_\sigma(\vec{r}) = \sum_{m,n} b^+_{mn\sigma} \varphi^+_{nm}(\vec{r}) \tag{3.42}$$

where $\varphi_{nm}(\vec{r})$ and $\varphi^+_{nm}(\vec{r})$ are wave functions of orbital m at the site n of the crystal, b and b^+ fermion annihilation and creation operators. The wave functions φ_{nm} constitute an orthogonal set. Equation (3.40) becomes

$$\hat{\mathcal{H}} = -\frac{1}{2} \sum \langle \varphi_{n_1 m_1 \sigma_1}(\vec{r}_1) \varphi_{n_2 m_2 \sigma_2}(\vec{r}_2) | \frac{e^2}{|\vec{r}_1-\vec{r}_2|} | \varphi_{n_3 m_3 \sigma_3}(\vec{r}_1) \varphi_{n_4 m_4 \sigma_4}(\vec{r}_2) \rangle$$

$$\times b^+_{n_1 m_1 \sigma_1} b^+_{n_2 m_2 \sigma_2} b_{n_3 m_3 \sigma_3} b_{n_4 m_4 \sigma_4} \tag{3.43}$$

where the sum runs over $(n_1, m_1, \sigma_1, \cdots, n_4, m_4, \sigma_4)$.

If $n_1 = n_2 = n_3 = n_4$, the interactions are between electrons of the same site. Equation (3.43) is the origin of the Hund's empirical rules. In addition, if $m_1 = m_2 = m_3 = m_4$, this equation is nothing but the Coulomb term in the Hubbard Hamiltonian which will be shown below. For simplicity, we suppose one electron per site and one orbital per electron in the following.

If $n_1 = n_3$ and $n_2 = n_4$, the Coulomb term is given by

$$\hat{\mathcal{H}}_c = -\frac{1}{2} \sum_{n_1, n_2} \langle n_1 n_2 | \frac{e^2}{|r_{12}|} | n_2 n_1 \rangle \sum_{\sigma_1, \sigma_2} b_{n_1 \sigma_1}^+ b_{n_2 \sigma_2}^+ b_{n_1 \sigma_1} b_{n_2 \sigma_2}.$$

If $n_1 = n_4$ and $n_2 = n_3$ (by consequence, $\sigma_1 = \sigma_2$), the exchange term becomes

$$\begin{aligned}
\hat{\mathcal{H}}_{ex} &= -\frac{1}{2} \sum_{n_1, n_2, \sigma_1 = \sigma_2} \langle n_1 n_2 | \frac{e^2}{|r_{12}|} | n_2 n_1 \rangle b_{n_1 \sigma_1}^+ b_{n_2 \sigma_2}^+ b_{n_1 \sigma_1} b_{n_2 \sigma_2} \\
&= -\frac{1}{2} \sum_{n_1, n_2, \sigma_1 = \sigma_2} \langle n_1 n_2 | \frac{e^2}{|r_{12}|} | n_2 n_1 \rangle b_{n_1 \sigma_1}^+ b_{n_1 \sigma_1} b_{n_2 \sigma_2}^+ b_{n_2 \sigma_2} \\
&= -\frac{1}{2} \sum_{n_1, n_2} J_{n_1 n_2} \sum_{\sigma_1, \sigma_2, \sigma_1 = \sigma_2} b_{n_1 \sigma_1}^+ b_{n_1 \sigma_1} b_{n_2 \sigma_2}^+ b_{n_2 \sigma_2} \quad (3.44)
\end{aligned}$$

where

$$J_{n_1 n_2} \equiv \langle n_1 n_2 | \frac{e^2}{|r_{12}|} | n_2 n_1 \rangle. \quad (3.45)$$

$$\begin{aligned}
\sum_{\sigma_1, \sigma_2, \sigma_1 = \sigma_2} b_{n_1 \sigma_1}^+ b_{n_1 \sigma_1} b_{n_2 \sigma_2}^+ b_{n_2 \sigma_2} &= b_{n_1 \uparrow}^+ b_{n_1 \uparrow} b_{n_2 \uparrow}^+ b_{n_2 \uparrow} + b_{n_1 \downarrow}^+ b_{n_1 \downarrow} b_{n_2 \downarrow}^+ b_{n_2 \downarrow} \\
&= \frac{1}{2} \left(b_{n_1 \uparrow}^+ b_{n_1 \uparrow} + b_{n_1 \downarrow}^+ b_{n_1 \downarrow} \right) \\
&\quad \times \left(b_{n_2 \uparrow}^+ b_{n_2 \uparrow} + b_{n_2 \downarrow}^+ b_{n_2 \downarrow} \right) \\
&\quad + \frac{1}{2} \left(b_{n_1 \uparrow}^+ b_{n_1 \uparrow} - b_{n_1 \downarrow}^+ b_{n_1 \downarrow} \right) \\
&\quad \times \left(b_{n_2 \uparrow}^+ b_{n_2 \uparrow} - b_{n_2 \downarrow}^+ b_{n_2 \downarrow} \right) \\
&\quad + b_{n_1 \uparrow}^+ b_{n_1 \downarrow} b_{n_2 \downarrow}^+ b_{n_2 \uparrow} \\
&\quad + b_{n_1 \downarrow}^+ b_{n_1 \uparrow} b_{n_2 \uparrow}^+ b_{n_2 \downarrow} \quad (3.46)
\end{aligned}$$

where the last two terms have been added. Note that these terms do not affect the result because their averages are zero in the diagonal representation: $\langle \hat{\psi} | b_{n_1 \uparrow}^+ b_{n_1 \downarrow} b_{n_2 \downarrow}^+ b_{n_2 \uparrow} | \hat{\psi} \rangle = \langle \hat{\psi} | b_{n_1 \downarrow}^+ b_{n_1 \uparrow} b_{n_2 \uparrow}^+ b_{n_2 \downarrow} | \hat{\psi} \rangle = 0$.

We define next the following spin operators

$$S_n^z = \frac{1}{2}(b_{n\uparrow}^+ b_{n\uparrow} - b_{n\downarrow}^+ b_{n\downarrow}) \tag{3.47}$$

$$S_n^+ \equiv S_n^x + iS_n^y = b_{n\uparrow}^+ b_{n\downarrow} \tag{3.48}$$

$$S_n^- \equiv S_n^x - iS_n^y = b_{n\downarrow}^+ b_{n\uparrow} \tag{3.49}$$

As we suppose one electron per site, we have

$$\left(b_{n_1\uparrow}^+ b_{n_1\uparrow} + b_{n_1\downarrow}^+ b_{n_1\downarrow}\right) = 1$$

$$\left(b_{n_2\uparrow}^+ b_{n_2\uparrow} + b_{n_2\downarrow}^+ b_{n_2\downarrow}\right) = 1$$

The right-hand side of (3.46) becomes

$$\frac{1}{2} + 2\vec{S}_{n_1}.\vec{S}_{n_2} \tag{3.50}$$

By using

$$\begin{aligned}
\vec{S}_{n_1}.\vec{S}_{n_2} &= S_{n_1}^z S_{n_2}^z + S_{n_1}^y S_{n_2}^y + S_{n_1}^x S_{n_2}^x \\
&= S_{n_1}^z S_{n_2}^z + \frac{1}{2}\left(S_{n_1}^+ S_{n_2}^- + S_{n_1}^- S_{n_2}^+\right).
\end{aligned}$$

we rewrite (3.44) as

$$\hat{\mathcal{H}}_{ex} = -\frac{1}{2}\sum_{n_1,n_2} J_{n_1 n_2}\left(\frac{1}{2} + 2\vec{S}_{n_1}.\vec{S}_{n_2}\right)$$

As the double sum $\sum_{n_1,n_2}$ is performed over the sites n_1 et n_2, we added a factor $\frac{1}{2}$ to remove the double counting of each pair (n_1, n_2). We have $\sum_{(n_1,n_2)}$ instead of $\frac{1}{2}\sum_{n_1,n_2}$ where (n_1, n_2) indicates *the pair* $(n_1 n_2)$.
Finally,

$$\hat{\mathcal{H}}_{ex} = -\sum_{(n_1 n_2)} J_{n_1 n_2}\left(\frac{1}{2} + 2\vec{S}_{n_1}.\vec{S}_{n_2}\right) \tag{3.51}$$

The first term does not depend on spins, while the second term does. This term is the Heisenberg model which shall be used later in (4.1).

Hamiltonian (3.51) is thus the origin of ferromagnetism observed in crystals such as CrO_2, $CrBr_3$, $\cdots$. In general, $J_{n_1 n_2}$ is positive:

$$J_{n_1 n_2} = \int \int \varphi_{n_1}^*(\vec{r}_1) \varphi_{n_2}^*(\vec{r}_2) \frac{e^2}{|\vec{r}_1 - \vec{r}_2|} \varphi_{n_1}(\vec{r}_1) \varphi_{n_2}(\vec{r}_2) d\vec{r}_1 d\vec{r}_2 \qquad (3.52)$$

Using

$$\frac{e^2}{|\vec{r}_1 - \vec{r}_2|} = \frac{1}{\Omega} \sum_{\vec{k}} \frac{4\pi e^2}{k^2} e^{i\vec{k}\cdot(\vec{r}_1 - \vec{r}_2)} \qquad (3.53)$$

we get

$$J_{n_1 n_2} = \frac{1}{\Omega} \sum_{k} \frac{4\pi e^2}{k^2} \int \varphi_{n_1}^*(\vec{r}_1) \varphi_{n_2}(\vec{r}_1) e^{-i\vec{k}\cdot\vec{r}_1} d\vec{r}_1 \int \varphi_{n_2}^*(\vec{r}_2) \varphi_{n_1}(\vec{r}_2) e^{i\vec{k}\cdot\vec{r}_2} d\vec{r}_2$$

$$= \frac{1}{\Omega} \sum_{\vec{k}} \frac{4\pi e^2}{k^2} I^2 \qquad (3.54)$$

where

$$I = \int_\Omega \varphi_{n_1}^*(\vec{r}_1) \varphi_{n_2}(\vec{r}_1) e^{-i\vec{k}\cdot\vec{r}_1} d\vec{r}_1 = \int_\Omega \varphi_{n_2}^*(\vec{r}_2) \varphi_{n_1}(\vec{r}_2) e^{-i\vec{k}\cdot\vec{r}_2} d\vec{r}_2$$

Note that these two integrals are identical because the indices and variables are dummy. $J_{n_1 n_2}$ is thus positive. From (3.51), we see that if $\vec{S}_{n_1}$ and $\vec{S}_{n_2}$ are parallel, then the energy is lower. The wave functions $\varphi_n(\vec{r})$ have been supposed to be orthogonal. They are Wannier wave functions constructed from linear combinations of Bloch wave functions.

3.2.2   *Hubbard model: Superexchange*

The Hubbard model has been introduced by Hubbard in 1963 [67] and almost at the same time by Gutzwiller [55] and Kanamori [71]. The Hubbard model is interesting because by changing a few parameters it can describe insulators, conductors and various magnetic states. In one dimension, it can be exactly solved [85]: at half-filling it gives an insulating state for $U > 0$, U being the on-site repulsive interaction defined below. Away from the half-filling, it yields a conducting state. Some simple cases are given as exercises in Problems 3 and 4 of section 3.3. The general Hubbard model cannot be solved without approximations. The reader is referred to reviews and books such as Refs. [45, 54, 97, 114]. In the following, we give some basic properties of the Hubbard model.

We have seen that the direct exchange interaction has an origin in a Coulomb interaction. In the Hubbard model, we suppose that the Coulomb interaction U is between the electrons of opposite spins which belong to the same ion. In addition, we introduce an interaction between the electrons of neighboring ions. The Hubbard Hamiltonian is written as

$$\hat{\mathcal{H}} = \sum_{n,n',\sigma} V_{nn'} b^+_{n'\sigma} b_{n\sigma} + U \sum_n b^+_{n\uparrow} b_{n\uparrow} b^+_{n\downarrow} b_{n\downarrow} \qquad (3.55)$$

where

$$V_{nn'} \equiv \int \varphi^*_{n'}(\vec{r}) \mathcal{H}_{cryst} \varphi_n(\vec{r}) d\vec{r} \qquad (\mathcal{H}_{cryst} = \text{crystalline field}) \quad (3.56)$$

The Hubbard model is often used for d electrons which are not completely localized: they are more or less *itinerant* because of the interaction $V_{nn'}$ between electrons of neighboring ions. In general, U is positive [demonstrations analogous to that for (3.54)]. We deduce

- If $U \gg V$: the crystal is an insulator because electrons prefer to stay on different sites to reduce their interaction energy; the second term in (3.55) is then equal to zero.
- If $U \ll V$: there is a possibility that electrons move from site to site under the action of V, the crystal is then a conductor.

We are interested in the first case where V is considered as a perturbation to U. We show below that the second-order perturbation in V gives rise to a magnetically ordered state. The Hamiltonian of the second order in V is written as

$$\mathcal{H}_2 = \sum_{n,n',\sigma,\sigma'} \frac{|V_{nn'}|^2}{E_0 - E_n} b^+_{n\sigma'} b_{n'\sigma'} b^+_{n'\sigma} b_{n\sigma} \qquad (3.57)$$

where E_0 is the energy of the initial state in which each spin occupies a site and E_n the energy of the excited state where two spins occupy the same site. We thus have $E_0 = V$ and $E_n = U$.

It follows that

$$\mathcal{H}_2 \cong \sum_{n,n',\sigma,\sigma'} \frac{|V_{nn'}|^2}{U} b^+_{n\sigma'} b_{n'\sigma'} b^+_{n'\sigma} b_{n\sigma} \quad \text{with V} \ll \text{U} \qquad (3.58)$$

The spin exchange due to the interaction (3.58) is schematically displayed in Fig. 3.1.

Obviously, the exchange mechanism is possible only if the spins of the sites n and n' are initially antiparallel. Otherwise, the intermediate state is

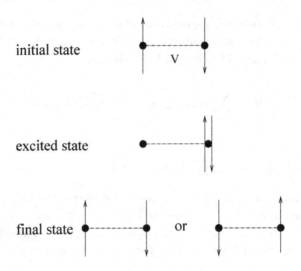

initial state

excited state

final state

Fig. 3.1   Spin exchange due to the second-order perturbation is schematically shown.

forbidden by the Pauli's principle. The sum on the spins in (3.54) clearly shows that fact:

$$\sum_{\sigma,\sigma'} b^+_{n\sigma'} b_{n'\sigma'} b^+_{n'\sigma} b_{n\sigma} = b^+_{n\uparrow} b_{n'\uparrow} b^+_{n'\uparrow} b_{n\uparrow} + b^+_{n\downarrow} b_{n'\downarrow} b^+_{n'\downarrow} b_{n\downarrow}$$

$$+ b^+_{n\uparrow} b_{n'\uparrow} b^+_{n'\downarrow} b_{n\downarrow} + b^+_{n\downarrow} b_{n'\downarrow} b^+_{n'\uparrow} b_{n\uparrow}$$

$$= b^+_{n\uparrow}(1 - b^+_{n'\uparrow} b_{n'\uparrow}) b_{n\uparrow} + b^+_{n\downarrow}(1 - b^+_{n'\downarrow} b_{n'\downarrow}) b_{n\downarrow}$$

$$- b^+_{n\uparrow} b^+_{n'\downarrow} b_{n'\uparrow} b_{n\downarrow} - b^+_{n\downarrow} b^+_{n'\uparrow} b_{n'\downarrow} b_{n\uparrow}$$

$$= (b^+_{n\uparrow} b_{n\uparrow} + b^+_{n\downarrow} b_{n\downarrow})$$

$$- \frac{1}{2}(b^+_{n\uparrow} b_{n\uparrow} + b^+_{n\downarrow} b_{n\downarrow})(b^+_{n'\uparrow} b_{n'\uparrow} + b^+_{n'\downarrow} b_{n'\downarrow})$$

$$- \frac{1}{2}(b^+_{n\uparrow} b_{n\uparrow} - b^+_{n\downarrow} b_{n\downarrow})(b^+_{n'\uparrow} b_{n'\uparrow} - b^+_{n'\downarrow} b_{n'\downarrow})$$

$$- b^+_{n\uparrow} b_{n\downarrow} b^+_{n'\downarrow} b_{n'\uparrow} - b^+_{n\downarrow} b_{n\uparrow} b^+_{n'\uparrow} b_{n'\downarrow} \qquad (3.59)$$

Since there is only one electron per site, one has

$$b^+_{n\uparrow} b_{n\uparrow} + b^+_{n\downarrow} b_{n\downarrow} = b^+_{n'\uparrow} b_{n'\uparrow} + b^+_{n'\downarrow} b_{n'\downarrow} = 1,$$

Equation (3.59) becomes, using (3.47)-(3.51),

$$\sum_{\sigma,\sigma'} b^+_{n\sigma'} b_{n'\sigma'} b^+_{n'\sigma} b_{n\sigma} = \frac{1}{2} - 2\vec{S}_n . \vec{S}_{n'} \qquad (3.60)$$

The Hamiltonian (3.58) is finally rewritten as

$$\mathcal{H}_2 = -\sum_{n,n'} \frac{|V_{nn'}|^2}{U} \left(\frac{1}{2} - 2\vec{S}_n.\vec{S}_{n'} \right) \tag{3.61}$$

We see here that if $\vec{S}_n$ and $\vec{S}_{n'}$ are parallel, namely $\vec{S}_n.\vec{S}_{n'} = \frac{1}{4}$, the energy given by (3.61) is indeed zero. If $\vec{S}_n$ and $\vec{S}_{n'}$ are antiparallel, this energy is minimum. Therefore, the stable state is an antiferromagnetic state. The Hamiltonian (3.61) is called "kinetic exchange interaction".

In the same manner, we can show that the interaction of the fourth order in V is given by

$$\mathcal{H}_4 \to -\frac{|V_{n'n}|^4}{U^3} (\vec{S}_n.\vec{S}_{n'})^2 \tag{3.62}$$

This interaction is called "biquadratic exchange interaction" which is at the origin of the ferromagnetic state observed in some materials such as MnO.

In the case where there are many electrons per site, we can show that the direct and kinetic exchange interactions are given by (see demonstration below)

$$\mathcal{H}_{ex} = -2 \sum_{n_1,n_2} J^{eff}_{n_1 n_2} \vec{S}_{n_1}.\vec{S}_{n_2} \tag{3.63}$$

where the sum is performed on the different pairs (n_1, n_2), and $J^{eff}_{n_1 n_2}$ is defined by

$$J^{eff}_{n_1 n_2} = \frac{2}{(2S^2)} \sum_{m,m'} \left[-\frac{|V^{mm'}_{n_1 n_2}|^2}{U} + J^{mm'}_{n_1 n_2} \right] \tag{3.64}$$

with

$$V^{mm'}_{n_1 n_2} \equiv \int \varphi_m^*(\vec{r}, n_1) \mathcal{H}_{cryst} \varphi_{m'}(\vec{r}, n_2) d\vec{r} \tag{3.65}$$

$$J^{mm'}_{n_1 n_2} \equiv \int \int \varphi_m^*(\vec{r}_1, n_1) \varphi_{m'}^*(\vec{r}_2, n_2) \frac{e^2}{|\vec{r}_1 - \vec{r}_2|} \varphi_m(\vec{r}_2, n_2) \varphi_{m'}(\vec{r}_1, n_1) d\vec{r}_1 d\vec{r}_2 \tag{3.66}$$

where $S = |\vec{S}|$ is the total spin per site. $J^{eff}_{n_1 n_2}$ is called "effective exchange integral". The sums on m and m' are made on different pairs (m, m') of occupied orbitals in the case of a less-than-half-filled shell. In the case of a more-than-half-filled shell, the sums are made on non occupied orbitals.

3.2.3   *Many electrons per site*

In the case where there is an arbitrary number of electrons per site, the direct and kinetic exchange interactions are written as, for sites n_1 and n_2,

$$\hat{\mathcal{H}}_{direct}(n_1, n_2) = -2 \sum_{m,m'} J_{n_1 n_2}^{mm'} \vec{S}_{n1}^m \cdot \vec{S}_{n2}^{m'} \tag{3.67}$$

$$\hat{\mathcal{H}}_{kinetic}(n_1, n_2) = 2 \sum_{m,m'} \frac{|V_{n_1 n_2}^{mm'}|^2}{U} \vec{S}_{n1}^m \cdot \vec{S}_{n2}^{m'} \tag{3.68}$$

where $V_{n_1 n_2}^{mm'}$ and $J_{n_1 n_2}^{mm'}$ are given by (3.65) and (3.66). We see that the denominator of (3.68) is U in spite of the fact that there are ν electrons per site. We can explain this by the following argument: if we suppose that the interaction U is identical for all electron pairs in the excited state, then

- at site n_1 : $\nu - 1$ electrons
- at site n_2 : $\nu + 1$ electrons

hence

- the number of U couplings in the excited state:
  - at site n_1 : $\frac{(\nu-1)!}{2!(\nu-3)!} = \frac{(\nu-1)(\nu-2)}{2}$
  - at site n_2 : $\frac{(\nu+1)!}{2!(\nu-1)!} = \frac{(\nu+1)\nu}{2}$
- the number of U couplings in the initial state for n_1 and n_2:

$$\frac{\nu!}{2!(\nu-2)!} = \frac{\nu(\nu-1)}{2}$$

We deduce

$$E_{excited} - E_{initial} = U \left[\frac{(\nu-1)(\nu-2)}{2} + \frac{\nu(\nu+1)}{2} - 2\frac{\nu(\nu-1)}{2} \right]$$

$$= U$$

The denominator of (3.68) is thus U. The sums in (3.67) and (3.68) give the effective exchange interaction:

$$\hat{\mathcal{H}}_{eff}(n_1, n_2) = -2 \sum_{m,m'} \left[J_{n_1 n_2}^{mm'} - \frac{|V_{n_1 n_2}^{mm'}|^2}{U} \right] \vec{S}_{n1}^m \cdot \vec{S}_{n2}^{m'} \tag{3.69}$$

Note that the spin $\vec{S}_n^m$ of the orbital m at the site n reads

$$\vec{S}_n^m = \frac{1}{2S}\vec{S}_n \tag{3.70}$$

where S is the spin amplitude of the site n, namely $S = |\vec{S}_n|$. It is noted that the spin distribution in the orbitals of the site n is given by the first Hund's rule: if the valence shell at the site n is less-than-half-filled, the spins in the orbitals of this shell are parallel:

$$S = |\vec{S}_n| = |\sum_m \vec{S}_n^m| = \sum_m |\vec{S}_n^m|$$

Example 1: The 3d shell having 3 electrons is shown in Fig. 3.2(a).

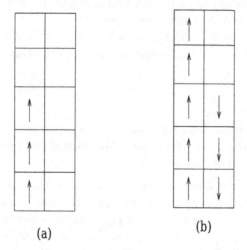

(a)   (b)

Fig. 3.2   3d shell filled with (a) 3 spins (b) 8 spins.

In the case of a more-than-half-filled shell, the total spin of the shell is given by the total spin of unoccupied orbitals:

$$S = |\vec{S}_n| = |\sum_{m(unocc.)} \vec{S}_n^m| = \sum_{m(unocc.)} |\vec{S}_n^m|$$

This method is used to facilitate the calculation because in both cases we have only "parallel" spins in the orbitals (each unoccupied orbital has a spin $\frac{1}{2}$).

Example 2: 3d shell with 8 electrons is shown in Fig. 3.2(b).

$$S = |\vec{S}_n| = \vec{S}_n^{m_1} + \vec{S}_n^{m_2} = \frac{1}{2} + \frac{1}{2} = 1$$

This is equivalent to the sum of spins in occupied orbitals.

Replacing (3.70) in (3.69) and performing the sum on pairs (n_1, n_2) we obtain (3.63).

3.3   Problems

Problem 1. Study properties of a free electron gas with the second quantization

Problem 2. Calculate the energy of an interacting electron gas at the first-order of perturbation with the second quantization

Problem 3. Hubbard model: one-site case

The Hubbard model has been described in subsection 3.2.2. We consider here a simple version defined by the Hamiltonian:

$$\hat{\mathcal{H}} = -t \sum_{i,i',\sigma} b_{i'\sigma}^+ b_{i\sigma} + U \sum_i b_{i\uparrow}^+ b_{i\uparrow} b_{i\downarrow}^+ b_{i\downarrow} - \mu \sum_i (b_{i\uparrow}^+ b_{i\uparrow} + b_{i\downarrow}^+ b_{i\downarrow})$$

$$(3.71)$$

where t is the hopping term between lattice sites i and i', U the on-site Coulomb interaction. We have added here the chemical potential term.

a) Consider the one-site case, namely $t = 0$ (no hopping). Give all possible microstates of the site. Calculate the grand partition function. Deduce the average energy. Calculate the occupation number. Show that it is given by $< n >= 2[\exp(\beta\mu) + \exp(2\beta\mu - \beta U)][1 + 2\exp(\beta\mu) + \exp(2\beta\mu - \beta U)]^{-1}$. Plot $< n >$ versus μ for $U = 4$, $T = 2, 0.5, 0.2, 0.1$. Comment.

b) Consider the case of no interaction, namely $U = 0$, in 1D. Use the Fourier transform of creation and annihilation operators to show that

$$\hat{\mathcal{H}} = \sum_{k,\sigma} (\epsilon_k - \mu) b_{k,\sigma}^+ b_{k,\sigma} \qquad (3.72)$$

where $\epsilon_k = -2t \cos(ka)$, k being the wave vector, a the lattice constant. This result is the same as the energy obtained from the tight-binding approximation. What is the result for the square lattice? Return to 1D. Calculate the grand partition function and the average energy $< E >$. Calculate the average occupation number $< n >$. Plot $< E >$ versus T for $t = 1$ at half-filling ($\mu = U/2$).

Problem 4. Hubbard model on a two-site system

We consider a system of two sites with the Hamiltonian given by (3.71).

a) Specify all possible configurations of the system using notations such as $a|0, \uparrow>$ to indicate the vector configuration in which there is no electron at the first site and an up spin at the second site, a being a complex number which represents the weight of the state vector.

b) Construct the state vectors each of which is a combination of connected configurations. Show that there are six such states. Write the Schrödinger equation for these states. Show that

$$
\begin{pmatrix}
0 & -t & 0 & 0 & 0 & 0 \\
-t & 0 & 0 & 0 & 0 & 0 \\
0 & 0 & U & -t & -t & 0 \\
0 & 0 & -t & 0 & 0 & -t \\
0 & 0 & -t & 0 & 0 & -t \\
0 & 0 & 0 & -t & -t & U
\end{pmatrix}
\begin{pmatrix}
a \\ b \\ c \\ d \\ e \\ f
\end{pmatrix}
= E
\begin{pmatrix}
a \\ b \\ c \\ d \\ e \\ f
\end{pmatrix}
\tag{3.73}
$$

where a, b, c, d, e and f are the weights of the six states. Solve these equations.

Problem 5. Show that $[\hat{\mathcal{H}}, \hat{N}] = 0$ where $\hat{N}$ is the field operator of occupation number defined in (3.36) and $\hat{\mathcal{H}}$ the Hamiltonian in the second quantization (3.35).

Problem 6. Show that $\hat{\Psi}(\vec{r})\hat{N} = (\hat{N}+1)\hat{\Psi}(\vec{r})$ for both boson and fermion cases.

Problem 7. Show that $\hat{\Psi}^+(\vec{r})|\text{vac} >$ ("vac" stands for vacuum) is a state in which there is a particle localized at $\vec{r}$.

Problem 8. Using the equation of motion $i\hbar\frac{d\hat{\Psi}(\vec{r})}{dt} = -[\hat{\mathcal{H}}, \hat{\Psi}(\vec{r})]$ where $\hat{\mathcal{H}}$ is the Hamiltonian in the second quantization of a system of fermions, show that we can obtain the Hartree-Fock equation by taking a first approximation of the right-hand side (linearization).

Problem 9. Bardeen-Cooper-Schrieffer theory of supraconductivity:

We consider a gas of N electrons with the reduced Hamiltonian in the superconducting regime:

$$
\mathcal{H} = \sum_{\vec{k}} \epsilon_{\vec{k}}(c_{\vec{k}}^+ c_{\vec{k}} + c_{-\vec{k}}^+ c_{-\vec{k}}) - V \sum_{\vec{k},\vec{k}'} c_{\vec{k}'}^+ c_{-\vec{k}'}^+ c_{-\vec{k}} c_{\vec{k}}
\tag{3.74}
$$

where V is an attractive interaction between electrons in states near the Fermi level (this attractive interaction is due to a coupling with phonons), $\epsilon_{\vec{k}}$ is the kinetic energy of the state $\vec{k}$ or $-\vec{k}$, c and c^+ are annihilation and creation operators. To simplify the writing of (3.74), the spins are implicitly supposed as follows: state $\vec{k}$ has a spin ↑ and state $-\vec{k}$ has a spin ↓.

a) Equation of motion: show that

$$i\hbar\dot{c}_{\vec{k}} = \epsilon_{\vec{k}}c_{\vec{k}} - c^+_{-\vec{k}}V\sum_{\vec{k}'}c_{-\vec{k}'}c_{\vec{k}'} \tag{3.75}$$

$$i\hbar\dot{c}^+_{-\vec{k}} = -\epsilon_{\vec{k}}c^+_{-\vec{k}} - c_{\vec{k}}V\sum_{\vec{k}'}c^+_{\vec{k}'}c^+_{-\vec{k}'} \tag{3.76}$$

b) To linearize the above equations we replace the products of operators by their average values, namely

$$\Delta_{\vec{k}} = V\sum_{\vec{k}'} <\phi^0_N|c_{-\vec{k}'}c_{\vec{k}'}|\phi^0_{N+2}> \tag{3.77}$$

$$\Delta^*_{\vec{k}} = V\sum_{\vec{k}'} <\phi^0_N|c^+_{\vec{k}'}c^+_{-\vec{k}'}|\phi^0_{N-2}> \tag{3.78}$$

where $|\phi^0_{N\pm2}>$ is the ground state having $N \pm 2$ particles. Show that a solution of the form $\exp(-i\omega_{\vec{k}}t)$ leads to

$$\omega_{\vec{k}} = (\epsilon^2_{\vec{k}} + \Delta^2)^{1/2} \tag{3.79}$$

where $\Delta^2 = \Delta_{\vec{k}}\Delta^*_{\vec{k}}$ (Δ is the gap in the excitation spectrum).

c) We use the following transform

$$a_{\vec{k}} = u_{\vec{k}}c_{\vec{k}} - v_{\vec{k}}c^+_{-\vec{k}} \tag{3.80}$$

$$a^+_{\vec{k}} = u_{\vec{k}}c^+_{\vec{k}} - v_{\vec{k}}c_{-\vec{k}} \tag{3.81}$$

where $u_{\vec{k}} = u_{-\vec{k}}$ and $v_{\vec{k}} = -v_{-\vec{k}}$ (real) and $u^2_{\vec{k}} + v^2_{\vec{k}} = 1$.

- Show that the operators $a_{\vec{k}}$ and $a^+_{\vec{k}}$ obey the anticommutation relations.
- Show that (3.75) with operator chains replaced by $\Delta_{\vec{k}}$ leads to

$$\omega_{\vec{k}}u_{\vec{k}} = \epsilon_{\vec{k}}u_{\vec{k}} + \Delta_{\vec{k}}v_{\vec{k}} \tag{3.82}$$

Putting this relation into square and using (3.79), show that

$$\tan\theta_{\vec{k}} = \Delta_{\vec{k}}/\epsilon_{\vec{k}} \tag{3.83}$$

where we have used the notations $u_{\vec{k}} = \cos(\theta_{\vec{k}}/2)$ and $v_{\vec{k}} = \sin(\theta_{\vec{k}}/2)$.

d) Wave function of the ground state:

- Show that the ground state of the superconducting regime is given by

$$\phi^0 = \prod_{\vec{k}} (u_{\vec{k}} + v_{\vec{k}} c^+_{\vec{k}} c^+_{-\vec{k}}) |\text{vac} > \tag{3.84}$$

- Show that ϕ^0 is normalized.
- Calculate $< \phi^0 | c^+_{\vec{k}'} c_{\vec{k}'} | \phi^0 >$ and $< \phi^0 | c^+_{\vec{k}'} c^+_{-\vec{k}'} c_{-\vec{k}''} c_{\vec{k}''} | \phi^0 >$. Find that the energy of the ground state in the superconducting regime is

$$E_g = -\sum_{\vec{k}} \epsilon_{\vec{k}} \cos \theta_{\vec{k}} - (\Delta^2/V) \tag{3.85}$$

Problem 10. Magnon-phonon interaction:

We consider the following Hamiltonian describing the interaction between magnon and phonon:

$$\mathcal{H} = \sum_{\vec{k}} \left[\omega^m_{\vec{k}} a^+_{\vec{k}} a_{\vec{k}} + \omega^p_{\vec{k}} b^+_{\vec{k}} b_{\vec{k}} + V_{\vec{k}} (a_{\vec{k}} b^+_{\vec{k}} + a^+_{\vec{k}} b_{\vec{k}}) \right] \tag{3.86}$$

where $V_{\vec{k}}$ is the coupling constant, $\omega^m_{\vec{k}}$ and $\omega^p_{\vec{k}}$ are eigenfrequencies of magnon and phonon, respectively, a and a^+ denote annihilation and creation operators of magnon, while b and b^+ denote those of phonon.

a) Using the following transformation

$$a_{\vec{k}} = \cos \theta_{\vec{k}} c_{\vec{k}} + \sin \theta_{\vec{k}} d_{\vec{k}}$$
$$b_{\vec{k}} = \cos \theta_{\vec{k}} d_{\vec{k}} - \sin \theta_{\vec{k}} c_{\vec{k}}$$

where $\theta_{\vec{k}}$ is real, show that operators c, c^+ and d, d^+ obey the commutation relations.

b) Show that $\mathcal{H}$ is diagonal in $c^+_{\vec{k}} c_{\vec{k}}$ and $d^+_{\vec{k}} d_{\vec{k}}$ if

$$\tan 2\theta_{\vec{k}} = 2V_{\vec{k}}/(\omega^p_{\vec{k}} - \omega^m_{\vec{k}}) \tag{3.87}$$

Show that when $\omega^p_{\vec{k}} = \omega^m_{\vec{k}} = \omega$ ("cross-over"), $\mathcal{H}$ is given by

$$\mathcal{H} = \sum_{\vec{k}} \left[(\omega - V_{\vec{k}}) c^+_{\vec{k}} c_{\vec{k}} + (\omega + V_{\vec{k}}) d^+_{\vec{k}} d_{\vec{k}} \right] \tag{3.88}$$

Comment.

Chapter 4

Magnetism: Mean-Field Theory

This chapter is devoted to the mean-field theory as applied to ferromagnetism and antiferromagnetism. A short treatment of a simple model of ferrimagnetism is also included. The origin of the ferromagnetism and other exchange interactions have been shown in chapter 3. The mean-field theory, sometimes called molecular-field approximation, is the simplest approximation for systems of interacting spins. Though simple, its results bear essential physical properties in high dimensions as will be seen below.

4.1 Systems of interacting spins

4.1.1 *Heisenberg model*

The Heisenberg model for the interaction between two spins localized at the lattice sites i and j is given by

$$-2J_{ij}\vec{S}_i \cdot \vec{S}_j \tag{4.1}$$

where J_{ij} is the exchange integral resulting from the Coulomb interaction between two electrons of spins $\vec{S}_i$ and $\vec{S}_j$, localized at $\vec{r}_i$ and $\vec{r}_j$. We have demonstrated (4.1) in chapter 3. In general, J_{ij} depends on the distance between the spins and on the orientation of $\vec{r}_j - \vec{r}_i$ with respect to the crystalline axes.

In the quantum model, $\vec{S}_i$ is a quantum spin whose components obey the spin commutation relations. For example, in the case of spin one-half its components are the Pauli matrices (1.3)-(1.5). In the classical model, $\vec{S}_i$ is considered as a vector.

We consider hereafter the simplest case where the exchange interaction is limited between nearest neighbors and this interaction is identical and equal to J for all pairs of nearest neighbors regardless of their relative positions with respect to the crystalline axes. In this case, the Hamiltonian reads

$$\mathcal{H} = -2J \sum_{<i,j>} \vec{S}_i \cdot \vec{S}_j \qquad (4.2)$$

where the sum is made over all pairs of nearest neighbors. We see that if $J > 0$, $\mathcal{H}$ is minimum when all spins are parallel. This spin configuration corresponds to the ferromagnetic ground state.

In the case where $J < 0$, we have to distinguish classical and quantum spin models:

(i) In the classical model, the spins are vectors. Except for lattices composed of equilateral triangular faces in their elementary cell such as the two-dimensional triangular lattice, the face-centered cubic lattice and the hexagonal-close-pack lattice, $\mathcal{H}$ is minimum in the other lattices when all nearest neighbors are antiparallel throughout the crystal. This spin configuration is called "classical antiferromagnetic ground state" or "Néel state".

(ii) In the quantum model, the spins can be decomposed into spin operators [see (3.47)-(3.49)]: $(S_i^z, S_i^+ \equiv S_i^x + iS_i^y, S_i^- \equiv S_i^x - iS_i^y)$. These spin operators obey the spin commutation relations

$$[S_l^+, S_m^-] = 2S_l^z \delta_{l,m} \quad \text{and} \quad [S_l^z, S_m^\pm] = \pm S_l^\pm \delta_{l,m} \qquad (4.3)$$

It is known that the Néel state is not the quantum ground state for antiferromagnets. As will be shown in chapter 5, at zero temperature the so-called zero-point quantum fluctuations make the spins contracted from its full length. The zero-point spin contraction is of the order of a few percents depending on the lattice structure. The real quantum ground state of antiferromagnets is not known, though we know that it is not far from the classical Néel state, except in low dimensions where strong quantum fluctuations can destroy the Néel order. Note that in the case of ferromagnets, the perfect parallel spin configuration is the real ground state in both classical and quantum spin models.

When the geometry of the lattice does not allow us to satisfy all the interactions between a spin with its neighbors, the system is frustrated. This happens when the elementary lattice cell is composed of equilateral triangles such as in the triangular lattice. There are many spectacular effects due to the frustration. We will outline some remarkable properties of frustrated systems in section 7.5.

4.1.2 *Ising and XY models*

Besides the Heisenberg model, there are three other spin models which are very popular in magnetism and in statistical physics:

- Ising model: The Ising model is defined by

$$\mathcal{H} = -\sum_{(i,j)} J_{ij}\sigma_i\sigma_j \tag{4.4}$$

where σ_i (σ_j) is the spin at lattice site i (j) with two possible values $+1$ and -1. Such a spin is called "Ising spin". The Ising model can be used to study spin systems with a strong uniaxial anisotropy (see section 4.3). It is also used as a simple model for phase transition investigation (see chapter 7). Any system of interacting entities where each entity has two individual states can be mapped into a system of Ising spins. For instance, in a binary alloy where each lattice site can be occupied by an A atom or a B atom, we can assign an up spin ($+1$) for an A atom and a down spin (-1) for a B atom. We can then study for example the structure of the alloy by studying the Ising model as a function of its composition.

- XY model: The XY model is defined by two-component spins. It is sometimes called "model of plane rotators". The Hamiltonian is given by

$$\mathcal{H} = -\sum_{(i,j)} J_{ij}(S_{ix}S_{jx} + S_{iy}S_{jy}) \tag{4.5}$$

The XY model is used to study spin systems with a strong planar anisotropy called "easy-plane anisotropy". It is used not only to describe some magnetic materials but also in statistical physics. A very interesting ferromagnetic XY model in two dimensions has been extensively studied because it gives rise to a very special phase transition known as the "Kosterlitz-Thouless" transition discovered in the 70's [80, 7, 26, 150]. A discussion on this phase transition is given in chapter 7.

- Potts model: The q-state Potts model is defined by

$$\mathcal{H} = -\sum_{(i,j)} J_{ij}\delta_{\sigma_i,\sigma_j} \tag{4.6}$$

where σ_i (σ_j) is a parameter taking q values, $\sigma_i = 1, 2, ..., q$ for example, and $\delta_{\sigma_i, \sigma_j}$ denotes the Kronecker symbol. The sum is often made over pairs of nearest neighbors. If interaction $J_{ij} = J > 0$ for nearest neighbors (i, j), then in the ground state there is only one value of q: it is ferromagnetic. Note that if $q = 2$, the model is equivalent to the Ising model. We define the Potts order parameter Q by

$$Q = \frac{[q \, \max(Q_1, Q_2, ..., Q_q) - 1]}{q - 1} \tag{4.7}$$

where Q_n is the spatial average defined by

$$Q_n = \frac{1}{N} \sum_{j=1}^{N} \delta_{\sigma_j, n} \tag{4.8}$$

where $n = 1, ..., q$, the sum runs over all lattice sites, and N is the total site number. From this definition we see that the ground state containing only one kind of spin has $Q = 1$, while in the disordered state q kinds of spin are equally present in the system, namely $Q_1 = Q_2 = ... = Q_q = 1/q$, so that $Q = 0$.

The q-state Potts model is used to study systems of interacting particles where each particle has q individual states. Exact methods to treat the Potts models in two dimensions are shown in a book by Baxter [16].

4.2 Ferromagnetism in mean-field theory

In this section, we present the mean-field theory by using the Heisenberg model for ferromagnets. The extension of the method to other spin models can be done without difficulty.

We consider the following Hamiltonian

$$\mathcal{H} = -2 \sum_{(i,j)} J_{ij} \vec{S}_i \cdot \vec{S}_j - g\mu_B \sum_i \vec{H}_0 \cdot \vec{S}_i \tag{4.9}$$

where $\vec{H}_0$ is a magnetic field applied in the z direction, g the Landé factor and μ_B the Bohr magneton. The first sum is performed over spin pairs $(\vec{S}_i, \vec{S}_j)$ occupying lattice sites i and j. For simplicity, we suppose in the following only interactions between nearest neighbors are not zero. Note that this hypothesis is not a hypothesis of the mean-field theory. We use it for presentation commodity. The mean-field theory can be applied to systems with far-neighbor interactions as seen in Problem 8 at the end of the chapter.

We consider the spin at the site i. The interactions with its nearest neighbors and with the magnetic field are written as

$$\mathcal{H}_i = -2J \sum_{\vec{\rho}} \vec{S}_i \cdot \vec{S}_{i+\vec{\rho}} - g\mu_B H_0 S_i^z \tag{4.10}$$

where $\vec{\rho}$ are vectors connecting the site i to its nearest neighbors and J the exchange integral.

4.2.1 *Mean-field equation*

The first assumption of the mean-field theory is to replace all neighboring spins by an average value $< \vec{S}_{i+\vec{\rho}} >$ which is the same for all nearest neighbors. This value is to be computed in the following. We choose the z axis as the spin quantization axis. The average values of the x and y spin components are then zero since the spin precesses circularly around the z axis:

$$< S_{i+\vec{\rho}}^x >=< S_{i+\vec{\rho}}^y >= 0.$$

For the z component, we have

$$< S_{i+\vec{\rho}}^z >=< S^z > + < \Delta S^z >$$

where $< S^z >$ is the average value in the absence of the magnetic field, and $< \Delta S^z >$ is the variation induced by the latter. We have

$$< \Delta S^z > \propto H_0$$

The spontaneous magnetization of the crystal is given by

$$M = Ng\mu_B < S^z > \tag{4.11}$$

where N is the total number of spins. $< S^z >$ depends on the temperature as seen below. Equation (4.10) becomes

$$\mathcal{H}_i \cong -2CJ \left[S_i^z \left(< S^z > + < \Delta S^z > \right) \right] - g\mu_B H_0 S_i^z \tag{4.12}$$

where C is the coordination number (number of nearest neighbors). We can express $\mathcal{H}_i$ as a function of the "molecular field" $\overline{H}$ defined by

$$\mathcal{H}_i = -g\mu_B \overline{H} S_i^z \tag{4.13}$$

where

$$\overline{H} = \frac{2CJ\left[< S^z > + < \Delta S^z >\right]}{g\mu_B} + H_0 \tag{4.14}$$

The average value $< S^z > + < \Delta S^z >$ is calculated by the canonical description (see Appendix A) as follows

$$< S^z > + < \Delta S^z > = \frac{\sum_{S_i^z=-S}^{S} S_i^z e^{-\beta \mathcal{H}_i}}{Z_i} \tag{4.15}$$

where $\beta = \frac{1}{k_B T}$ and Z_i the partition function defined by

$$
\begin{aligned}
Z_i &= \sum_{S_i^z=-S}^{S} \exp(\beta g\mu_B \overline{H} S_i^z) \\
&= \frac{\sinh[\beta g\mu_B \overline{H}(S + \frac{1}{2})]}{\sinh[\frac{1}{2}\beta g\mu_B \overline{H}]}
\end{aligned} \tag{4.16}
$$

where $S = |\vec{S}_i|$. We obtain

$$
\begin{aligned}
\sum_{S_i^z=-S}^{S} S_i^z e^{-\beta \mathcal{H}_i} &= \frac{\partial}{\partial \alpha} \sum_{S_i^z=-S}^{S} e^{\alpha S_i^z} \quad (\alpha \equiv \beta g\mu_B \overline{H}) \\
&= \frac{\partial}{\partial \alpha} Z_i \\
&= \frac{(S + \frac{1}{2})\cosh(S + \frac{1}{2})\alpha \sinh\frac{\alpha}{2} - \frac{1}{2}\sinh(S + \frac{1}{2})\alpha \cosh\frac{\alpha}{2}}{\sinh^2\frac{\alpha}{2}}
\end{aligned}
$$

whence

$$< S^z > + < \Delta S^z > = S B_S(x) \tag{4.17}$$

$B_S(x)$ is the Brillouin function defined by

$$B_S(x) = \frac{2S+1}{2S} \coth\frac{(2S+1)x}{2S} - \frac{1}{2S} \coth\frac{x}{2S} \tag{4.18}$$

where

$$x = \beta g\mu_B S\overline{H} = \beta[2CJS(< S^z > + < \Delta S^z >) + g\mu_B S H_0] \tag{4.19}$$

Equation (4.17) is called "mean-field equation". Since the argument x of $B_S(x)$ contains $< S^z > + < \Delta S^z >$, (4.17) is an implicit equation of $[< S^z > + < \Delta S^z >]$ which depends on the temperature. In the case of spin one-half, $S = \frac{1}{2}$, the Brillouin function is

$$B_{\frac{1}{2}}(x) = \tanh x.$$

In the case where $S \to \infty$, we have

$$B_\infty(x) = \coth x - \frac{1}{x} \equiv \quad \text{Langevin function.}$$

If H_0 is very weak, $< \Delta S^z >$ is then very small. In such a case, we can expand the Brillouin function near $x_0 = \beta 2CJS < S^z >$. By identifying the second terms of the two sides of (4.17), we have

$$< \Delta S^z >= SB'_S(x_0) \left[\frac{1}{k_b T}(g\mu_B SH_0 + 2CJS < \Delta S^z >) \right] \qquad (4.20)$$

where $B'_S(x_0)$ is the derivative of $B_S(x)$ with respect to x taken at x_0.

In zero applied field, (4.17) becomes

$$< S^z >= SB_S(x_0) \qquad (4.21)$$

4.2.2 Critical temperature

At high temperatures, $\beta < S^z > \ll 1$, we obtain from (4.18)

$$B_S(x) \simeq \frac{S+1}{3S}x - \frac{[S^2 + (S+1)^2](S+1)}{90S^3}x^3 + O(x^5) \qquad (4.22)$$

Equation (4.21) becomes

$$< S^z > \left[\frac{2CJS(S+1)}{3k_B T} - 1 \right] = \frac{S(S+1)[S^2 + (S+1)^2]}{90} \left(\frac{2CJ}{k_B T} \right)^3 < S^z >^3$$
$$+ O(x^5) \qquad (4.23)$$

This equation has a solution $< S^z > \neq 0$ only if

$$\left[\frac{2}{3}CS(S+1) - \frac{k_B T}{J} \right] > 0$$

namely

$$T < \frac{2CJS(S+1)}{3k_B} \equiv T_c \qquad (4.24)$$

Once this condition is satisfied, $< S^z >$ is given by

$$< S^z >^2 \simeq \frac{10}{3} \frac{S^2(S+1)^2}{[S^2 + (S+1)^2]} \left[\frac{T_c - T}{T_c} \right] \qquad (4.25)$$

T_c is called "critical temperature". When $T \geq T_c$, the solution of (4.23) is $< S^z >= 0$.

At low temperatures, $2CJS < S^z >$ is much larger than $k_B T$, the expansion of (4.18) gives

$$B_S(x) \simeq 1 - \frac{1}{S} e^{(-x/S)} + \cdots \qquad (4.26)$$

whence

$$< S^z >= SB_S(x) \simeq S - e^{-2JCS/k_B T} + \cdots \qquad (4.27)$$

If $T = 0$, we have $< S^z >= S$.

4.2.3   *Graphical solution*

In general, we solve (4.21) by a graphical method: we look for the intersection of the two curves $y_1 = \frac{<S^z>}{S} = \frac{k_B T}{2CJS^2} x$ and $y_2 = B_S(x)$ which represent the two sides of (4.21). The first curve, y_1, is a straight line with a slope proportional to the temperature. For a given value of T, there are two symmetric intersections at $\pm M$ as shown in Fig. 4.1. It is obvious that if the slope of y_1 is larger than the slope of y_2 at $x = 0$, there is no intersection other than the one at $x = 0$. The solution is then $< S^z >= 0$. The slope of y_2 at $x = 0$ thus determines the critical temperature T_c.

For each temperature, we note the graphical solutions. We represent the solutions of $< S^z >$ so collected versus T in Fig. 4.2.

4.2.4   *Specific heat*

The average energy of a spin when $H_0 = 0$ is calculated by

$$\overline{E}_i = -\frac{\partial \ln Z_i}{\partial \beta} \simeq -2CJ < S^z >^2 \qquad (4.28)$$

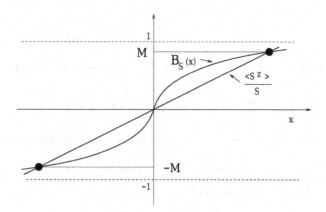

Fig. 4.1   Graphical solutions of Eq. (4.21).

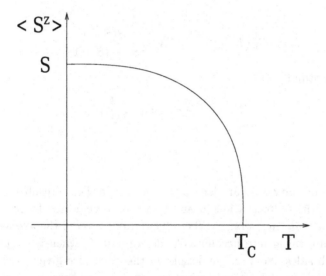

Fig. 4.2   Thermal average $< S^z >$ versus T.

The total ferromagnetic energy of the crystal is

$$\overline{E} = \frac{1}{2}N\overline{E}_i \qquad (4.29)$$

where the factor $\frac{1}{2}$ is added in order to count each interaction just once. The specific heat is

$$C_V = \left(\frac{\partial \overline{E}}{\partial T}\right)_V = -2NCJ < S^z > \frac{\partial < S^z >}{\partial T} \qquad (4.30)$$

At low temperatures, $< S^z > \cong S - e^{-2CJS/k_BT}$ [see (4.27)] , we have

$$C_V(T \simeq 0) \simeq 4Nk_B \left[\frac{CJS}{k_BT}\right]^2 e^{-2CJS/k_BT} \qquad (4.31)$$

When $T \to 0$, we have $C_V \simeq 0$. The third thermodynamic principle is satisfied.

For $T > T_c$, we have $\overline{E} = 0$, therefore $C_V = 0$. Let us calculate C_V when $T \to T_c^-$. We have from (4.25)

$$\lim_{T \to T_c^-} < S^z > \frac{\partial < S^z >}{\partial T} = -\frac{5}{2}\frac{k_B}{JC}\frac{S(S+1)}{[S^2 + (S+1)^2]} \qquad (4.32)$$

so that

$$C_V(T \to T_c^-) = 5Nk_B\frac{S(S+1)}{S^2 + (S+1)^2} \qquad (4.33)$$

The discontinuity of C_V at T_c is thus

$$\text{For } \ S = \frac{1}{2} \Rightarrow \Delta C_V = \frac{3}{2}Nk_B$$

$$\text{For } \ S = \infty \Rightarrow \Delta C_V = \frac{5}{2}Nk_B$$

This discontinuity is an artefact of the mean-field theory resulting from the fact that critical fluctuations near T_c have been neglected by replacing all spins by a uniform average. When fluctuations around the average values of spins are taken into account, C_V diverges at T_c when we approach T_c from both sides. Some more details on this point are given in chapter 7. We show in Fig. 4.3 C_V calculated by the mean-field theory as a function of T.

4.2.5 *Susceptibility*

By definition, the susceptibility is written as

$$\chi_\| = \left(\frac{\partial M}{\partial H_0}\right)_{H_0=0} = Ng\mu_B \left(\frac{\partial < S^z >}{\partial H_0}\right)_{H_0=0} \qquad (4.34)$$

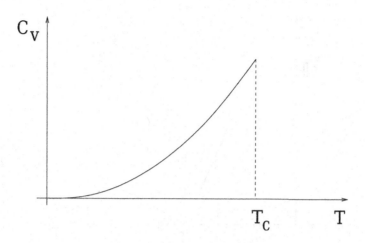

Fig. 4.3 C_V calculated by the mean-field theory versus T.

Equation (4.20) gives

$$< \Delta S^z > = \frac{SB'_S(x)\frac{g\mu_B S}{k_B T}H_0}{1 - SB'_S(x)\left(\frac{2CJS}{k_B T}\right)} \qquad (4.35)$$

so that

$$\chi_\parallel = \frac{N(g\mu_B)^2 S^2 B'_S(x)}{k_B T - 2CJS^2 B'_S(x)} \qquad (4.36)$$

where $x = \frac{2CJS<S^z>}{k_B T}$.

When $T \geq T_c$, we have $< S^z > = 0$ and $B'_x(0) = \frac{S+1}{3S}$. We get

$$\chi_\parallel(T \geq T_c) = \frac{N(g\mu_B)^2 S(S+1)}{3k_B(T - T_c)} \qquad \text{Curie-Weiss law} \qquad (4.37)$$

When $T \lesssim T_c$, we have $< S^z > \to 0$. Expanding $B'_S(x)$ with respect to $< S^z >$, we obtain

$$\chi_\parallel(T \lesssim T_c) = \frac{N(g\mu_B)^2 S(S+1)}{6k_B(T_c - T)} \qquad (4.38)$$

It is noted that the coefficient in this case is twice smaller than that in (4.37). When $T \to 0$, $\chi_\parallel \to 0$ because $M \to$ constant. The inverse of the susceptibility is schematically shown as a function of T in Fig. 4.4.

Note that $\chi_\parallel$ calculated above corresponds to a perfect ferromagnetic crystal where all spins are parallel to a crystalline direction. In reality, a

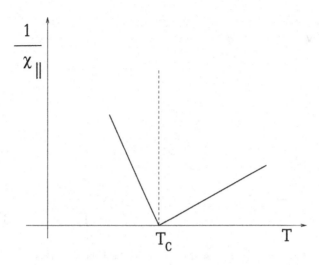

Fig. 4.4   Inverse of the susceptibility obtained by mean-field theory versus T.

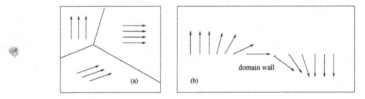

Fig. 4.5   (a) Ferromagnetic domains in an imperfect crystal (b) Example of a domain wall spin structure.

ferromagnetic crystal can have several ferromagnetic domains with spins pointing in different directions. This is due to the presence of defects, dislocations, imperfections, ... during the formation of the crystal. The region between two magnetic domains is called "domain wall" in which the matching of two spin orientations is progressively realized. We show schematically magnetic domains and a domain wall spin configuration in Fig. 4.5. The presence of domain walls makes it difficult to compare calculated and experimental susceptibilities.

4.2.6   *Validity of mean-field theory*

The mean-field theory neglects instantaneous fluctuations of spins. Due to this approximation, it overestimates the critical temperature T_c. Fluctuations favor disorder, so when taken into account, fluctuations cause a transition at a temperature lower than T_c given by (4.24). This point is treated in section 7.2 with the Landau-Ginzburg theory.

Another related problem of the mean-field theory is that it results in a phase transition at a finite temperature in spin systems in any space dimension. This is not correct because we know that in dimensions $d = 1$ and $d = 2$, fluctuations are so strong that they destroy magnetic long-range order at any finite temperature in many systems. The mean-field theory, however, becomes exact for dimension $d > 4$ (see chapter 7 for more details).

4.3 Magnetic anisotropy

The Heisenberg model (4.1) is isotropic because the energy is invariant with respect to an arbitrary global rotation of the spins in the space. The energy depends only on the relative orientation of the two spins, namely the angle formed by $\vec{S}_i$ and $\vec{S}_j$. In a real crystal, there are many factors that break the isotropic character of the model. We can mention a few examples such as spin-orbit interaction, crystalline field and lattice symmetry. There are several types of magnetic anisotropy:

- Anisotropy of exchange interaction J_{ij}: exchange interaction results from interaction (overlap) between neighboring orbitals, therefore it can be different in different directions depending on the spatial form of orbitals such as non spherical p or d orbitals.
- Single-ion magnetic anisotropy: an anisotropy of this category is of the form

$$-D \sum_i (S_i^z)^2 \qquad (4.39)$$

where D is a constant resulting from for example a spin-orbit interaction of second order perturbation. In the case of ferromagnets, if $D > 0$ spins prefer to align themselves in the $+z$ or $-z$ directions to reduce their energy. For this reason, we call (4.39) "uniaxial anisotropy". The z axis is called "easy-magnetization axis". If $D < 0$, spins prefer to reduce their z components and stay in the xy plane. We have here the case of an "easy-magnetization plane". When $D \to +\infty$, we deal with the Ising model and when $D \to -\infty$ we have the XY model.

Without entering into details of a phenomenological theory on magnetic anisotropy [62, 91, 94], we write the following expression for the anisotropy in a crystal of cubic symmetry

$$E_A = K_1(\alpha_1^2\alpha_2^2 + \alpha_2^2\alpha_3^2 + \alpha_3^2\alpha_1^2) + K_2\alpha_1^2\alpha_2^2\alpha_3^2 + \dots \qquad (4.40)$$

where $(\alpha_1, \alpha_2, \alpha_3)$ are the direction cosines of the spin $\vec{S}_i$ with respect to the three principal crystal axes. K_1 and K_2 denote anisotropy constants.

In the case of a hexagonal symmetry such as that of cobalt, we write

$$E_A = K_1 \cos^2\theta + K_2 \cos^4\theta + \dots \qquad (4.41)$$

where θ is the angle between $\vec{S}_i$ and the hexagonal axis.

In real crystals, because of different kinds of anisotropy, we often observe a preferential spin orientation. It is important to include in the Hamiltonian a correct anisotropy term describing such an orientational preference.

4.4 Antiferromagnetism in mean-field theory

In chapter 3 we have seen that depending on the sign of the exchange interaction a spin system can have an antiferromagnetic order at zero and low temperatures. We study here properties of antiferromagnets by the mean-field theory. This case requires some extra precautions with respect to the case of ferromagnets.

We consider a system of Heisenberg spins interacting with each other via the Hamiltonian

$$\mathcal{H} = \sum_{(i,j)} J_{ij}\vec{S}_i \cdot \vec{S}_j - g\mu_B \sum_i \vec{H}_0 \cdot \vec{S}_i \qquad (4.42)$$

where g and μ_B are the Landé factor and the Bohr magneton, respectively. $\vec{H}_0$ is a magnetic field applied along the z axis. To simplify the presentation, we suppose that the exchange interaction J_{ij} is limited to the nearest neighbors with $J_{ij} = J$. We have

$$\mathcal{H} = J \sum_{(i,j)} \vec{S}_i \cdot \vec{S}_j - g\mu_B \sum_i \vec{H}_0 \cdot \vec{S}_i \qquad (4.43)$$

Note that we have defined the Hamiltonian with a positive sign so that the antiferromagnetic interaction corresponds to $J > 0$. In zero applied field, the neighboring spins are antiparallel, except in geometrically frustrated systems (see chapter 7). A few antiferromagnetic systems are displayed in Fig. 4.6.

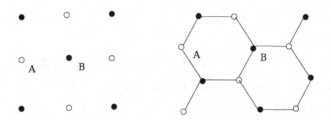

Fig. 4.6 Antiferromagnetic ordering: black and white circles denote ↑ and ↓ spins, respectively.

4.4.1 *Mean-field theory*

For commodity, we divide the antiferromagnetic lattice into two sublattices: sublattice of ↑ spins and sublattice of ↓ spins, indicated by indices l and m respectively.

The mean-field theory applied above to ferromagnets can be applied in the same manner to antiferromagnets. We write two coupled mean-field equations for two sublattices. We have the following energies of l and m spins

$$H_l = CJ < S^z_- > S^z_l + [CJ < \Delta S_- > -g\mu_B H_0] S^z_l \qquad (4.44)$$

$$H_m = CJ < S^z_+ > S^z_m + [CJ < \Delta S_+ > -g\mu_B H_0] S^z_m \qquad (4.45)$$

where C is the coordination number, $< S^z_l >=< S^z_+ > + < \Delta S_+ >$ denotes the average value of S^z_l, and $< \Delta S_+ >$ the spin variation induced by the applied field. The amplitude of $\vec{H}_0$ is supposed to be very small hereafter. With H_l, we calculate $< S^z_l >$ as follows

$$< S^z_l > = < S^z_+ > + < \Delta S_+ >= \frac{\mathrm{Tr} S^z_l e^{-\beta H_l}}{\mathrm{Tr} e^{-\beta H_l}}$$

$$= S B_S(x) \qquad (4.46)$$

where $B_S(x)$ is the Brillouin function given by

$$B_S(x) = \frac{2S+1}{2S} \coth \frac{(2S+1)x}{2S} - \frac{1}{2S} \coth \frac{x}{2S} \qquad (4.47)$$

with

$$x = \beta[-CJS(< S^z_- > + < \Delta S_- >) + g\mu_B S H_0] \qquad (4.48)$$

For weak fields, we expand the function $B_S(x)$ around

$$x_0 = -\beta C J S < S^z_- > .$$

We then obtain

$$< S_+^z > + < \Delta S_+ >$$
$$\simeq S B_S(-\beta C J S < S_-^z >) - \beta[C J S^2 < \Delta S_- >$$
$$-g\mu_B S^2 H_0]B_S'(x_0)$$

therefore

$$< S_+^z > \simeq S B_S(-\beta C J S < S_-^z >) \tag{4.49}$$
$$< \Delta S_+ > \simeq -\beta[C J S^2 < \Delta S_- > -g\mu_B S^2 H_0]B_S'(x_0) \tag{4.50}$$

$B_S'(x_0)$ being the derivative of $B_S(x_0)$ with respect to x taken at x_0. In the same manner, we obtain for down sublattice spin $< S_m^z >$

$$< S_-^z > \simeq S B_S(-\beta C J S < S_+^z >) \tag{4.51}$$
$$< \Delta S_- > \simeq -\beta[C J S^2 < \Delta S_+ > -g\mu_B S^2 H_0]B_S'(x_0^-) \tag{4.52}$$

with $x_0^- = -\beta C J S < S_+^z >$.

If the two sublattices are symmetric, namely

$$< S_+^z >= - < S_-^z >\equiv< S^z >$$

Equations (4.49) and (4.51) are equivalent because the Brillouin function is an odd function. We then have only one implicit equation for $< S^z >$ to solve

$$< S^z >= S B_S(\beta C J S < S^z >) \tag{4.53}$$

This mean-field equation for a sublattice spin is the same as the mean-field equation for ferromagnets. We have thus the same result on the temperature dependence and on the critical temperature. Note that the critical temperature for antiferromagnets is called "Néel temperature" and denoted by T_N:

$$\frac{k_B T_N}{J} = \frac{C S(S + 1)}{3} \tag{4.54}$$

Note that we did not use the factor 2 in the Hamiltonian (4.42). So, this result is the same as (4.24) for ferromagnets.

For $< \Delta S_\pm >$, we have $< \Delta S_+ >=< \Delta S_- >\equiv< \Delta S >$. $B_S'(x)$ is an even function of x, therefore

$$< \Delta S >= -\beta[C J S^2 < \Delta S > -g\mu_B S^2 H_0]B_S'(\beta C J S < S^z >) \tag{4.55}$$

The susceptibility is given by

$$\chi_\| = (\frac{\partial M}{\partial H_0})_{H_0=0} = \frac{Ng\mu_B <\Delta S>}{H_0}$$
$$= \frac{N(g\mu_B S)^2 B'_S(\beta CJS <S^z>)}{k_B T + CJS^2 B'_S(\beta CJS <S^z>)} \tag{4.56}$$

When $T \to 0$, $B'_S(...)$ tends to 0 faster than T. We deduce that $\chi_\| = 0$. On the contrary, for $T \geq T_N$, $B'_S(...) \simeq \frac{S+1}{3S}$, we get

$$\chi_\| = \frac{N(g\mu_B)^2 S(S+1)}{3k_B(T+T_N)} \tag{4.57}$$

where we notice the $+$ sign in front of T_N, in contrast to the ferromagnetic case. There is thus no divergence of the susceptibility at the phase transition for an antiferromagnet.

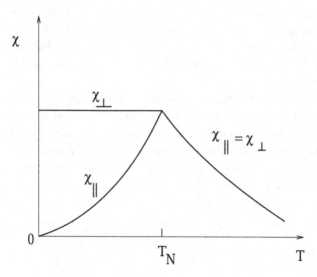

Fig. 4.7 Susceptibility $\chi_\|$ and $\chi_\perp$ of an antiferromagnet versus T.

In the case where the applied field is also weak but perpendicular to the z axis, for example $\vec{H}_0 \parallel \vec{Ox}$, we modify (4.44) and (4.45) to obtain

$$\chi_\perp(T \geq T_N) = \frac{N(g\mu_B)^2 S(S+1)}{3k_B(T+T_N)} = \chi_\|(T \geq T_N) \tag{4.58}$$

and

$$\chi_\perp(T \leq T_N) = \frac{N(g\mu_B)^2}{4CJ} = \text{constant.} \tag{4.59}$$

We show in Fig. 4.7 $\chi_\parallel$ and $\chi_\perp$ versus T.

In materials which have magnetic domains or in powdered systems, experimental susceptibility at $T \leq T_N$ is an average with spatial weight coefficients $1/3$ and $2/3$:

$$\chi(T \leq T_N) = \frac{1}{3}\chi_\parallel + \frac{2}{3}\chi_\perp \qquad (4.60)$$

4.4.2   *Spin orientation in a strong applied magnetic field*

The results shown above have been calculated with the assumption of weak field. When $\vec{H}_0$ is sufficiently strong, the results will be different as seen below.

We suppose that $\vec{H}_0$ is parallel to the z axis. The $\uparrow$ spins have their energy lowered by the Zeeman effect $-g\mu_B S_l^z H_0$ while the $\downarrow$ spins have their energy increased by $-g\mu_B S_m^z H_0 > 0$ ($S_m^z < 0$). Contrary to the weak field case where the spins remain approximately antiparallel because of the dominant J, in the case of strong field the competition between the Zeeman effect and the exchange interaction determines the stable spin configuration as seen below.

We consider the general case where we add a uniaxial anisotropy term to the Hamiltonian (4.43) to fix the easy-magnetization axis. We suppose that $\vec{H}_0$ is applied in the direction which forms an angle ζ with respect to the easy-magnetization axis. The competition between the Zeeman effect and J gives rise to a configuration of the two sublattices shown in Fig. 4.8 where θ is the angle of $\vec{H}_0$ with respect to the $+z$ axis.

The exchange energy is written as

$$E_e = \lambda M^2 \cos(\pi - 2\phi) \qquad (4.61)$$

where

$$\lambda = \frac{2CJ}{N(g\mu_B)^2} \qquad (4.62)$$

with M being the magnetization modulus.

The anisotropy energy is written for sublattices $\vec{M}_+$ and $\vec{M}_-$ (see Fig. 4.8) as

$$E_a = \frac{K}{2}\left[\cos^2(\zeta - \theta - \phi) + \cos^2(\zeta - \theta + \phi)\right] \qquad (4.63)$$

where K is the anisotropy constant.

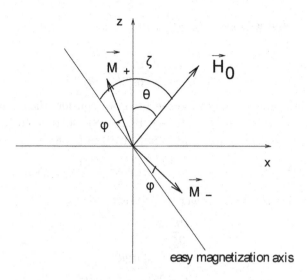

Fig. 4.8   Spin orientation with respect to the direction of $\vec{H}_0$.

The Zeeman energy is

$$E_Z = -H_0 M \left[\cos(\zeta - \phi) + \cos(\pi - \zeta - \phi) \right] \tag{4.64}$$

The energy induced by the variation of M under the applied field is

$$E_i = -\frac{1}{2} \chi_\parallel (H_0 \cos \zeta)^2 \tag{4.65}$$

because, by definition,

$$dM = \chi_\parallel dH \cos \zeta \tag{4.66}$$

$$\Rightarrow \quad E_i = -\int H \cos \zeta dM = -\chi_\parallel \int_0^{H_0} H dH \cos^2 \zeta$$

$$= -\frac{1}{2} \chi_\parallel (H_0 \cos \zeta)^2 \tag{4.67}$$

The total energy is thus

$$E = E_e + E_a + E_Z + E_i \tag{4.68}$$

By minimizing E of (4.68) with respect to ϕ and ζ we obtain

$$0 = \frac{\partial E}{\partial \phi}$$

$$= \cos \phi \{ 4\lambda M^2 \sin \phi + 2K \cos[2(\zeta - \theta)] \sin \phi$$

$$-2H_0 M \sin \zeta \}, \quad \text{therefore,}$$

$$\sin \phi = \frac{H_0 M \sin \zeta}{2\lambda M^2 + K \cos[2(\zeta - \theta)]} \tag{4.69}$$

Since, by definition, $\chi_\perp = \frac{2M \sin \phi}{H_0 \sin \zeta}$, we get

$$\chi_\perp = \frac{2M^2}{2\lambda M^2 + K \cos[2(\zeta - \theta)]} \tag{4.70}$$

Before minimizing (4.68) with respect to ζ, we consider the regime where $\lambda \gg K, H_0$. In this case $\phi \simeq 0$ (see Fig. 4.8), so that $\cos(\zeta - \phi) \simeq \cos(\pi - \zeta - \phi)$. Replacing this and (4.69)-(4.70) in (4.68), we arrive at

$$E = -\lambda M^2 - K \cos^2(\zeta - \phi) - \frac{1}{2}\chi_\| H_0^2 \cos^2 \zeta - \frac{1}{2}\chi_\perp H_0^2 \sin^2 \zeta \tag{4.71}$$

The minimization with respect to ζ leads to

$$0 = \frac{\partial E}{\partial \zeta}$$

$$\tan(2\zeta) = \frac{\sin(2\theta)}{\cos(2\theta) - \frac{\chi_\perp - \chi_\|}{2K} H_0^2} \tag{4.72}$$

We examine a particular case where $\theta = 0$ ($\vec{H}_0 \parallel \vec{Oz}$). The solutions are

- $\zeta = 0$ if $H_0 < H_c$,
- $\zeta \simeq \pi$ if $H_0 > H_c$ where

$$H_c(\text{critical field}) = \sqrt{\frac{2K}{\chi_\perp - \chi_\|}} \tag{4.73}$$

The spin configurations corresponding to these two solutions are displayed in Fig. 4.9. The transition between these phases when $H_0 = H_c$ is called "spin-flop transition": the spins are approximately perpendicular to $\vec{H}_0$ for $H_0 > H_c$. H_c est called "critical field".

This result has been obtained with the hypothesis $\lambda \gg K, H_0$. In the case where H_0 is larger than the local field acting on a spin (due to exchange interaction with its neighbors), all spins will turn into the direction of $\vec{H}_0$.

4.4.3   *Phase transition in an applied magnetic field*

The results shown above were obtained at $T = 0$. We discuss now the effect of T in an antiferromagnet under a strong applied field.

To simplify the description, let us consider the Ising spin model. The field $\vec{H}_0$ is supposed to be parallel to $\vec{Oz}$. If $H_0 < H_c$ where H_c is the critical field which is to be determined for the Ising model (see Problem 11

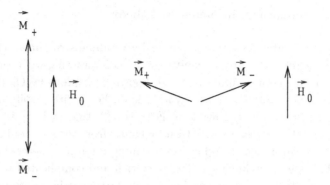

Fig. 4.9   Spin orientation with respect to $\vec{H}_0$ when $H_0 < H_c$ (left) and $H_0 > H_c$ (right).

below), the spins remain antiparallel between them. If $H_0 > H_c$, all spins are parallel to $\vec{H}_0$: there is no spin-flop phase for Ising spins.

In the case of ferromagnets in a field, the magnetization is never zero, so a phase transition is impossible at any temperature. In the case of antiferromagnets, when $H_0 < H_c$ there is a possibility that the antiferromagnetic order is broken with increasing T: at high temperatures, spins excited by the temperature finish by turning themselves parallel to the field at a temperature T_c. Of course, T_c depends on H_0 (see Problem 8 of chapter 7). We display schematically a phase diagram in Fig. 4.10. More details on the phase transition are given in chapter 7.

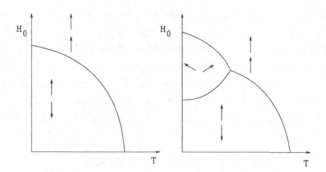

Fig. 4.10   Left: Phase diagram of an antiferromagnet with Ising spins under an applied field of amplitude H_0. The line separates the antiferromagnetic and paramagnetic phases under field. Right: Phase diagram in the case of Heisenberg spins, there is a spin-flop phase.

4.5   Ferrimagnetism in mean-field theory

Ferrimagnetic materials have complicated crystalline structures. There are often many sublattices of non equivalent spins interacting with each other via antiferromagnetic couplings. A well-known example is Fe_2O_3 (ferrites) which has three sublattices: Fe^{3+} with $\uparrow$ spin, Fe^{3+} with $\downarrow$ spin and Fe^{2+} with $\uparrow$ spin. The spin amplitude of Fe^{3+} is 5/2, that of Fe^{2+} is 2. As a consequence, the resulting magnetization comes from the spins of Fe^{2+}. For ferrimagnets, we should introduce several interaction parameters intra and extra sublattices. Ferrimagnets have very rich and complicated properties which are used in numerous applications in particular in recording industries thanks to their very high critical temperatures of the order of 500-800 Celcius degrees.

We introduce hereafter a very simple model to illustrate some remarkable properties of ferrimagnets. This does not replace the reading of more realistic models in the literature. We consider a system of Heisenberg spins which is composed of two sublattices, sublattice A containing $\uparrow$ spins of amplitude S_A and sublattice B containing $\downarrow$ spins of amplitude S_B. The Hamiltonian is written as

$$\mathcal{H} = J_1 \sum_{(l,m)} \vec{S}_l \cdot \vec{S}_m + J_2^A \sum_{(l,l')} \vec{S}_l \cdot \vec{S}_{l'}$$

$$+ J_2^B \sum_{(m,m')} \vec{S}_m \cdot \vec{S}_{m'} \tag{4.74}$$

where (l,l') and (m,m') indicate the sites of A and B, respectively. The interactions are J_1 between inter-sublattice nearest neighbors, J_2^A between the neighbors belonging to the sublattice A and J_2^B between neighbors belonging to the sublattice B. We suppose $J_1 > 0$ (antiferromagnetic). The signs of J_2^A and J_2^B can be arbitrary. The above Hamiltonian will be used in several examples later. But for now, we assume $J_2^A = J_2^B = 0$. We can start with equations (4.49) and (4.51) for two sublattices in zero applied field:

$$< S_A^z > = S_A B_{S_A}(-\beta C J_1 S_A < S_B^z >) \tag{4.75}$$

$$< S_B^z > = S_B B_{S_B}(-\beta C J_1 S_B < S_A^z >) \tag{4.76}$$

where C is the coordination number. Since the sublattices are not equivalent because $S_A \neq S_B$, we have to solve these two coupled equations by iteration. We put $M_A = < S_A^z >$ and $M_B = < S_B^z >$. At $T = 0$, the expansion of the functions $B_{S_A}(...)$ and $B_{S_B}(...)$ gives $M_A = S_A$ and $M_B = S_B$ (see section 4.2). At low temperatures, we can obtain the solution for M_A and M_B by solving graphically Eqs. (4.75)-(4.76). However, it is more complicated to calculate the critical temperature. The high-temperature expansion similar to (4.22) gives two equations containing M_A and M_B of the form $M_A = a(S_A, T)M_B + b(S_A, T)M_B^3 + ...$ and $M_B = c(S_B, T)M_A + d(S_B, T)M_A^3 + ...$ where $a(S_A, T)$, $b(S_A, T)$, $c(S_B, T)$ and $d(S_B, T)$ are coefficients depending on S_A, S_B and T. An explicit expression of the critical temperature T_N can be obtained (see Problem 6 below). We have

$$k_B T_N = \frac{CJ_1}{3} \sqrt{S_A(S_A + 1)S_B(S_B + 1)} \qquad (4.77)$$

This result is equivalent to (4.54) for antiferromagnets if $S_A = S_B$. Let us give a qualitative argument. We suppose that $S_A > S_B$. When M_B becomes very small M_A is still large. It induces a local field on its B neighbors, keeping them from going to zero. As long as M_A is not zero, M_B is maintained at a non zero value. However, fluctuations of M_B affect in turn M_A. Therefore, the critical temperature is somewhere between the two temperatures where the respective sublattices become disordered when each occupies the entire lattice, namely

$$\frac{k_B T_A}{J_1} = \frac{CS_A(S_A + 1)}{3} > \frac{k_B T_N}{J_1} > \frac{k_B T_B}{J_1} = \frac{CS_B(S_B + 1)}{3}.$$

For $S_A = 2$, $S_B = 1$ and $C = 8$ (body-centered cubic lattice), we have $k_B T_A/J_1 = 16$, $k_B T_B/J_1 = 16/3 \simeq 5.33$, and $k_B T_N/J_1 = 8\sqrt{12}/3 \simeq 9.2376$. The numerical solution of (4.75) and (4.76) for the above values of S_A, S_B and C is shown as a function of T in Fig. 4.11.

4.6 Conclusion

In this chapter, we have presented the mean-field theory with application to ferromagnetism, antiferromagnetism and ferrimagnetism. This theory is the first approximation to study interacting spin systems: low-temperature properties and the phase transition are described with physical quantities such as the order parameter, the critical temperature, the susceptibility and the specific heat. The theory gives incorrect results in low dimensions but it can be improved by higher-order approximations such as the Landau-Ginzburg theory (see section 7.2) and the cluster-variation method (CVM).

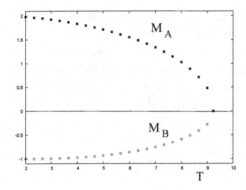

Fig. 4.11   Numerical solution of (4.75) and (4.76) is shown as a function of T for $S_A = 2$, $S_B = 1$, $C = 8$ and $k_B/J_1 = 1$. See text for comments.

The CVM consists in an exact treatment of a cluster of small size embedded in a crystal: the interaction of the cluster with the crystal is treated with the mean-field theory. The larger the cluster size is, the better the results become. However, the calculation is very complicated for large sizes. The method with smallest clusters is sometimes called "Bethe-Peierls-Weiss" (see Problem 7). To obtain more precise properties of interacting spin systems, we can use other methods such as the spin-wave theory (chapter 5), the Green's function theory (chapter 6), the renormalization group (chapter 7) or the Monte Carlo simulation (chapter 8).

4.7   Problems

Problem 1. Define the order parameter of an antiferromagnetic lattice of Ising spins.

Problem 2. Consider the q-state Potts model defined by the Hamiltonian (4.6) on a square lattice.

    a) Define the order parameter of the q-state Potts model.

    b) Describe the ground state and its degeneracy when $J > 0$.

    c) If $J < 0$, what is the ground state for $q = 2$ and $q = 3$? For $q = 3$, find ways to construct some ground states and give comments.

    d) Show that the Potts model is equivalent to the Ising model when $q = 2$.

Problem 3. Domain walls:

    In magnetic materials, due to several reasons, we may have magnetic domains schematically illustrated in Fig. 4.5. The spins at the interface between two neighboring domains should arrange themselves in a smooth configuration in order to make a gradual change from one domain to the other. An example of such a "domain wall" is shown in that figure. Calculate the energy of a wall of thickness of N spins.

Problem 4. Bragg-Williams approximation:

The mean-field theory presented in this chapter can be demonstrated by the Bragg-Williams approximation described in this problem.

Consider a crystal of N sites each occupied by an Ising spin at a given temperature T. The coordination number (number of nearest neighbors) is z. One supposes the periodic boundary conditions. Let N_+ and N_- be the number of up and down spins, respectively. The energy of a pair of parallel spins is $-J$ ($J > 0$) and that of a pair of antiparallel spins is $+J$. Let X be defined by $N_+ = N(1+X)/2$. One has then $N_- = N(1 - X)/2$.

a) Calculate the entropy S.
b) Calculate the probability to have an up spin at a lattice site. Deduce the numbers of up-up, down-down and up-down spin pairs, as functions of X.
c) Calculate the energy of the crystal as a functions of z, J and X.
d) Calculate the free energy F. Deduce the expression of X at thermal equilibrium, namely at the minimum of F. Show that this expression is equivalent to the mean-field equation (4.17) with $S = \pm 1$.
e) Give the mean-field solution for the critical temperature T_c. Calculate the entropy for $T > T_c$.

Problem 5. Binary alloys by spin language:

Consider a lattice where there are two kinds of site such as the one shown in Fig. 4.12: sites of type I (white circles) and sites of type II (black circles). There are two kinds of atoms A and B occupying the lattice sites. The number of each atom type is $N/2$. The interaction energy between two neighbors of the same kind is ϵ, that between two neighbors of different kinds is ϕ. One supposes $\epsilon > \phi$.

In the disordered phase, half of A atoms are on the white sites and the other half on the black sites. The same situation is for B atoms. We can study the ordering structure of this binary alloy by map-

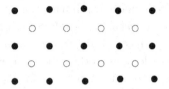

Fig. 4.12   Binary alloy (see Problem 5): white and black circles represent sites of type *I* and *II*, respectively.

ping the problem into a spin language: an A atom is represented by an up Ising spin and a B atom by a down Ising spin. The $A - B$ attractive interaction is replaced by an antiferromagnetic interaction.

a) Describe the ground state.

b) The system energy is E. Let $N_{\uparrow,I}$ be the number of $\uparrow$-spins occupying sites of the type I. We define x by

$$N_{\uparrow,I} = N(1+x)/4 \qquad (4.78)$$

- What is the value domain of x ? Which state does $x = 0$ correspond to? Calculate as a function of x the number of $\uparrow$-spins occupying sites of type II ($N_{\uparrow,II}$). The same question is for $\downarrow$-spins. One considers $x > 0$ hereafter.
- Calculate the probabilities as functions of x for a $\uparrow$-spin at a site of type I and at a site of type II, supposing that all probabilities are independent. The same question is for a $\downarrow$-spin.
- Let $N_{\uparrow,\uparrow}$, $N_{\downarrow,\downarrow}$, and $N_{\uparrow,\downarrow}$ be the numbers of $\uparrow\uparrow$, $\downarrow\downarrow$ and $\uparrow\downarrow$ spin pairs, respectively. Calculate these quantities as functions of x. Show that $N_{\uparrow,\uparrow} = N(1 - x^2)/2$, $N_{\downarrow,\downarrow} = N(1 - x^2)/2$, $N_{\uparrow,\downarrow} = N(1 + x^2)$.
- Calculate E as functions of x, ϵ and ϕ. Show that E can be written as

$$E = N(\epsilon + \phi) - N(\epsilon - \phi)x^2 \qquad (4.79)$$

- Calculate $\Omega(E)$ the number of microscopic states of energy E. Deduce the entropy S.
- Calculate temperature T. Show that

$$x = \tanh[2(\epsilon - \phi)x/(k_B T)] \qquad (4.80)$$

Show that x tends to 1 at low T, and that $x=0$ when $T > T_c = 2(\epsilon - \phi)/k_B$.

Problem 6. Critical temperature of ferrimagnet:

Using the mean-field theory, calculate the critical temperature T_N of the simple model for a ferrimagnet described in section 4.5.

Problem 7. Improvement of mean-field theory:

a) Two-spin problem:

Consider the following Hamiltonian

$$\mathcal{H} = -2J\vec{S}_1 \cdot \vec{S}_2 - D[(S_1^z)^2 + (S_2^z)^2] - B(S_1^z + S_2^z) \qquad (4.81)$$

where J (exchange interaction) and D (magnetic anisotropy) are positive constants and B magnitude of an applied magnetic field in the z direction. Find the eigenvalues and eigenvectors of $\mathcal{H}$ for spin one-half.

b) Improved mean-field theory:

Consider the Heisenberg spin model:

$$\mathcal{H} = -2J \sum_{(i,j)} \vec{S}_i \cdot \vec{S}_j$$

In the first step, we treat exactly the interaction of two neighboring spins. In the second step, we use the mean-field theory to treat the interaction of the two-spin cluster embedded in the crystal. Explicitly, consider two spins $\vec{S}_i$ and $\vec{S}_j$ embedded in a crystal. The Hamiltonian is given by

$$\mathcal{H}_{ij} = -2J\vec{S}_i \cdot \vec{S}_j - 2(Z-1)J < S^z > (S_i^z + S_j^z) \qquad (4.82)$$

where Z is the coordination number. Show that the critical temperature T_c for $S = 1/2$ is given by

$$e^{-2J/k_B T_c} + 3 - 2(Z-1)J/k_B T = 0 \qquad (4.83)$$

Problem 8. Interaction between next-nearest neighbors in mean-field treatment:

Consider a centered cubic lattice where each site is occupied by an Ising spin with values ± 1. The Hamiltonian is given by

$$\mathcal{H} = -J_1 \sum_{(i,j)} \sigma_i \sigma_j - J_2 \sum_{(i,k)} \sigma_i \sigma_k \qquad (4.84)$$

where σ_i is the spin at site i, J_1 (> 0) exchange interaction between nearest neighbors and J_2 (> 0) interaction between next-nearest neighbors. The first and second sums are made over pairs of corresponding neighbors.

a) Describe the magnetic ordering at temperature $T = 0$.

b) Give briefly the hypothesis of the mean-field theory.

c) By a qualitative argument, show that the interaction between next-nearest neighbors, J_2, increases the critical temperature.

d) Using the mean-field theory, calculate the partition function of a spin at a given temperature T. Deduce an equation which allows us to calculate $< \sigma >$, mean value of a spin, at T.

e) Determine the critical temperature T_c as functions of J_1 and J_2.

f) In the case where J_2 is negative (antiferromagnetic interaction), the above result is no more valid beyond a critical value of $|J_2|$. Determine that critical value J_2^c. What is the magnetic ordering when $|J_2| \gg |J_2^c|$ at $T = 0$?

Problem 9. Improved mean-field theory: Bethe's approximation

We can improve the mean-field theory in the case of Ising spins by using the partition function for an approximate Hamiltonian constructed as follows. We separate the whole Hamiltonian into two parts: the exact interaction between a spin with all of its z neighbors and a mean-field interaction between each of these neighbors with its own neighbors outside of the cell. The Hamiltonian is written as

$$\mathcal{H} = -J \sum_{i=1}^{z} \sigma_0 \sigma_i - \mu_B B \sigma_0 - \mu_B (B + H) \sum_{i=1}^{z} \sigma_i \qquad (4.85)$$

where B is an applied magnetic field, H the molecular field acting on the neighboring spins of σ_0 from outside spins.

Calculate the critical temperature and make a comparison with the result from the elementary mean-field theory.

Problem 10. Repeat Problem 7 in the case of an antiferromagnet.

Problem 11. Calculate the critical field H_c in the following cases

a) a simple cubic lattice of Ising spins with antiferromagnetic interaction between nearest neighbors

b) a square lattice of Ising spins with antiferromagnetic interaction J_1 between nearest neighbors and ferromagnetic interaction J_2 between next-nearest neighbors.

Chapter 5

Theory of Magnons

Elementary excitations of a system of interacting spins have a wave nature as in many other systems such as collective vibrations of coupled atoms in crystals (phonons), waves of charge density in plasmas, etc. In spin systems, the collective excitations are called spin waves or magnons when they are quantized. Spin waves propagate in magnetically ordered systems in the way phonons do in crystalline solids. At finite temperatures, as long as the magnetic order exists, i. e. $T < T_c$, spin waves are the only physical process that determines low-T magnetic properties of the system.

This chapter is devoted to spin waves in ferromagnets, antiferromagnets and, to a lesser extent, ferrimagnets and helimagnets. A classical treatment is first introduced to give a simple picture of spin waves. A quantum method is next presented to study in a more efficient manner finite-temperature properties of spin systems.

5.1   Ferromagnetism

5.1.1   *Classical treatment*

In ferromagnets, spins are parallel in the ground state. One supposes that the spins are aligned along the Oz axis. One considers each spin as a vector of modulus S with three components. This is the classical Heisenberg model. As the temperature increases, the spins rotate each around its z axis in a collective manner as shown in Fig. 5.1. The energy brought about by the temperature is shared by the whole system.

Let us calculate the spin-wave energy. Consider the classical Heisenberg spin model on a lattice. The spins are supposed to interact with each other via a nearest-neighbor exchange interaction J. For the spin $\vec{S}_l$, the

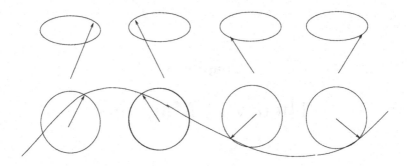

Fig. 5.1   A spin wave : side view (upper) and top view (lower).

interaction with its nearest neighbors is written as

$$\mathcal{H}_l = -2J\vec{S}_l \cdot \sum_{\vec{R}} \vec{S}_{l+\vec{R}} \tag{5.1}$$

where $\vec{R}$ is a vector connecting the spin $\vec{S}_l$ with one of its neighbors. The sum is performed over all nearest neighbors. Equation (5.1) can be rewritten as

$$\mathcal{H}_l = -\vec{M}_l \cdot \vec{H}_{ex} \tag{5.2}$$

where

$$\vec{M}_l = g\mu_B \vec{S}_l \tag{5.3}$$

and

$$\vec{H}_{ex} = \frac{2J}{g\mu_B} \sum_{\vec{R}} \vec{S}_{l+\vec{R}} \tag{5.4}$$

g and μ_B are the Landé factor and Bohr magneton. $\vec{M}_l$ and $\vec{H}_{ex}$ are the magnetic moment of spin $\vec{S}_l$ and the field created by the nearest neighbors acting on $\vec{S}_l$.

The equation of motion of the kinetic moment $\hbar\vec{S}_l$ is written as

$$\hbar\frac{d\vec{S}_l}{dt} = \left[\vec{M}_l \wedge \vec{H}_{ex}\right] = 2J\left[\vec{S}_l \wedge \sum_{\vec{R}} \vec{S}_{l+\vec{R}}\right] \tag{5.5}$$

If $\vec{S}_l$ is parallel to its neighbors then one has $\hbar\frac{d\vec{S}_l}{dt} = 0$, i.e. $\vec{S}_l$ is equal to a constant vector, thus there is no spin wave.

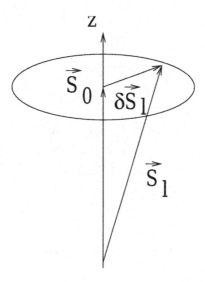

Fig. 5.2 Spin decomposition around the z axis.

In an excited state due to a spin rotation around the z axis, one can decompose $\vec{S}_l$ into components as shown in Fig. 5.2.

$$\vec{S}_l = \vec{S}_0 + \delta\vec{S}_l \tag{5.6}$$

where $\delta\vec{S}_l$ represents the deviation and $\vec{S}_0$ the z spin component. For a homogenous system and for a given spin-wave mode, it is natural to suppose that $\vec{S}_0$ is time-independent. Equation (5.5) thus becomes

$$\hbar\frac{d(\delta\vec{S}_l)}{dt} = 2J\sum_{\vec{R}}\left[(\vec{S}_0 + \delta\vec{S}_l)\wedge(\vec{S}_0 + \delta\vec{S}_{l+\vec{R}})\right]$$

$$\simeq 2J\sum_{\vec{R}}\left[(\delta\vec{S}_l - \delta\vec{S}_{l+\vec{R}})\wedge\vec{S}_0\right] \tag{5.7}$$

where $\delta\vec{S}_l$ and $\delta\vec{S}_{l+\vec{R}}$ are supposed to be small. This hypothesis is justified at low temperatures.

Let x and y be the two other Cartesian coordinates lying in the plane perpendicular to the z axis. One has $\vec{S}_0 \simeq S\hat{k}$, $(\delta\vec{S}_l)^x = S_l^x$ and $(\delta\vec{S}_l)^y = S_l^y$, $\hat{k}$ being the unit vector on the z axis (see Fig. 5.3).

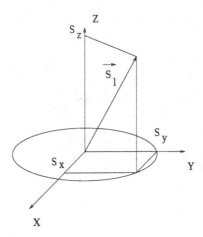

Fig. 5.3   Spin components.

Equation (5.7) reads

$$\hbar\frac{dS_l^x}{dt} = 2JS\sum_{\vec{R}}\left(S_l^y - S_{l+\vec{R}}^y\right) \tag{5.8}$$

$$\hbar\frac{dS_l^y}{dt} = -2JS\sum_{\vec{R}}\left(S_l^x - S_{l+\vec{R}}^x\right) \tag{5.9}$$

$$\hbar\frac{dS_l^z}{dt} = 0 \tag{5.10}$$

S_l^z is thus a constant of motion. One looks for the solutions of S_l^x and S_l^y of the form

$$S_l^x = Ue^{i(\vec{k}\cdot\vec{l}-\omega_{\vec{k}}t)} \tag{5.11}$$

$$S_l^y = Ve^{i(\vec{k}\cdot\vec{l}-\omega_{\vec{k}}t)} \tag{5.12}$$

where $\vec{k}$ and $\vec{l}$ are the wave vector and the position of $\vec{S}_l$, respectively. Replacing (5.11)-(5.12) in (5.8)-(5.9), one obtains

$$-i\hbar\omega_{\vec{k}}U = 2JSZ\left[1 - \frac{1}{Z}\sum_{\vec{R}}e^{i\vec{k}\cdot\vec{R}}\right]V \tag{5.13}$$

$$i\hbar\omega_{\vec{k}}V = 2JSZ\left[1 - \frac{1}{Z}\sum_{\vec{R}}e^{i\vec{k}\cdot\vec{R}}\right]U \tag{5.14}$$

where Z is the coordination number (number of nearest neighbors). The non trivial solutions of U and V verify

$$\begin{vmatrix} i\hbar\omega_{\vec{k}} & 2JSZ(1-\gamma_{\vec{k}}) \\ -2JSZ(1-\gamma_{\vec{k}}) & i\hbar\omega_{\vec{k}} \end{vmatrix} = 0$$

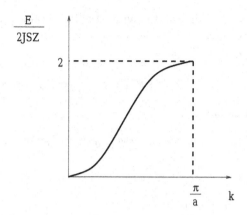

Fig. 5.4   Spin-wave dispersion relation of a ferromagnet in one dimension.

from which one has

$$\hbar\omega_{\vec{k}} = 2JSZ\left(1 - \gamma_{\vec{k}}\right) \tag{5.15}$$

where

$$\gamma_{\vec{k}} = \frac{1}{Z}\sum_{\vec{R}} e^{i\vec{k}\cdot\vec{R}} \tag{5.16}$$

With (5.16) and (5.15) one finds $V = -iU$. This relation indicates that the spin rotation has a circular precession around the z axis.

The relation (5.15) is called "dispersion relation" of spin waves in ferromagnets.

Example : In the case of a chain of spins one has

$$\gamma_k = \frac{1}{2}(e^{ika} + e^{-ika}) = \cos(ka) \tag{5.17}$$

where a is the lattice constant. Equation (5.15) becomes

$$\hbar\omega_k = 2JSZ\left[1 - \cos(ka)\right] \tag{5.18}$$

Figure 5.4 shows $\hbar\omega_k$ versus k in the first Brillouin zone. When $ka << 1$, with $Z = 2$ for one dimension, one has

$$\hbar\omega_k \simeq 2JS(ka)^2 \tag{5.19}$$

ω_k is thus proportional to k^2 for small k (long wave-length), in contrast to the case of phonons where ω_k is proportional to k. A consequence of this difference is that macroscopic physical properties which are averaged over these elementary excitations will show different temperature-dependent behaviors as seen below.

Remark: Effect of the crystal symmetry is contained in the factor $\gamma_{\vec{k}}$. Here are a few examples:

1) Square lattice:

$$\gamma_{\vec{k}} = \frac{1}{4}(e^{ik_x a} + e^{-ik_x a} + e^{ik_y a} + e^{-ik_y a}) = \frac{1}{2}[\cos(k_x a) + \cos(k_y a)] \quad (5.20)$$

2) Simple cubic lattice:

$$\gamma_{\vec{k}} = \frac{1}{6}(e^{ik_x a} + e^{-ik_x a} + e^{ik_y a} + e^{-ik_y a} + e^{ik_z a} + e^{-ik_z a})$$

$$= \frac{1}{3}[\cos(k_x a) + \cos(k_y a) + \cos(k_z a)] \quad (5.21)$$

3) Centered cubic lattice:

$$\gamma_{\vec{k}} = \cos(\frac{k_x a}{2})\cos(\frac{k_y a}{2})\cos(\frac{k_z a}{2}) \quad (5.22)$$

4) Face-centered cubic lattice:

$$\gamma_{\vec{k}} = \frac{1}{3}[\cos(\frac{k_x a}{2})\cos(\frac{k_y a}{2}) + \cos(\frac{k_y a}{2})\cos(\frac{k_z a}{2}) + \cos(\frac{k_z a}{2})\cos(\frac{k_x a}{2})]$$
$$(5.23)$$

5.1.2 *Quantum theory: Holstein-Primakoff approximation*

We consider now quantum spins. The spin $\vec{S}_l$ at the lattice site l can be decomposed into spin operators S_l^z and $S_l^{\pm} = S_l^x \pm iS_l^y$. The spin operators obey the following commutation relations:

$$[S_l^+, S_{l'}^-] = 2S_l^z \delta_{ll'} \quad (5.24)$$

$$[S_l^z, S_{l'}^{\pm}] = \pm S_l^{\pm} \delta_{ll'} \quad (5.25)$$

Let Θ_S^m be a spin function where S indicates the spin amplitude and m the spin component along the z axis. The quantum theory of angular momentum gives

$$S^+\Theta_S^m = \sqrt{(S-m)(S+m+1)}\Theta_S^{m+1} \quad (5.26)$$

$$S^-\Theta_S^m = \sqrt{(S+m)(S-m+1)}\Theta_S^{m-1} \quad (5.27)$$

$$S^z\Theta_S^m = m\Theta_S^m \quad (5.28)$$

In the particular case where $S = 1/2$, one has $\Theta_{1/2}^{1/2} = 1/2$, $\Theta_{1/2}^{-1/2} = -1/2$, $S^+\Theta_{1/2}^{1/2} = 0$, $S^-\Theta_{1/2}^{1/2} = -1/2$, $S^+\Theta_{1/2}^{-1/2} = 1/2$, $S^-\Theta_{1/2}^{-1/2} = 0$.

The eigenvalue of $S^+ = S^x + iS^y$ "increases" thus with m while that of $S^- = S^x - iS^y$ decreases. As m varies from $-S$ to S, it has $(2S+1)$ values.

In general, one writes $S^z = S - n$ where n is called the excited spin-wave "number".

Now, instead of Θ_S^m, one can use the following corresponding functions in n

$$\Theta_S^m \leftrightarrow F_S(n) \tag{5.29}$$
$$\Theta_S^{m+1} \leftrightarrow F_S(n-1) \tag{5.30}$$
$$\Theta_S^{m-1} \leftrightarrow F_S(n+1) \tag{5.31}$$

A decrease of m corresponds to an increase of n and vice-versa. Equations (5.26)-(5.28) become

$$S^+ F_S(n) = \sqrt{n(2S - n + 1)} F_S(n-1)$$
$$= \sqrt{2S}\sqrt{1 - \frac{n-1}{2S}} \sqrt{n} F_S(n-1) \tag{5.32}$$
$$S^- F_S(n) = \sqrt{(2S - n)(n+1)} F_S(n+1)$$
$$= \sqrt{2S}\sqrt{n+1}\sqrt{1 - \frac{n}{2S}} F_S(n+1) \tag{5.33}$$
$$S^z F_S(n) = (S - n) F(n) \tag{5.34}$$

The Holstein-Primakoff method consists in introducing the operators a and a^+ as follows

$$a F_S(n) = \sqrt{n} F_S(n-1) \tag{5.35}$$
$$a^+ F_S(n) = \sqrt{n+1} F_S(n+1) \tag{5.36}$$

These operators obey the commutation relations (see Problem 8 in section 5.6). Using Eqs. (5.35) and (5.36), one has

$$a^+ a F_S(n) = n F_S(n) \tag{5.37}$$

This relation shows that n is an eigenvalue of $a^+ a$. Thus, $a^+ a$ is called "spin-wave number operator".

From (5.35)-(5.37) one has

$$S^z = S - a^+ a \tag{5.38}$$
$$S^+ = \sqrt{2S} f(S) a \tag{5.39}$$
$$S^- = \sqrt{2S} a^+ f(S) \tag{5.40}$$

where

$$f(S) = \sqrt{1 - \frac{a^+ a}{2S}} \tag{5.41}$$

The transformations (5.38)-(5.41) are called "Holstein-Primakoff transformations".

Remark: One can apply S^z, S^+ and S^- given by (5.38)-(5.40) on the function $F_S(n)$ then make use of (5.35)-(5.37) to find again (5.32)-(5.34).

We consider now the following Heisenberg Hamiltonian

$$\mathcal{H} = -2J \sum_{(l,m)} \vec{S}_l \cdot \vec{S}_m - g\mu_B H \sum_l S_l^z \qquad (5.42)$$

where interactions are limited to nearest neighbor pairs (l, m) with an exchange integral $J > 0$ (ferromagnetic). H is the amplitude of a magnetic field applied along the z direction, g and μ_B are respectively the Landé factor and Bohr magneton.

Using $S^\pm = S^x \pm iS^y$ for $\vec{S}_l$ and $\vec{S}_m$ one rewrites $\mathcal{H}$ as

$$\mathcal{H} = -2J \sum_{(l,m)} \left[S_l^z S_m^z + \frac{1}{2}(S_l^+ S_m^- + S_l^- S_m^+) \right] - g\mu_B H \sum_l S_l^z \qquad (5.43)$$

Replacing $S^\pm$ and S^z by (5.38)-(5.40) while keeping position indices l and m of operators a and a^+ one obtains

$$\mathcal{H} = -2J \sum_{(l,m)} [S^2 + Sa_l^+ f_l(S)f_m(S)a_m + Sf_l(S)a_l a_m^+ f_m(S) - Sa_l^+ a_l$$

$$-Sa_m^+ a_m - a_l^+ a_l a_m^+ a_m] - g\mu_B H \sum_l (S - a_l^+ a_l) \qquad (5.44)$$

It is impossible to find a solution of this equation because of nonlinear terms such as $a_l^+ a_l a_m^+ a_m$, $f_l(S)$ and $f_m(S)$. In a first approximation, one can assume that the number of excited spin-waves n is small with respect to $2S$ (namely $a^+ a << 2S$) so that one can expand $f_l(S)$ and $f_m(S)$ as follows

$$f_l(S) \simeq 1 - \frac{a_l^+ a_l}{4S} + \dots \qquad (5.45)$$

$$f_m(S) \simeq 1 - \frac{a_m^+ a_m}{4S} + \dots \qquad (5.46)$$

Equation (5.44) becomes, to the quadratic order in a and a^+,

$$\mathcal{H} \simeq -ZJNS^2 - g\mu_B HNS + 4JS \sum_{(l,m)} (a_l^+ a_l - a_l^+ a_m) + g\mu_B H \sum_l a_l^+ a_l$$

$$\qquad (5.47)$$

where one has used the following relation

$$\sum_{(l,m)} 1 = \frac{1}{2} \sum_l \sum_{\vec{R}} 1 = \frac{Z}{2} \sum_l 1 = \frac{Z}{2} N \qquad (5.48)$$

$\vec{R}$ being the vector connecting the spin at l to one of its nearest neighbors, Z the coordination number and N the total number of spins.

The first term of (5.47) is the energy of the ground state where all spins are parallel and the second is a constant which will be omitted in the following. One introduces next the following Fourier transformations

$$a_l^+ = \frac{1}{\sqrt{N}} \sum_{\vec{k}} e^{-i\vec{k}\cdot\vec{l}} a_{\vec{k}}^+ \qquad (5.49)$$

$$a_l = \frac{1}{\sqrt{N}} \sum_{\vec{k}} e^{i\vec{k}\cdot\vec{l}} a_{\vec{k}} \qquad (5.50)$$

One can show that $a_{\vec{k}}$ and $a_{\vec{k}}^+$ obey the boson commutation relations just as real-space operators a_l and a_l^+ (see Problem 8 in section 5.6). Putting (5.50) in (5.47) one finds

$$\mathcal{H} = \sum_{\vec{k}} \left[2ZJS(1 - \gamma_{\vec{k}}) + g\mu_B H \right] a_{\vec{k}}^+ a_{\vec{k}} = \sum_{\vec{k}} \epsilon_{\vec{k}} a_{\vec{k}}^+ a_{\vec{k}} \qquad (5.51)$$

where

$$\epsilon_{\vec{k}} = 2ZJS(1 - \gamma_{\vec{k}}) + g\mu_B H \qquad (5.52)$$

One sees that in the case where $H = 0$ one finds again $\epsilon_{\vec{k}} = \hbar\omega_{\vec{k}}$, the magnon dispersion (5.15) obtained above by the classical treatment. The Holstein-Primakoff method allows however to go further by taking into account terms of order higher than quadratic in a^+ and a. By using expansions (5.45)-(5.46) one obtains terms of four operators, six operators, ... which represent interactions between spin waves. These terms play an important role when the temperature increases.

5.1.3 *Properties at low temperatures*

One studies here some low-temperature properties of spin waves using the dispersion relation (5.52).

5.1.3.1   *Magnetization*

One has seen above that a and a^+ are boson operators. The number of spin waves (or magnons) of $\vec{k}$ mode at temperature T is therefore given by the Bose-Einstein distribution

$$< n_{\vec{k}} >=< a_{\vec{k}}^+ a_{\vec{k}} >= \frac{1}{\exp[\beta\epsilon_{\vec{k}}] - 1} \qquad (5.53)$$

The magnetization is defined by

$$M = g\mu_B \sum_j < S_j^z >= g\mu_B \sum_{j=1}^N (S- < a_j^+ a_j >) \qquad (5.54)$$

where the sum is performed over all spins. One notes that the magnetization is defined as the magnetic moment per unit of volume. In the above expression, one takes thus the volume $\Omega = 1$.

With (5.50), Eq. (5.54) becomes

$$M = g\mu_B \sum_j < S_j^z >= g\mu_B N(S - \frac{1}{N} \sum_{\vec{k}} < n_{\vec{k}} >) \qquad (5.55)$$

where $< n_{\vec{k}} >$ is given by (5.53).

One shows here how to calculate M in the case of a simple cubic lattice and $H = 0$. The sum in (5.55) reads

$$\frac{1}{N} \sum_{\vec{k}} < n_{\vec{k}} >= \frac{\Omega}{(2\pi)^3} \frac{1}{N} \int \int \int \frac{dk_x dk_y dk_z}{\exp[\beta 2ZJS(1 - \gamma_{\vec{k}})] - 1} \qquad (5.56)$$

Using $\gamma_{\vec{k}}$ of (5.21), one has

$$1 - \gamma_{\vec{k}} = 1 - \frac{1}{3}[\cos(k_x a) + \cos(k_y a) + \cos(k_z a)]$$

$$\simeq \frac{1}{6}[(k_x a)^2 + (k_y a)^2 + (k_z a)^2 - O(k^4)] \qquad (5.57)$$

where one used an expansion for small $\vec{k}$ because at low temperatures (large β) the main contribution to the integral (5.56) comes from the region of small $\vec{k}$. With $\Omega = 1$, $Z = 6$ (simple cubic lattice) and (5.57), Eq. (5.56) becomes

$$\frac{1}{N} \sum_{\vec{k}} < n_{\vec{k}} > \simeq \frac{1}{(2\pi)^3} \int_0^\infty \frac{4\pi k^2 dk}{e^{\beta 2JS(ka)^2} - 1}$$

$$= \frac{1}{2\pi^2} \int_0^\infty \sum_{l=1}^\infty \exp[-l\beta 2JS(ka)^2] k^2 dk \qquad (5.58)$$

where the upper limit of the integral which is the border of the first Brillouin zone has been replaced by ∞ as justified by the fact that important contributions are due to small k. Changing the variable $x = l\beta 2JS(ka)^2$, one obtains

$$\frac{1}{N}\sum_{\vec{k}} <n_{\vec{k}}> \simeq \frac{1}{(2\pi)^2}(\frac{k_BT}{2JS})^{3/2}\sum_{l=1}^{\infty}\frac{1}{l^{3/2}}\int_0^{\infty}e^{-x}x^{1/2}dx \qquad (5.59)$$

One notes that the integral of the right-hand side is equal to $\frac{\sqrt{\pi}}{2}$ and that the sum on l is the Riemann's series $\zeta(3/2)$. Finally, one arrives at

$$\frac{M}{g\mu_B N} = S - \zeta(3/2)(\frac{k_B}{8\pi JS})^{3/2}T^{3/2} \qquad (5.60)$$

The magnetization decreases with increasing T by a term proportional to $T^{3/2}$. This is called the Bloch's law.

As T increases further one has to take into account higher-order terms in (5.57). In doing so, one obtains

$$\frac{M}{g\mu_B N} = S - \zeta(3/2)t^{3/2} - \frac{3\pi}{4}\zeta(5/2)t^{5/2} - \frac{33\pi^2}{32}\zeta(7/2)t^{7/2} - ... \qquad (5.61)$$

where $t = \frac{k_BT}{8\pi JS}$.

This result, exact at low temperatures, has been confirmed by experiments at least up to $T^{5/2}$.

Note: $\zeta(3/2) = 2.612$, $\zeta(5/2) = 1.341$, $\zeta(7/2) = 1.127$

5.1.3.2 *Energy and heat capacity*

The energy of a ferromagnet is calculated in the same manner. One obtains

$$E = -ZJNS^2 + \sum_{\vec{k}}\epsilon_{\vec{k}} <n_{\vec{k}}> \qquad (5.62)$$

$$E \simeq -ZJNS^2 + 12N\pi JS\zeta(5/2)t^{5/2} + ... \qquad (5.63)$$

where the first term is the ground-state energy and the second term the energy of excited magnons to the quadratic order.

The magnetic heat capacity C_V^m is thus

$$C_V^m = \frac{dE}{dT} = \frac{dE}{dt}\frac{dt}{dT} \simeq \frac{15}{4}Nk_B\zeta(5/2)t^{3/2} + ... \qquad (5.64)$$

One notes that at low temperatures the power of T in C_V^m is different from the heat capacity of an electron gas where C_V^e is proportional to T. It is also different from that of phonons where $C_V^p \simeq T^3$. This dependence of C_V^m on T has been experimentally confirmed.

5.2   Antiferromagnetism

5.2.1   *Dispersion relation*

Consider the following Heisenberg Hamiltonian

$$\mathcal{H} = J \sum_{(l,m)} \vec{S}_l \cdot \vec{S}_m - g\mu_B H \left(\sum_l S_l^z + \sum_m S_m^z \right)$$

$$= J \sum_{(l,m)} \left[S_l^z S_m^z + \frac{1}{2} \left(S_l^+ S_m^- + S_l^- S_m^+ \right) \right]$$

$$- g\mu_B H \left(\sum_l S_l^z + \sum_m S_m^z \right) \tag{5.65}$$

where interactions are limited to pairs of nearest neighbors (l, m) with an exchange integral $J > 0$ (antiferromagnetic). H is the amplitude of a magnetic field applied in the z direction. l and m indicate the sites belonging respectively to $\uparrow$ and $\downarrow$ sublattices. Note that sometimes we do not put the factor 2 in front of J in the Hamiltonian (5.65).

We use the Holstein-Primakoff method in the same manner as in the case of ferromagnets shown in (5.38)-(5.41) but with a distinction of up and down sublattices. For the up sublattice, one has

$$S_l^z = S - a_l^+ a_l \tag{5.66}$$

$$S_l^+ = \sqrt{2S} f_l(S) a_l \tag{5.67}$$

$$S_l^- = \sqrt{2S} a_l^+ f_l(S) \tag{5.68}$$

where

$$f_l(S) = \sqrt{1 - \frac{a_l^+ a_l}{2S}} \tag{5.69}$$

For the down sublattice, one defines

$$S_m^z = -S + b_m^+ b_m \tag{5.70}$$

$$S_m^+ = \sqrt{2S} b_m^+ f_m(S) \tag{5.71}$$

$$S_m^- = \sqrt{2S} f_m(S) b_m \tag{5.72}$$

where

$$f_l(S) = \sqrt{1 - \frac{b_m^+ b_m}{2S}} \tag{5.73}$$

The operators a, a^+, b and b^+ obey the commutation relations (Problem 8 in section 5.6). Replacing operators $S^\pm$ and S^z in (5.65) by these operators, one gets

$$
\mathcal{H} = J \sum_{(l,m)} [-S^2 + S f_l(S) a_l f_m(S) b_m + S a_l^+ f_l(S) b_m^+ f_m(S) + S a_l^+ a_l
$$
$$
+ S b_m^+ b_m - a_l^+ a_l b_m^+ b_m] - g\mu_B H [\sum_l (S - a_l^+ a_l)
$$
$$
- \sum_m (-S + b_m^+ b_m)] \tag{5.74}
$$

In a first approximation , one supposes that the number of excited spin waves n is small with respect to $2S$, namely $a_l^+ a_l << 2S$ and $b_m^+ b_m << 2S$, so that an expansion is possible for $f_l(S)$ and $f_m(S)$. One has then

$$
f_l(S) \simeq 1 - \frac{a_l^+ a_l}{4S} + ... \tag{5.75}
$$

$$
f_m(S) \simeq 1 - \frac{b_m^+ b_m}{4S} + ... \tag{5.76}
$$

Equation (5.74) becomes at the quadratic order

$$
\mathcal{H} \simeq -\frac{ZJNS^2}{2} + JS \sum_{(l,m)} \left(a_l^+ a_l + b_m^+ b_m + a_l^+ b_m^+ + a_l b_m \right)
$$
$$
+ g\mu_B H \left(\sum_l a_l^+ a_l - \sum_m b_m^+ b_m \right) \tag{5.77}
$$

where Z is the coordination number and the following relation has been used

$$
\sum_{(l,m)} 1 = \sum_l \sum_{\vec{R}} 1 = Z \sum_l 1 = Z \frac{N}{2} \tag{5.78}
$$

$\vec{R}$ is a vector connecting the spin at l to a nearest neighbor belonging to the other sublattice, $N/2$ the total number of spins in a sublattice.

The first term of (5.77) is the classical ground-state energy where neighboring spins are antiparallel (Néel ground state). One introduces now the following Fourier transformations

$$a_l^+ = \sqrt{\frac{2}{N}} \sum_{\vec{k}} e^{i\vec{k}\cdot\vec{l}} a_{\vec{k}}^+ \tag{5.79}$$

$$a_l = \sqrt{\frac{2}{N}} \sum_{\vec{k}} e^{-i\vec{k}\cdot\vec{l}} a_{\vec{k}} \tag{5.80}$$

$$b_m^+ = \sqrt{\frac{2}{N}} \sum_{\vec{k}} e^{-i\vec{k}\cdot\vec{m}} b_{\vec{k}}^+ \tag{5.81}$$

$$b_m = \sqrt{\frac{2}{N}} \sum_{\vec{k}} e^{i\vec{k}\cdot\vec{m}} b_{\vec{k}} \tag{5.82}$$

One can show that the Fourier components $a_{\vec{k}}$, $a_{\vec{k}}^+$, $b_{\vec{k}}$ and $b_{\vec{k}}^+$ obey the boson commutation relations. Putting these into (5.77), one gets

$$\mathcal{H} = -\frac{ZJNS^2}{2} + ZJS \sum_{\vec{k}} \left[(1+h)a_{\vec{k}}^+ a_{\vec{k}} + (1-h)b_{\vec{k}}^+ b_{\vec{k}} + \gamma_{\vec{k}}(a_{\vec{k}} b_{\vec{k}} + a_{\vec{k}}^+ b_{\vec{k}}^+) \right] \tag{5.83}$$

where

$$\gamma_{\vec{k}} = \frac{1}{Z} \sum_{\vec{R}} e^{i\vec{k}\cdot\vec{R}} \tag{5.84}$$

and

$$h = \frac{g\mu_B H}{ZJS} \tag{5.85}$$

One sees that $\mathcal{H}$ of Eq. (5.83) does not have a diagonal form of the "harmonic oscillator", namely $a_{\vec{k}}^+ a_{\vec{k}}$ and $b_{\vec{k}}^+ b_{\vec{k}}$. This is because of the existence of the term $a_{\vec{k}} b_{\vec{k}} + a_{\vec{k}}^+ b_{\vec{k}}^+$. One can diagonalize $\mathcal{H}$ with the following transformation

$$\alpha_{\vec{k}} = a_{\vec{k}} \cosh\theta_k + b_{\vec{k}}^+ \sinh\theta_k \tag{5.86}$$

$$\alpha_{\vec{k}}^+ = a_{\vec{k}}^+ \cosh\theta_k + b_{\vec{k}} \sinh\theta_k \tag{5.87}$$

$$\beta_{\vec{k}} = a_{\vec{k}}^+ \sinh\theta_k + b_{\vec{k}} \cosh\theta_k \tag{5.88}$$

$$\beta_{\vec{k}}^+ = a_{\vec{k}} \sinh\theta_k + b_{\vec{k}}^+ \cosh\theta_k \tag{5.89}$$

where θ_k is a variable to be determined as follows. The inverse transformation gives

$$a_{\vec{k}}^+ = \alpha_{\vec{k}}^+ \cosh\theta_k - \beta_{\vec{k}} \sinh\theta_k \tag{5.90}$$

$$a_{\vec{k}} = \alpha_{\vec{k}} \cosh\theta_k - \beta_{\vec{k}}^+ \sinh\theta_k \tag{5.91}$$

$$b_{\vec{k}}^+ = -\alpha_{\vec{k}} \sinh\theta_k + \beta_{\vec{k}}^+ \cosh\theta_k \tag{5.92}$$

$$b_{\vec{k}} = -\alpha_{\vec{k}}^+ \sinh\theta_k + \beta_{\vec{k}} \cosh\theta_k \tag{5.93}$$

One can verify that the new operators also obey the commutation relations (see Problem 9 of section 5.6). Replacing (5.90)-(5.93) in (5.83) one gets

$$\mathcal{H} = -\frac{ZJNS^2}{2} + ZJS\sum_{\vec{k}}\{\cosh(2\theta_k) - 1 - \gamma_{\vec{k}}\sinh(2\theta_k)$$

$$+[\cosh(2\theta_k) - \gamma_{\vec{k}}\sinh(2\theta_k) + h]\alpha_{\vec{k}}^+\alpha_{\vec{k}}$$

$$+[\cosh(2\theta_k) - \gamma_{\vec{k}}\sinh(2\theta_k) - h]\beta_{\vec{k}}^+\beta_{\vec{k}}$$

$$-[\sinh(2\theta_k) - \gamma_{\vec{k}}\cosh(2\theta_k)](\alpha_{\vec{k}}\beta_{\vec{k}} + \alpha_{\vec{k}}^+\beta_{\vec{k}}^+)\} \qquad (5.94)$$

In order that $\mathcal{H}$ is diagonal, the coefficient before the term $\alpha_{\vec{k}}\beta_{\vec{k}} + \alpha_{\vec{k}}^+\beta_{\vec{k}}^+$ should be zero. This requirement allows us to determine the variable θ_k. One has

$$\tanh(2\theta_k) = \gamma_{\vec{k}} \qquad (5.95)$$

Replacing $\sinh(2\theta_k)$ and $\cosh(2\theta_k)$ as functions of $\tanh(2\theta_k) = \gamma_{\vec{k}}$ one obtains

$$\mathcal{H} = -\frac{ZJNS^2}{2} + ZJS\sum_{\vec{k}}\left[\sqrt{1 - \gamma_{\vec{k}}^2} - 1\right]$$

$$+ZJS\sum_{\vec{k}}\left\{\left[\sqrt{1 - \gamma_{\vec{k}}^2} + h\right]\alpha_{\vec{k}}^+\alpha_{\vec{k}} + \left[\sqrt{1 - \gamma_{\vec{k}}^2} - h\right]\beta_{\vec{k}}^+\beta_{\vec{k}}\right\} \qquad (5.96)$$

One recognizes that for a given wave vector $\vec{k}$, there are two modes corresponding to

$$\epsilon_{\vec{k}}^{\pm} = ZJS\left[\sqrt{1 - \gamma_{\vec{k}}^2} \pm h\right] \qquad (5.97)$$

Without an applied field, these modes are degenerate. It is important to note that for small k, $\gamma_{\vec{k}}^2 \simeq (1 + ak^2 + ...)$ which leads to

$$\epsilon_{\vec{k}}^{\pm} \propto k \qquad (5.98)$$

This result for antiferromagnets is different from $\epsilon_{\vec{k}}^{\pm} \propto k^2$ obtained for ferromagnets. One expects therefore that thermodynamic properties are different for the two cases in particular at low temperatures where small k modes dominate. This will be indeed seen below.

5.2.2 *Properties at low temperatures*

If one knows the dispersion relation $\epsilon_{\vec{k}}$ one can in principle use formulas of statistical mechanics to study properties of a system as a function of the temperature (see Appendix A). One writes the partition function as follows [see Eq. (A.9) with a change of the notation to avoid a confusion with Z the coordination number used above]:

$$\Xi = \text{Tr}e^{-\beta\mathcal{H}}$$

$$= \sum_{n_k=0}^{\infty} \sum_{n'_k=0}^{\infty} \exp\left\{-\beta\left[E_0 + \sum_{\vec{k}}(n_k\epsilon_{\vec{k}}^+ + n'_k\epsilon_{\vec{k}}^-)\right]\right\}$$

$$= e^{-\beta E_0} \prod_{\vec{k}}\left[\frac{1}{1-e^{-\beta\epsilon_{\vec{k}}^+}}\frac{1}{1-e^{-\beta\epsilon_{\vec{k}}^-}}\right] \tag{5.99}$$

where one has used

$$n_k = \alpha_{\vec{k}}^+ \alpha_{\vec{k}} \tag{5.100}$$

$$n'_k = \beta_{\vec{k}}^+ \beta_{\vec{k}} \tag{5.101}$$

$$E_0 = -\frac{ZJNS^2}{2} + ZJS\sum_{\vec{k}}\left[\sqrt{1-\gamma_{\vec{k}}^2} - 1\right] \tag{5.102}$$

The free energy is written as

$$F = -k_BT\ln\Xi = E_0 + k_BT\sum_{\vec{k}}\left[\ln(1-\epsilon_{\vec{k}}^+) + \ln(1-\epsilon_{\vec{k}}^-)\right]$$

$$= E_0 + 2k_BT\sum_{\vec{k}}\ln(1-\epsilon_{\vec{k}}) \quad \text{if } h = 0 \tag{5.103}$$

To calculate various thermodynamic properties, one uses the above expression of F as seen below.

5.2.2.1 *Energy*

For $h = 0$, one has

$$\mathcal{H} = -\frac{ZJNS^2}{2} + ZJS\frac{N}{2} + \sum_{\vec{k}}\epsilon_{\vec{k}}(\alpha_{\vec{k}}^+\alpha_{\vec{k}} + \beta_{\vec{k}}^+\beta_{\vec{k}} + 1)$$

$$= -\frac{ZJNS(S+1)}{2} + 2\sum_{\vec{k}}\epsilon_{\vec{k}}(n_k + \frac{1}{2}) \tag{5.104}$$

where one has used $\sum_{\vec{k}}1 = \frac{N}{2} = $ number of microscopic states in the first Brillouin zone which is equal the number of spins in each sublattice. At $T = 0$, $n_k = n'_k = 0$ one obtains

$$E(T=0) = -\frac{ZJNS(S+1)}{2} + \sum_{\vec{k}}\epsilon_{\vec{k}} \tag{5.105}$$

The second term is a correction to the classical ground-state energy $-\frac{ZJNS^2}{2}$ (Néel state). This correction is due to quantum fluctuations in analogy with the zero-point phonon energy.

At low temperatures, one calculates the magnon energy by the use of a low-temperature expansion. One gets

$$< E >= -T^2 \frac{\partial}{\partial T}(\frac{F}{T}) \simeq E(T = 0) + aT^4 \qquad (5.106)$$

where a is a coefficient proportional to ZJS. One notes that the power of T is different from that of the ferromagnetic case [Eq. (5.63)]. This is a consequence of $\epsilon_{\vec{k}} \propto k$ for small k.

5.2.2.2 *Magnetization at low temperatures*

To calculate the sublattice magnetization, one writes for the ↑ sublattice

$$
\begin{aligned}
M &= \sum_l (S- < a_l^+ a_l >) \\
&= \frac{N}{2}S - \frac{2}{N} \sum_{\vec{k}} \sum_{\vec{k}'} < a_{\vec{k}}^+ a_{\vec{k}'} > \sum_{\vec{l}} e^{i(\vec{k}-\vec{k}')\cdot\vec{l}} \\
&= \frac{N}{2}S - \frac{2}{N} \sum_{\vec{k}} \sum_{\vec{k}'} < a_{\vec{k}}^+ a_{\vec{k}'} > \frac{N}{2}\delta_{\vec{k},\vec{k}'} \\
&= \frac{N}{2}S - \sum_{\vec{k}} < a_{\vec{k}}^+ a_{\vec{k}} > \\
&= \frac{N}{2}S - \sum_{\vec{k}} < \cosh^2 \theta_k \alpha_{\vec{k}}^+ \alpha_{\vec{k}} + \sinh^2 \theta_k \beta_{\vec{k}}^+ \beta_{\vec{k}} + \sinh^2 \theta_k >
\end{aligned}
$$

$$(5.107)$$

where one has used successively the Fourier transformation and relations (5.90)-(5.93). One expresses now $\cosh^2 \theta_k$ and $\sinh^2 \theta_k$ in terms of $\gamma_{\vec{k}}$ using (5.95), then one uses

$$< \alpha_{\vec{k}}^+ \alpha_{\vec{k}} >=< \beta_{\vec{k}}^+ \beta_{\vec{k}} >= \frac{1}{e^{\beta \epsilon_{\vec{k}}} - 1} \qquad (5.108)$$

to obtain M. At low temperatures, using an expansion of $\frac{1}{e^{\beta \epsilon_{\vec{k}}}-1}$ for small k with $\epsilon_{\vec{k}} \propto k$, one gets

$$M \simeq \frac{N}{2}(S - \Delta S - AT^2) \qquad (5.109)$$

where $\Delta S = \sum_{\vec{k}} \sinh^2 \theta_k$ is independent of T, and A a coefficient.

One sees that at $T = 0$, the magnetization is $S - \Delta S$ which is smaller than the spin magnitude S. ΔS is called the zero-point spin contraction. ΔS depends on the lattice: $\Delta S \simeq 0.197$ for an antiferrromagnetic square lattice, $\Delta S \simeq 0.078$ for a cubic antiferromagnet of NaCl type.

Note that the sublattice magnetization of an antiferromagnet depends on T^2 while the ferromagnetic magnetization depends on $T^{3/2}$ [Eq. (5.60)].

5.3   Ferrimagnetism

In this section, one calculates the magnon dispersion relation in the case of a ferrimagnet. In principle, one can use the Holstein-Primakoff method as described above. However, the purpose here is to obtain the dispersion relation in a simplest manner. So, one will use the method of equation of motion as described hereafter.

One considers here a simple model of ferrimagnet which is composed of two sublattices of A and B Heisenberg spins described by the Hamiltonian (4.74): the A sublattice contains $\uparrow$ spins of amplitude S_A and the B sublattice contains $\downarrow$ spins of amplitude S_B. One rewrites (4.74) as

$$
\mathcal{H} = J_1 \sum_{(l,m)} \left[S_l^z S_m^z + \frac{1}{2}(S_l^+ S_m^- + S_l^- S_m^+) \right]
$$
$$
+ J_2^A \sum_{(l,l')} \left[S_l^z S_{l'}^z + \frac{1}{2}\left(S_l^+ S_{l'}^- + S_l^- S_{l'}^+\right) \right]
$$
$$
+ J_2^B \sum_{(m,m')} \left[S_m^z S_{m'}^z + \frac{1}{2}(S_m^+ S_{m'}^- + S_m^- S_{m'}^+) \right] \tag{5.110}
$$

To simplify the presentation, one supposes in the following $J_2^A = J_2^B = J_2$.

The Heisenberg equation of motion for the operator $S_l^-(t)$ reads

$$
i\hbar \frac{dS_l^-}{dt} = [S_l^-, \mathcal{H}] \simeq J_1 \left[\sum_m < S_m^z > S_l^- - < S_l^z > S_m^+ \right]
$$
$$
+ J_2 \left[\sum_{l'} < S_{l'}^z > S_l^- - < S_l^z > S_{l'}^- \right] \tag{5.111}
$$

The equation of motion for S_m^+ can be obtained from this equation by exchanging $l \leftrightarrow m$, $l' \leftrightarrow m'$ and $S^+ \leftrightarrow S^-$. It is noted that in this equation, one has replaced, in the mean-field spirit, the operators S_m^z and S_l^z by their averaged values $< S_m^z >$ and $< S_l^z >$. Using for equations of

S_l^- and S_m^+ the Fourier transformations

$$S_l^- = \frac{1}{(2\pi)^3} \int_{-\infty}^{-\infty} d\omega e^{-i\omega t} \int_{BZ} d\vec{k} e^{i\vec{k}\cdot\vec{l}} U_A(\vec{k},\omega) \qquad (5.112)$$

$$S_m^+ = \frac{1}{(2\pi)^3} \int_{-\infty}^{-\infty} d\omega e^{-i\omega t} \int_{BZ} d\vec{k} e^{i\vec{k}\cdot\vec{l}} U_B(\vec{k},\omega) \qquad (5.113)$$

where BZ indicates the first Brillouin zone, one obtains for a body-centered cubic lattice the following coupled equations

$$(E + E_A)U_A = -8(1-\alpha)\gamma_1(\vec{k})U_B \qquad (5.114)$$

$$(E - E_B)U_B = 8(1+\alpha)\gamma_1(\vec{k})U_A \qquad (5.115)$$

where one has used

$$< S_l^z > = S_A = (1-\alpha)S \qquad (5.116)$$

$$< S_m^z > = S_B = -(1+\alpha)S \qquad (5.117)$$

$$E = \frac{\hbar\omega}{J_1 S} \qquad (5.118)$$

$$\epsilon = \frac{J_2}{J_1} \qquad (5.119)$$

$$E_A = 8(1+\alpha) - 6\epsilon(1-\alpha)[1-\gamma_2(\vec{k})] \qquad (5.120)$$

$$E_B = 8(1-\alpha) - 6\epsilon(1+\alpha)[1-\gamma_2(\vec{k})] \qquad (5.121)$$

$$\gamma_1(\vec{k}) = \frac{1}{C_1} \sum_{\vec{\rho}_1 \in NN} e^{i\vec{k}\cdot\vec{\rho}_1} = \cos(\frac{k_x a}{2})\cos(\frac{k_y a}{2})\cos(\frac{k_z a}{2}) \qquad (5.122)$$

$$\gamma_2(\vec{k}) = \frac{1}{C_2} \sum_{\vec{\rho}_2 \in NNN} e^{i\vec{k}\cdot\vec{\rho}_2}$$

$$= \frac{1}{3}[\cos(k_x a) + \cos(k_y a) + \cos(k_z a)] \qquad (5.123)$$

a being the lattice constant, $C_1 = 8$ and $C_2 = 6$ the numbers of nearest and next nearest neighbors. One recognizes here that $\vec{\rho}_1$ and $\vec{\rho}_2$ connect a site to its nearest neighbors (NN) and next nearest neighbors (NNN), respectively.

For a nontrivial solution of (5.114) and (5.115), one imposes the secular equation

$$(E + E_A)(E - E_B) = -64(1-\alpha^2)\gamma_1^2(\vec{k}) \qquad (5.124)$$

from which, one obtains

$$E_\pm = \frac{1}{2}\left[E_B - E_A \pm \sqrt{(E_A - E_B)^2 + 4[E_A E_B - 64(1-\alpha^2)\gamma_1^2(\vec{k})]} \right]$$

$$(5.125)$$

Let us examine some particular cases:

- For $k = 0$, one has $E_+ = 0$ and $E_- = -16\alpha$, independent of ϵ, namely J_2
- For $k_x a = k_y a = k_z a = \pi$, one has $E_+ = 8(1-\alpha) - 12\epsilon(1+\alpha)$ and $E_- = -8(1+\alpha) + 12\epsilon(1-\alpha)$.
- If $\epsilon = 0$, then for $k_x a = k_y a = k_z a = \pi$, one has $E_+ = 8(1-\alpha)$ and $E_- = -8(1+\alpha)$.

These results show a gap in the magnon spectrum at $\vec{k} = 0$ with a width proportional to α. Figure 5.5 shows the magnon spectrum versus $k_x = k_y$ for $k_z = 0$.

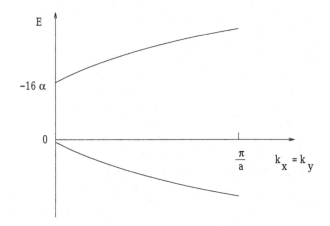

Fig. 5.5   Magnon spectrum of a ferrimagnet versus $k_x = k_y$ with $k_z = 0$, $\alpha = -1/3$, $\epsilon = 0$.

The reader is recommended to use the method of the preceding section to calculate the energy, the heat capacity and the magnetization at low temperatures.

5.4   Helimagnetism

Helimagnets are a family of materials in which the spins are not collinear in the low-T ordered phase as in other systems considered above. Due to a competition between various kinds of interaction, the neighboring spins make an angle different from 0 and π. Helimagnets are thus frustrated systems which present many spectacular properties [39, 40].

Consider here a simplest example of helimagnet which is a chain with a ferromagnetic interaction $J_1(> 0)$ between nearest neighbors and an antiferromagnetic interaction $J_2(<0)$ between next nearest neighbors. When $\varepsilon = |J_2|/J_1$ is larger than a critical value ε_c, the spin configuration of the ground state becomes non collinear. One shows that the helical configuration displayed in Fig. 5.6 is obtained by minimizing the following interaction energy

$$E = -J_1 \sum_i \mathbf{S}_i \cdot \mathbf{S}_{i+1} + |J_2| \sum_i \mathbf{S}_i \cdot \mathbf{S}_{i+2}$$

$$= S^2 \left[-J_1 \cos\theta + |J_2| \cos(2\theta) \right] \sum_i 1 \tag{5.126}$$

$$\frac{\partial E}{\partial \theta} = S^2 \left[J_1 \sin\theta - 2|J_2| \sin(2\theta) \right] \sum_i 1 = 0$$

$$= S^2 \left[J_1 \sin\theta - 4|J_2| \sin\theta \cos\theta \right] \sum_i 1 = 0, \tag{5.127}$$

where one has supposed that the angle between nearest neighbors is θ.

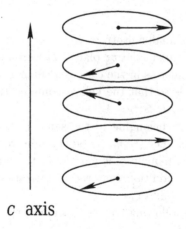

c axis

Fig. 5.6   Helical configuration when $\varepsilon = |J_2|/J_1 > \varepsilon_c = 1/4$ ($J_1 > 0$, $J_2 < 0$).

The solutions are

- Ferromagnetic and antiferromagnetic configurations:

$$\sin\theta = 0 \longrightarrow \theta = 0, \pi$$

- Helical configuration:

$$\cos\theta = \frac{J_1}{4|J_2|} \longrightarrow \theta = \pm \arccos\left(\frac{J_1}{4|J_2|} \right). \tag{5.128}$$

The last solution is possible if $-1 \leq \cos\theta \leq 1$, i.e. $J_1/(|J_2|) \leq 4$ or $|J_2|/J_1 \geq 1/4 \equiv \varepsilon_c$. There are two degenerate configurations corresponding to clockwise and counter-clockwise turning angles. Replacing the above solutions into Eq. (5.126), we see that the antiferromagnetic solution $(\theta = \pi)$ corresponds to the maximum of E. It is to be discarded. The ferromagnetic solution has an energy lower than that of the helical solution for $|J_2|/J_1 < 1/4$. It is therefore more stable in this range of parameters. For $|J_2|/J_1 > 1/4$, the reverse is true.

For the magnon spectrum, let us consider a three-dimensional version of the helical chain considered above which has a body-centered cubic lattice with Heisenberg spins interacting with each other via i) a ferromagnetic interaction $J_1 > 0$ between nearest neighbors, ii) an antiferromagnetic interaction $J_2 < 0$ between next nearest neighbors along the y axis. The Hamiltonian is given by

$$\mathcal{H} = -J_1 \sum_{<i,j>} \vec{S}_i \cdot \vec{S}_j - J_2 \sum_{<i,l>} \vec{S}_i \cdot \vec{S}_l + D \sum_i (S_i^y)^2 \qquad (5.129)$$

where $D > 0$ is a very small anisotropy of the type "easy-plane anisotropy" which stabilizes the spins in the xz plane. The ground state can be calculated in the same manner as in the case of a chain given above. The result shows that $\cos\theta = \frac{J_1}{|J_2|}$ so that the helical configuration in the y direction is stable when $\varepsilon = |J_2|/J_1 > \varepsilon_c = 1$.

Note that the spins belonging to the same xz plane perpendicular to the y axis are parallel. Let $(\vec{\xi}_i, \vec{\eta}_i, \vec{\zeta}_i)$ be the unit vectors making a direct trihedron at the site i, namely $\vec{\eta}_i$ is parallel to the y axis as shown in Fig. 5.7. One supposes in addition that the quantization axis of the spin $\vec{S}_i$ coincides with the local axis $\vec{\zeta}_i$.

One uses now the following transformation in the local coordinates associated with $\vec{S}_i$ and $\vec{S}_j$

$$\vec{\eta}_j = \vec{\eta}_i \qquad (5.130)$$

$$\vec{\zeta}_j = \cos Q \vec{\zeta}_i + \sin Q \vec{\xi}_i \qquad (5.131)$$

$$\vec{\xi}_j = -\sin Q \vec{\zeta}_i + \cos Q \vec{\xi}_i \qquad (5.132)$$

One writes

$$\vec{S}_i = S_i^x \vec{\xi}_i + S_i^y \vec{\eta}_i + S_i^z \vec{\zeta}_i \qquad (5.133)$$

$$\vec{S}_j = S_j^x \vec{\xi}_j + S_j^y \vec{\eta}_j + S_j^z \vec{\zeta}_j \qquad (5.134)$$

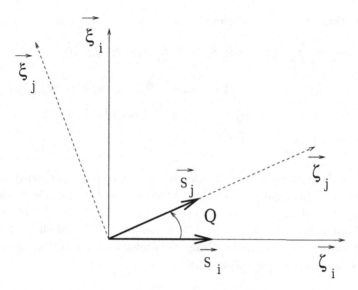

Fig. 5.7 Local coordinates defined for two spins $\vec{S}_i$ and $\vec{S}_j$. The axis $\vec{\eta}$ is common for the two spins.

Their scalar product becomes

$$
\begin{aligned}
\vec{S}_i \cdot \vec{S}_j &= [S_i^x \vec{\xi}_i + S_i^y \vec{\eta}_i + S_i^z \vec{\zeta}_i] \cdot [S_j^x (-\sin Q \vec{\zeta}_i + \cos Q \vec{\xi}_i) \\
&\quad + S_j^y \vec{\eta}_i + S_j^z (\cos Q \vec{\zeta}_i + \sin Q \vec{\xi}_i)] \\
&= S_i^z (-\sin Q S_j^x + \cos Q S_j^z) + S_i^y S_j^y \\
&\quad + S_i^x (\cos Q S_j^x + \sin Q S_j^z) \\
&= \cos Q S_i^z S_j^z - \sin Q S_i^z \left[\frac{S_j^+ + S_j^-}{2} \right] - \frac{(S_i^+ - S_i^-)(S_j^+ - S_j^-)}{4} \\
&\quad + \cos Q \frac{(S_i^+ + S_i^-)(S_j^+ + S_j^-)}{4} + \sin Q \left[\frac{S_i^+ + S_i^-}{2} \right] S_j^z \quad (5.135)
\end{aligned}
$$

To be general, the angle Q should depend on positions of $\vec{S}_i$ and $\vec{S}_j$. One defines $\cos Q = \cos(\vec{Q} \cdot \vec{R}_{ij})$ where $\vec{Q}$ is the vector of modulus Q, perpendicular to the plane of the angle Q, namely plane $(\vec{\zeta}, \vec{\xi})$, and $\vec{R}_{ij}$ the vector connecting the positions of $\vec{S}_i$ and $\vec{S}_j$. One shall keep in the following $J(\vec{R}_{ij})$ as interaction between $\vec{S}_i$ and $\vec{S}_j$ which will be replaced by J_1 and J_2 depending on $\vec{R}_{ij}$ at the end of the calculation.

Equation (5.129) is rewritten as

$$\mathcal{H} = -\frac{1}{4} \sum_{(i,j)} J(\vec{R}_{ij})\{(S_i^+ S_j^- + S_i^- S_j^+)[1 + \cos(\vec{Q} \cdot \vec{R}_{ij})]$$

$$-(S_i^+ S_j^+ + S_i^- S_j^-)[1 - \cos(\vec{Q} \cdot \vec{R}_{ij})] + 4S_i^z S_j^z \cos(\vec{Q} \cdot \vec{R}_{ij})$$

$$+2[(S_i^+ + S_i^-)S_j^z - S_i^z(S_j^+ + S_j^-)] \sin(\vec{Q} \cdot \vec{R}_{ij})\}$$

$$-\frac{D}{4} \sum_i (S_i^+ - S_i^-)^2 \tag{5.136}$$

With this Hamiltonian, one can choose an appropriate method to calculate the magnon dispersion relation. One can use the Holstein-Primakoff method by replacing the operators $S^{\pm}$ and S^z by (5.38)-(5.41), or the Green's function method (chapter 6) or simply by the method of equation of motion (section 5.3). Using the Holstein-Primakoff method, one obtains the magnon dispersion relation

$$\mathcal{H} = -NSJ(\vec{Q}) + \frac{S}{2} \sum_{\vec{k}} [A(\vec{k}, \vec{Q})(a_{\vec{k}} a_{\vec{k}}^+ + a_{\vec{k}}^+ a_{\vec{k}})$$

$$+B(\vec{k}, \vec{Q})(a_{\vec{k}} a_{-\vec{k}} + a_{\vec{k}}^+ a_{-\vec{k}}^+)] \tag{5.137}$$

where

$$J(\vec{k}) = \sum_{\vec{R}_{ij}} J(\vec{R}_{ij}) \exp(\vec{k} \cdot \vec{R}_{ij}) \quad (\text{sum on } \vec{R}_{ij}) \tag{5.138}$$

$$A(\vec{k}, Q) = \left[2J(\vec{Q}) - J(\vec{k}) - \frac{1}{2}[J(\vec{k} + \vec{Q}) + J(\vec{k} - \vec{Q})] + D \right] \tag{5.139}$$

$$B(\vec{k}, \vec{Q}) = \left[J(\vec{k}) - \frac{1}{2}[J(\vec{k} + \vec{Q}) + J(\vec{k} - \vec{Q})] - D \right] \tag{5.140}$$

The Hamiltonian (5.137) can be diagonalized by introducing the new operators $\alpha_{\vec{k}}$ and $\alpha_{\vec{k}}^+$ just as in the antiferromagnetic case studied above

$$a_{\vec{k}} = \alpha_{\vec{k}} \cosh\theta_k - \alpha_{-\vec{k}}^+ \sinh\theta_k \tag{5.141}$$

$$a_{\vec{k}}^+ = \alpha_{\vec{k}}^+ \cosh\theta_k - \alpha_{-\vec{k}} \sinh\theta_k \tag{5.142}$$

where $\alpha_{\vec{k}}$ and $\alpha_{\vec{k}}^+$ obey the boson commutation relations [see similar transformation in Eqs. (5.86)-(5.89)]. Hamiltonian (5.137) is diagonal if one takes

$$\tanh(2\theta_k) = \frac{B(\vec{k}, \vec{Q})}{A(\vec{k}, \vec{Q})} \tag{5.143}$$

One then has

$$\mathcal{H} = \frac{S}{2} \sum_{\vec{k}} \hbar \omega_{\vec{k}} [\alpha_{\vec{k}}^{+} \alpha_{\vec{k}} + \alpha_{\vec{k}} \alpha_{\vec{k}}^{+}] \qquad (5.144)$$

where the energy of the magnon of mode $\vec{k}$ is

$$\hbar \omega_{\vec{k}} = \sqrt{A(\vec{k}, \vec{Q})^2 - B(\vec{k}, \vec{Q})^2} \qquad (5.145)$$

In the case of the body-centered cubic lattice, one has

$$J(\vec{k}) = 8J_1 \cos(k_x a/2) \cos(k_y a/2) \cos(k_z c/2) + 2J_2 \cos(k_z c)$$
$$= 2J_2 \left[-4 \cos Q \cos(k_x a/2) \cos(k_y a/2) \cos(k_z c/2) + \cos(k_z c) \right]$$

where $\epsilon_c = 1$ has been used, a and c being the lattice constants (one uses c for the helical axis). Figure 5.8 shows the magnon spectrum for J_2/J_1 corresponding to $Q = \pi/3$. One observes that the magnon frequency is zero not only at $k = 0$ but also at $k_z = Q$.

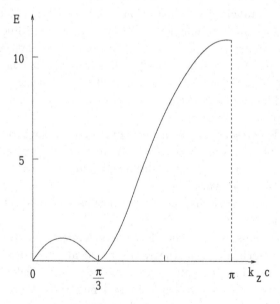

Fig. 5.8 Magnon spectrum $E = \hbar \omega_{\vec{k}}$, Eq. (5.145), versus $k_z c$ in a helimagnet defined by the Hamiltonian (5.129), with $Q = \pi/3$, $k_x = k_y = 0$.

5.5   Conclusion

The theory of magnons presented in this chapter allows us to calculate the spin-wave dispersion relation which is used to study thermodynamic properties of magnetic systems at low temperatures in a precise manner. When the temperature increases, it is necessary to take into account higher-order terms in the Hamiltonian which represent interactions between magnons. The calculation then becomes more complicated and needs other approximations to decouple chains of operators to renormalize the harmonic, or free-magnon, spectrum. One can also use the Green's function method which allows us to include implicitly magnon-magnon interactions up to the transition temperature (see chapter 6). The method involves however some decoupling schemes which make the results near the transition less precise.

Let us summarize some main points of this chapter. First, on the dispersion relation, results of antiferromagnets are quite different from those of ferromagnets. For instance, at small $\vec{k}$, one has for antiferromagnets $\epsilon_{\vec{k}} \propto k$, while for ferromagnets $\epsilon_{\vec{k}} \propto k^2$. This difference yields different temperature-dependence of physical properties. One notes also that the completely antiparallel spin configuration (Néel state) is not the real ground state of quantum antiferromagnets. Zero-point quantum fluctuations reduce the spin amplitude at $T = 0$ in antiferromagnets. For ferrimagnets, only a simple example has been used to illustrate the magnon spectrum. A gap due to the difference of spin magnitudes is frequently found. However, it should be emphasized that real ferrimagnets often have much more complicated lattice structures.

We have also presented some aspects of helimagnets where the magnon spectrum has been shown. It is important to note that systems with non collinear spin configurations have been and still are subject of intensive investigations since 30 years [39, 40].

Finally, to close this chapter, let us outline some aspects which are important to know:

1) *Magnetic anisotropy:* The Heisenberg model is isotropic. It does not tell us in which direction the spins should align themselves. It is a habit to suppose that the spins are on the z axis for calculation. But all other directions are equivalent. In real materials, there often exists a preferential direction which is called "easy-magnetization axis". This magnetic "anisotropy" stems from complicated microscopic origins such as spin-orbit interaction, dipole-dipole interaction, etc. [62].

2) *Long-range interactions*: In this chapter, one has supposed for simplicity that interactions are limited to nearest neighbors. The calculations can be of course extended to interactions up to second, third, ... nearest neighbors. However for infinite-range interaction specific methods should be used.

3) *Low-dimensional systems*: In the case of one dimension, a system of spins with short-range interactions are disordered for nonzero T whatever the spin model is. In the case of two dimensions, systems of discrete spins such as Ising and Potts models have a phase transition at a finite temperature [16]. Systems of vector spins such as Heisenberg and XY models are not ordered at $T \neq 0$. This has been rigorously shown by a theorem of Mermin-Wagner [95]. One can see this in an approximative manner: in two dimensions one replaces $4\pi k^2 dk$ in (5.58) by $2\pi k dk$, the integral then becomes divergent at the lower bound ($k \rightarrow 0$), causing undefined M except when $T = 0$.

4) There exist other methods to study temperature-dependent properties of magnetic systems such as low- and high-temperature series expansions [42], renormalization-group analysis and numerical simulations. These are shown in chapters 7 and 8.

5.6 Problems

Problem 1. Prove (5.63)-(5.64).

Problem 2. Chain of Heisenberg spins:

a) Calculate the magnon spectrum $\epsilon(k)$ for a chain of Heisenberg spins of lattice constant a with ferromagnetic interactions J_1 between nearest neighbors and J_2 between next-nearest-neighbors. Plot $\epsilon(k)$ versus k within the first Brillouin zone (BZ).

b) The spectrum $\epsilon(k)$ obtained in the previous question is supposed to be valid when J_2 becomes antiferromagnetic as long as $\epsilon(k) \geq 0$. Show that the ferromagnetic order becomes unstable when $|J_2|$ is larger than a critical value. Determine this value and compare to ε_c given below Eq. (5.128).

Problem 3. Heisenberg spin systems in two dimensions:

Consider the Heisenberg spin model on a two-dimensional lattice with a ferromagnetic interaction J between nearest neighbors.

a) Calculate the magnon spectrum $\epsilon(\vec{k})$ as a function of $\vec{k}$. Check that $\epsilon(\vec{k}) \propto k^2$ as $k \rightarrow 0$.

b) Write down the formal expression connecting the magnetization M to the temperature T. Show that M is undefined as soon as T becomes non zero. Comments.

Problem 4. Prove Eqs. (5.143)-(5.145).

Problem 5. Consider the Ising spin model on a 'Union-Jack' lattice, namely the square lattice in which one square out of every two has a centered site. Define sublattice 1 containing the centered sites, and sublattice 2 containing the remaining sites (namely the cornered sites). Let J_1 be the interaction between a centered spin and its nearest neighbors, J_2 and J_3 the interactions between two nearest spins on the y and x axes of the sublattice 2, respectively. Determine the phase diagram of the ground state in the space (J_1, J_2, J_3). Indicate the phases where the centered spins are undefined (partial disorder).

Problem 6. Using the method described in section 5.4, determine the ground-state spin configuration of a triangular lattice with XY spins interacting with each other via an antiferromagnetic exchange J_1 between nearest neighbors.

Problem 7. Uniaxial anisotropy

a) Show that if one includes in the Heisenberg Hamiltonian for ferromagnets the following anisotropy term $-D\sum_i(S_i^z)^2$ where D is a positive constant and the sum is performed over all spins, one obtains the following magnon spectrum
$\epsilon_{\vec{k}} = 2ZJS(1 - \gamma_{\vec{k}} + d)$
where $d \equiv \frac{D}{2ZJS}$ [see the definitions of other notations in Eq. (5.52)].

b) Is it possible to have a long-range magnetic ordering at finite temperature in two dimensions ? (cf. Problem 3).

Problem 8. Show that the operators a^+ and a defined in the Holstein-Primakoff approximation, Eqs. (5.35) and (5.36), respect rigorously the commutation relations between the spin operators.

Problem 9. Show that the operators defined in Eqs. (5.86)-(5.89) obey the commutation relations.

Problem 10. Show that the magnon spectrum (5.125) becomes unstable when the interaction between next-nearest neighbors defined in ϵ, Eq. (5.119), is larger than a critical constant.

Chapter 6

Green's Function Method in Magnetism

The Green's function is a very general method in quantum field theory. It can be used for various problems in many areas such as condensed matter, nuclear physics and elementary-particle physics [50]. The general formulation is rather complicated, abstract, not suitable for a quick application. For our purpose, we present in this chapter a simplified version which, from the beginning, aims at an application to systems of interacting spins. This formulation does not need much of basic knowledge in quantum field theory. A basic level in quantum mechanics is enough to understand and to master the method.

As seen below, the Green's function method completes the theory of magnons presented in chapter 5: while the theory of magnons is valid only at low temperatures, the Green's function method can treat, with some precision, the whole temperature region going from low-temperature phase up to the transition temperature T_c. In order to take into account magnon-magnon interactions in the theory of magnons, we have to treat terms of higher orders in the Holstein-Primakoff expansion at finite temperature to modify the harmonic magnon spectrum. This is a formidable task. The advantage of the Green's function method is even at the lowest level (see the Tyablikov decoupling scheme below) it implicitly includes spin-spin correlations so that one can follow the evolution of the magnon spectrum with varying temperature. However, since all correlations are not hierarchically included, the question of its validity is still an open question. Applications of the Green's function method in surface magnetism are shown in chapter 9.

6.1   Definition of the Green's function

Let $A(t)$ and $B(t')$ be two operators in the Heisenberg representation at times t and t', respectively. We define the retarded Green's function by

$$G^r_{AB}(t - t') = <<A(t); B(t')>>^r = -i\theta(t - t') < [A(t), B(t')] > \quad (6.1)$$

and the advanced Green's function by

$$G^a_{AB}(t - t') = <<A(t); B(t')>>^a = i\theta(t' - t) < [A(t), B(t')] > \quad (6.2)$$

where

$$\theta(t - t') = 1 \qquad\qquad \text{if } t > t' \qquad\qquad (6.3)$$
$$\theta(t - t') = 0 \qquad\qquad \text{if } t < t' \qquad\qquad (6.4)$$

$[A(t), B(t')]$ being the commutation relation and $< ... >$ denoting the thermal average. In spite of a complicated notation in its definition, the Green's function is just a thermal average of a commutation relation between operators which is connected, as will be seen below, to physical properties of the system. Therefore, the choice of the operators $A(t)$ and $B(t')$ depends on what we want to study. They should be chosen to allow us to calculate desired physical quantities. For instance, to study an electron gas we can choose $A(t) = a_i^+(t)$ and $B(t') = a_i(t')$ where $a_i^+(t)$ and $a_i(t')$ are creation and annihilation operators of electron state i (cf. chapter 3), to study phonon excitations we can choose $A(t)$ and $B(t')$ as phonon creation and annihilation operators, and for a spin system we can choose $A(t) = S_i^+(t)$ and $B(t') = S_i^-(t')$ where $S_i^+(t)$ and $S_i^-(t')$ are spin operators (cf. chapter 5).

The Green's function contains information on the system, in particular on the elementary excitation spectrum. In what follows we present the formulation of the method as it is applied to a spin system so that the reader can associate mathematical tools presented here to physical meanings of a phenomenon that has been seen in chapter 5.

6.2   Formulation

We consider the following functions

$$F_{AB}(t - t') = < A(t)B(t') > \qquad\qquad (6.5)$$
$$F_{BA}(t' - t) = < B(t')A(t) > \qquad\qquad (6.6)$$

The time Fourier transformation of $F_{BA}(t' - t)$ is written as

$$F_{BA}(t' - t) = \int_{-\infty}^{\infty} I(\omega)e^{i\omega(t'-t)}d\omega \tag{6.7}$$

We show that

$$< A(t)B(t') > = < B(t')A(t + i\beta) > \tag{6.8}$$

where $\beta = (k_B T)^{-1}$.

Demonstration: We use essentially the circular permutation properties of operators in Tr[...] and $A(t) = e^{\beta \mathcal{H} t}Ae^{-\beta \mathcal{H} t}$:

$$< A(t)B(t') > = \frac{\text{Tr}A(t)B(t')e^{-\beta \mathcal{H}}}{\text{Tre}^{-\beta \mathcal{H}}} = \frac{\text{Tre}^{-\beta \mathcal{H}}A(t)B(t')}{\text{Tre}^{-\beta \mathcal{H}}}$$

$$= \frac{\text{Tr}B(t')e^{-\beta \mathcal{H}}A(t)}{\text{Tre}^{-\beta \mathcal{H}}}$$

$$= \frac{\text{Tre}^{\beta \mathcal{H} t'}Be^{-\beta \mathcal{H} t'}e^{-\beta \mathcal{H}}e^{\beta \mathcal{H} t}Ae^{-\beta \mathcal{H} t}}{\text{Tre}^{-\beta \mathcal{H}}}$$

$$= \frac{\text{Tre}^{-\beta \mathcal{H}}e^{\beta \mathcal{H} t'}Be^{-\beta \mathcal{H} t'}e^{-\beta \mathcal{H}}e^{\beta \mathcal{H} t}Ae^{-\beta \mathcal{H} t}e^{\beta \mathcal{H}}}{\text{Tre}^{-\beta \mathcal{H}}}$$

$$= \frac{\text{Tre}^{-\beta \mathcal{H}}e^{\beta \mathcal{H} t'}Be^{-\beta \mathcal{H} t'}e^{i\mathcal{H}(t+i\beta)}Ae^{-i\mathcal{H}(t+i\beta)}}{\text{Tre}^{-\beta \mathcal{H}}}$$

$$= < B(t')A(t + i\beta) >$$

We have thus

$$F_{AB}(t - t') = F_{BA}(t' - t - i\beta) \tag{6.9}$$

The Fourier transformation of $F_{AB}(t - t')$ is

$$F_{AB}(t - t') = F_{BA}(t' - t - i\beta)$$

$$= \int_{-\infty}^{\infty} I(\omega)e^{-i\omega(t-t')}e^{\beta\omega}d\omega \tag{6.10}$$

Using this formula for the Green's function $G_{AB}^r(t - t')$

$$G_{AB}^r(t - t') = -i\theta(t - t') < A(t)B(t') - B(t')A(t) > \tag{6.11}$$

we obtain

$$G_{AB}^r(\omega) = \frac{1}{(2\pi)^2} \int \int \int_{-\infty}^{\infty} \frac{I(\omega')(e^{\beta\omega'} - 1)e^{i(\omega-\omega'-x)(t-t')}}{x + i\epsilon}d\omega'd(t-t')dx \tag{6.12}$$

where we have used the following expression

$$\theta(t - t') = \lim_{\epsilon \to 0^+} \frac{i}{2\pi} \int_{-\infty}^{\infty} \frac{e^{-ix(t-t')}}{x + i\epsilon}dx \tag{6.13}$$

Integrating on $(t - t')$ and using the formula

$$\delta(x) = \frac{1}{2\pi i}\left(\frac{1}{x - i\epsilon} - \frac{1}{x + i\epsilon}\right) \tag{6.14}$$

we arrive at

$$G^r_{AB}(\omega) = \frac{1}{2\pi}\int_{-\infty}^{\infty}\frac{I(\omega')(e^{\beta\omega'} - 1)}{\omega - \omega' + i\epsilon}d\omega' \tag{6.15}$$

In the same manner, we obtain

$$G^a_{AB}(\omega) = \frac{1}{2\pi}\int_{-\infty}^{\infty}\frac{I(\omega')(e^{\beta\omega'} - 1)}{\omega - \omega' - i\epsilon}d\omega' \tag{6.16}$$

The difference between the retarded and advanced Green's functions is the sign in front of $i\epsilon$ in the denominator. We write from now on these functions without superscripts r and a but we distingush them by the sign in their arguments. Combining these functions we write

$$\begin{aligned}
G^r_{AB}(\omega) - G^a_{AB}(\omega) &= G_{AB}(\omega + i\epsilon) - G_{AB}(\omega - i\epsilon) \\
&= \frac{1}{2\pi}\int_{-\infty}^{\infty}d\omega' I(\omega')(e^{\beta\omega'} - 1) \\
&\quad \times \left[\frac{1}{\omega - \omega' + i\epsilon} - \frac{1}{\omega - \omega' - i\epsilon}\right] \\
&= \frac{1}{2\pi}\int_{-\infty}^{\infty}d\omega' I(\omega')(e^{\beta\omega'} - 1)\left[-2\pi i\delta(\omega - \omega')\right] \\
&= -iI(\omega)(e^{\beta\omega} - 1) \tag{6.17}
\end{aligned}$$

We shall use this relation to calculate the spin-spin correlation in the following sections..

6.3 Ferromagnetism by Green's function method

We consider the Heisenberg spins with ferromagnetic interaction between nearest neighbors. The Hamiltonian is given by

$$\begin{aligned}
\mathcal{H} &= -J\sum_{(l,m)}\vec{S}_l \cdot \vec{S}_m \\
&= -J\sum_{(l,m)}\left[S^z_l S^z_m + \frac{1}{2}(S^+_l S^-_m + S^-_l S^+_m)\right] \tag{6.18}
\end{aligned}$$

where the spin operators obey the commutation relations

$$[S_l^+, S_m^-] = 2S_l^z \delta_{lm} \tag{6.19}$$

$$[S_m^z, S_l^\pm] = \pm S_l^\pm \delta_{lm} \tag{6.20}$$

We define the following Green's function

$$G_{lm}(t-t') = << S_l^+(t); S_m^-(t') >> = -i\theta(t-t') < [S_l^+(t), S_m^-(t')] > \tag{6.21}$$

We set $t' = 0$ hereafter to simplify the writing.

6.3.1 *Equation of motion*

The equation of motion for $G_{lm}(t)$ is written as

$$i\hbar \frac{dG_{lm}(t)}{dt} = 2 < S_l^z > \delta_{lm}\delta(t) - << [\mathcal{H}, S_l^+](t); S_m^- >>$$

$$= 2 < S_l^z > \delta_{lm}\delta(t)$$

$$-J \sum_{\vec{\rho}} << S_l^z(t)S_{l+\vec{\rho}}^+(t) - S_l^+(t)S_{l+\vec{\rho}}^z(t); S_m^- >> \tag{6.22}$$

where the sum on $\vec{\rho}$ is performed on the vectors connecting spin $\vec{S}_l$ to its nearest neighbors.

Remark: We have used the identity $[AB, C] = [A, C]B + A[B, C]$ then (6.19) and (6.20) to calculate $[\mathcal{H}, S_l^+]$ (see Problem 2 in section 6.7).

We see that the right-hand side of Eq. (6.22) contains Green's functions of higher order with three operators. Writing the equation of motion for these functions will generate functions of five operators. In a first approximation, we reduce higher-order Green's functions by using the so-called Tyablikov decoupling scheme

$$<< S_l^z(t)S_{l+\vec{\rho}}^+(t); S_m^- >> \simeq < S_l^z(t) > << S_{l+\vec{\rho}}^+(t); S_m^- >> \tag{6.23}$$

$$<< S_l^+(t)S_{l+\vec{\rho}}^z(t); S_m^- >> \simeq < S_{l+\vec{\rho}}^z(t) > << S_l^+(t); S_m^- >> \tag{6.24}$$

We obtain on the right-hand side of Eq. (6.22) Green's functions of the same order as the initial one defined in (6.21). This decoupling has the same spirit as in the mean-field theory (chapter 4): replacing an operator in a product by its average value, i. e. by a "c-number". This approximation is called sometimes "random-phase approximation" (RPA) . The RPA is hierarchically higher than the mean-field theory in the sense that one operator in a three-operator product is replaced by its average value in the RPA while in the mean-field theory one operator in a two-operator product is replaced.

We notice that S_l^z is a constant of motion. Therefore, in a homogeneous system, we can suppose that

$$< S_l^z(t) >=< S_{l+\vec{\rho}}^z(t) >=< S^z > \qquad \text{site-independent} \qquad (6.25)$$

We obtain then

$$i\hbar\frac{dG_{lm}(t)}{dt} = 2 < S^z > \delta_{lm}\delta(t)$$
$$-J\sum_{\vec{\rho}} < S^z > [<< S_{l+\vec{\rho}}^+(t); S_m^- >>$$
$$- << S_l^+(t); S_m^- >>]$$
$$= 2 < S^z > \delta_{lm}\delta(t)$$
$$-J < S^z > \sum_{\vec{\rho}}[G_{l+\vec{\rho},m}(t) - G_{lm}(t)] \qquad (6.26)$$

6.3.2   Dispersion relation

The time Fourier transformation gives

$$\hbar\omega G_{lm}(\omega) = \frac{1}{\pi} < S^z > \delta_{lm} - J < S^z > \sum_{\vec{\rho}}[G_{l+\vec{\rho},m}(\omega) - G_{lm}(\omega)] \quad (6.27)$$

We use next the following spatial Fourier transformation

$$G_{lm}(\omega) = \frac{1}{N}\sum_{\vec{k}} G_{\vec{k}}(\omega)e^{i\vec{k}\cdot(\vec{l}-\vec{m})} \qquad (6.28)$$

where the sum on the wave vector $\vec{k}$ is performed in the first Brillouin zone of N states. We obtain

$$\hbar\omega G_{\vec{k}}(\omega) = \frac{< S^z >}{\pi} - J < S^z > [G_{\vec{k}}(\omega)\sum_{\vec{\rho}} e^{i\vec{k}\cdot\vec{\rho}} - ZG_{\vec{k}}(\omega)]$$
$$= \frac{< S^z >}{\pi} + J < S^z > ZG_{\vec{k}}(\omega)[1 - \frac{1}{Z}\sum_{\vec{\rho}} e^{i\vec{k}\cdot\vec{\rho}}] \quad (6.29)$$

where Z is the coordination number. We get

$$G_{\vec{k}}(\omega) = \frac{< S^z >}{\pi}\frac{1}{\hbar\omega - ZJ < S^z > (1 - \gamma_{\vec{k}})}$$
$$= \frac{< S^z >}{\pi}\frac{1}{\hbar\omega - \epsilon_{\vec{k}}} \qquad (6.30)$$

where

$$\gamma_{\vec{k}} = \frac{1}{Z} \sum_{\vec{\rho}} e^{i\vec{k}\cdot\vec{\rho}} \tag{6.31}$$

$$\epsilon_{\vec{k}} = ZJ < S^z > (1 - \gamma_{\vec{k}}) \tag{6.32}$$

We note that the singularity of the Green's function is at $\hbar\omega = \epsilon_{\vec{k}}$. This determines the eigen-energy $\hbar\omega_{\vec{k}}$ of the magnon of wave vector $\vec{k}$:

$$\hbar\omega_{\vec{k}} = ZJ < S^z > (1 - \gamma_{\vec{k}}) \tag{6.33}$$

This is the ferromagnetic dispersion relation which is to be compared to (5.52) obtained by the theory of magnons: $\epsilon_{\vec{k}} = 2ZJS(1 - \gamma_{\vec{k}})$. We see that apart from the factor 2 due to the model defined by (5.42) the only difference is that $< S^z >$ appears in (6.33) instead of S. As $< S^z >$ depends on the temperature, the magnon spectrum varies with T. We now calculate $< S^z >$ in the following.

6.3.3 *Magnetization and critical temperature*

We write

$$\vec{S}_l \cdot \vec{S}_l = S(S+1) = (S_l^z)^2 + \frac{1}{2}(S_l^+ S_l^- + S_l^- S_l^+)$$
$$= (S_l^z)^2 + (S_l^- S_l^+ + S_l^z) \tag{6.34}$$

where we have used (6.19). For $S = 1/2$, we have $(S_l^z)^2 = 1/4$ (see Pauli's matrices in chapter 1). We get then

$$\frac{3}{4} = \frac{1}{4} + S_l^- S_l^+ + S_l^z \tag{6.35}$$

from which we have

$$< S_l^z > = \frac{1}{2} - < S_l^- S_l^+ > \tag{6.36}$$

We now calculate $< S_l^- S_l^+ >$. We have

$$< S_m^- S_l^+ > = \frac{1}{N} \sum_{\vec{k}} \int_{-\infty}^{\infty} I_{\vec{k}} e^{i\vec{k}\cdot(\vec{l}-\vec{m})} e^{-i\omega t} d\omega \tag{6.37}$$

from which, for $t = 0$,

$$< S_l^- S_l^+ > = \frac{1}{N} \sum_{\vec{k}} \int_{-\infty}^{\infty} I_{\vec{k}} d\omega \tag{6.38}$$

Using (6.17) we write

$$< S_l^- S_l^+ > = \frac{1}{N} \sum_{\vec{k}} \int_{-\infty}^{\infty} d\omega \left[\frac{G_{\vec{k}}(\omega + i\epsilon) - G_{\vec{k}}(\omega - i\epsilon)}{-i(e^{\beta\omega} - 1)} \right] \tag{6.39}$$

Replacing $G_{\vec{k}}(\omega + i\epsilon)$ and $G_{\vec{k}}(\omega - i\epsilon)$ by (6.30) and using (6.14), we obtain

$$< S_l^- S_l^+ > = -\frac{1}{N} \frac{< S^z >}{i\pi} \sum_{\vec{k}} \int_{-\infty}^{\infty} \frac{d\omega}{e^{\beta\omega} - 1}$$

$$\times \left[\frac{1}{\hbar\omega - \epsilon_{\vec{k}} + i\epsilon} - \frac{1}{\hbar\omega - \epsilon_{\vec{k}} - i\epsilon} \right]$$

$$= -\frac{1}{N} \frac{< S^z >}{i\pi} \sum_{\vec{k}} \int_{-\infty}^{\infty} \frac{d\omega}{e^{\beta\omega} - 1} \left[-2\pi i \delta(\hbar\omega - \epsilon_{\vec{k}}) \right]$$

$$= \frac{2}{N} < S^z > \sum_{\vec{k}} \frac{1}{e^{\beta\hbar\omega_{\vec{k}}} - 1} \tag{6.40}$$

Equation (6.36) becomes

$$< S_l^z > = \frac{1}{2} - \frac{2}{N} < S^z > \sum_{\vec{k}} \frac{1}{e^{\beta\hbar\omega_{\vec{k}}} - 1} \tag{6.41}$$

This is an implicit equation for $< S_l^z >$: the right-hand side indeed contains $< S_l^z >$ in $\omega_{\vec{k}}$. Therefore, we have to solve this equation by iteration to get $< S_l^z >$.

At low temperatures, we follow the same method as that used for (5.58). We obtain the following result

$$< S_l^z > \simeq \frac{1}{2} - a_1 T^{3/2} - a_2 T^{5/2} - a_3 T^3 - a_4 T^{7/2} - \dots \tag{6.42}$$

where $a_i (i = 1, 2, 3, \dots)$ are constants. We notice that the T^3 term does not exist in the low-temperature expansion (5.61) of the theory of magnons. The Tyablikov decoupling (6.23)-(6.24), which is the only approximation used here, can be improved. The reader who wishes to go further in this method is referred to the references [128, 91].

At high temperatures, $< S_l^z > \to 0$. An expansion of $e^{\beta\hbar\omega_{\vec{k}}}$ yields

$$< S_l^z > \simeq \frac{1}{2} - \frac{2}{N} < S^z > \sum_{\vec{k}} \frac{1}{1 + \beta\hbar\omega_{\vec{k}} + \dots - 1}$$

$$\simeq \frac{1}{2} - \frac{2}{N} < S^z > \sum_{\vec{k}} \frac{1}{\beta Z J < S^z > (1 - \gamma_{\vec{k}})}$$

$$\simeq \frac{1}{2} - \frac{2}{N} \sum_{\vec{k}} \frac{1}{\beta Z J (1 - \gamma_{\vec{k}})} \tag{6.43}$$

At $T = T_c$, $< S_l^z >= 0$. Equation (6.43) gives then

$$(\frac{k_B T_c}{J})^{-1} = \frac{4}{ZN} \sum_{\vec{k}} \frac{1}{1 - \gamma_{\vec{k}}} \qquad (6.44)$$

In the same manner, for spins of amplitude $S \neq 1/2$, one has

$$(\frac{k_B T_c}{J})^{-1} = \frac{3}{ZS(S+1)N} \sum_{\vec{k}} \frac{1}{1 - \gamma_{\vec{k}}} \qquad (6.45)$$

It is noted that when transforming the sum on $\vec{k}$ in (6.44) into an integral, we clearly see that the integral diverges at small k for dimensions $d = 1$ and 2 (see the demonstration and the discussion in the conclusion of chapter 5). As a consequence, $T_c = 0$ for these low dimensions, in agreement with exact results (see chapter 7). For $d = 3$, the values of T_c numerically calculated by (6.44) are

- Simple cubic lattice: $k_B T_c/J \simeq 0.994$
- Body-centered cubic lattice: $k_B T_c/J \simeq 1.436$
- Face-centered cubic lattice: $k_B T_c/J \simeq 2.231$

These values are very low with respect to those, overestimated, given by the mean-field theory: 3, 4 and 6, respectively. The present results are better than those of the mean-field theory, due to the fact that in the Green's function method, fluctuations due to spin-spin correlations are taken into account.

6.4   Antiferromagnetism by Green's function method

In a spin system with an antiferromagnetic interaction, we have to define two Green's functions, the first one relates spins of the same sublattice and the second one connects spins between sublattices.

We consider a system of Heisenberg spins with an antiferromagnetic interaction between nearest neighbors

$$\mathcal{H} = J \sum_{(l,m)} \left[S_l^z S_m^z + \frac{1}{2}(S_l^+ S_m^- + S_l^- S_m^+) \right] \qquad (6.46)$$

where the spin operators satisfy the commutation relations (6.19) and

(6.20). We define the following Green's functions

$$G_{ll'}(t - t') = <\!\!< S_l^+(t); S_{l'}^-(t') >\!\!>$$
$$= -i\theta(t - t') < [S_l^+(t), S_{l'}^-(t')] > \qquad (6.47)$$
$$F_{ml'}(t - t') = <\!\!< S_m^+(t); S_{l'}^-(t') >\!\!>$$
$$= -i\theta(t - t') < [S_m^+(t), S_{l'}^-(t')] > \qquad (6.48)$$

where l, l' belong to sublattice of $\uparrow$ spins and m belongs to sublattice of $\downarrow$ spins. For writing simplicity, we take $t' = 0$.

The equations of motion for $G_{ll'}(t)$ and $F_{ml'}(t)$ are written as, after the Tyablikov decoupling,

$$i\hbar \frac{dG_{ll'}(t)}{dt} = 2 < S^z > \delta_{ll'}\delta(t)$$
$$+ J < S^z > \sum_{\vec{\rho}}[F_{l+\vec{\rho},l'}(t) + G_{ll'}(t)] \qquad (6.49)$$

$$i\hbar \frac{dF_{ml'}(t)}{dt} = -J < S^z > \sum_{\vec{\rho}}[G_{m+\vec{\rho},l'}(t) + F_{ml'}(t)] \qquad (6.50)$$

where we have used the fact that $l + \vec{\rho}$ is a site m of the $\downarrow$ sublattice and that $m + \vec{\rho}$ is a site l of the $\uparrow$ sublattice. As in the ferromagnetic case we have supposed $< S_l^z > = < S_{m+\vec{\rho}}^z > = - < S_m^z > = - < S_{l+\vec{\rho}}^z > = < S^z >$, independent of the lattice site.

We use the Fourier transformations

$$G_{ll'}(t) = \frac{2}{N} \int_{-\infty}^{\infty} \sum_{\vec{k}} G_{\vec{k}}(\omega) e^{i[\vec{k}\cdot(\vec{l}-\vec{l'})-\omega t]} d\omega \qquad (6.51)$$

$$F_{ml'}(t) = \frac{2}{N} \int_{-\infty}^{\infty} \sum_{\vec{k}} F_{\vec{k}}(\omega) e^{i[\vec{k}\cdot(\vec{m}-\vec{l'})-\omega t]} d\omega \qquad (6.52)$$

where the factor $2/N$ comes from the fact that each sublattice has $N/2$ sites. The equations (6.49)-(6.50) become

$$(\hbar\omega - A_{\vec{k}})G_{\vec{k}}(\omega) - B_{\vec{k}}F_{\vec{k}}(\omega) = \frac{< S^z >}{\pi} \qquad (6.53)$$

$$B_{\vec{k}}G_{\vec{k}}(\omega) + (\hbar\omega + A_{\vec{k}})F_{\vec{k}}(\omega) = 0 \qquad (6.54)$$

where $A_{\vec{k}} = ZJ < S^z >$ and $B_{\vec{k}} = ZJ < S^z > \gamma_{\vec{k}}$. The solution of (6.53)-(6.54) is

$$G_{\vec{k}}(\omega) = \frac{<S^z>}{2\pi} \left[\frac{1 + A_{\vec{k}}/\epsilon_{\vec{k}}}{\hbar\omega - \epsilon_{\vec{k}}} + \frac{1 - A_{\vec{k}}/\epsilon_{\vec{k}}}{\hbar\omega + \epsilon_{\vec{k}}} \right] \tag{6.55}$$

$$F_{\vec{k}}(\omega) = -\frac{<S^z> B_{\vec{k}}}{2\pi\epsilon_{\vec{k}}} \left[\frac{1}{\hbar\omega - \epsilon_{\vec{k}}} - \frac{1}{\hbar\omega + \epsilon_{\vec{k}}} \right] \tag{6.56}$$

where

$$\epsilon_{\vec{k}} = \sqrt{A_{\vec{k}}^2 - B_{\vec{k}}^2} = ZJ <S^z> \sqrt{1 - \gamma_{\vec{k}}^2} \tag{6.57}$$

The singularities of these functions are $\pm\epsilon_{\vec{k}}$. The magnon mode $\vec{k}$ has thus two opposite precessions. This degeneracy comes from the reversal symmetry of the two sublattices. The antiferromagnetic dispersion relation is

$$\hbar\omega_{\vec{k}} = \pm\epsilon_{\vec{k}} = \pm\sqrt{A_{\vec{k}}^2 - B_{\vec{k}}^2} = \pm ZJ <S^z> \sqrt{1 - \gamma_{\vec{k}}^2} \tag{6.58}$$

It is noted that

- for small k, one has $\epsilon_{\vec{k}} \propto k$ instead of k^2 of the ferromagnetic case (cf. previous paragraph and chapter 5)
- as in the ferromagnetic case, the magnon spectrum depends on the temperature via $<S^z>$ in $\hbar\omega_{\vec{k}}$.

To calculate the magnetization we follow the same method used in the ferromagnetic case. We have

$$<S_{l'}^- S_l^+> = \frac{2}{N} \int_{-\infty}^{\infty} \sum_{\vec{k}} I_{\vec{k}}(\omega) e^{i[\vec{k}\cdot(\vec{l}-\vec{l}')-\omega t]} d\omega \tag{6.59}$$

$$<S_{l'}^- S_m^+> = \frac{2}{N} \int_{-\infty}^{\infty} \sum_{\vec{k}} K_{\vec{k}}(\omega) e^{i[\vec{k}\cdot(\vec{m}-\vec{l}')-\omega t]} d\omega \tag{6.60}$$

Replacing $I_{\vec{k}}(\omega)$ and $K_{\vec{k}}(\omega)$ each with its corresponding Green's function by using (6.17), we obtain

$$I_{\vec{k}}(\omega) = <S^z> \frac{(1 + A_{\vec{k}}/\epsilon_{\vec{k}})\delta(\hbar\omega - \epsilon_{\vec{k}}) + (1 - A_{\vec{k}}/\epsilon_{\vec{k}})\delta(\hbar\omega + \epsilon_{\vec{k}})}{e^{\beta\hbar\omega} - 1} \tag{6.61}$$

$$K_{\vec{k}}(\omega) = -<S^z> \frac{B_{\vec{k}}}{\epsilon_{\vec{k}}} \frac{\delta(\hbar\omega - \epsilon_{\vec{k}}) - \delta(\hbar\omega + \epsilon_{\vec{k}})}{e^{\beta\hbar\omega} - 1} \tag{6.62}$$

Using these relations in (6.59)-(6.60), we have

$$<S_l^- S_l^+> = \frac{2}{N} <S^z> \sum_{\vec{k}} \left[-1 + \frac{A_{\vec{k}}}{\epsilon_{\vec{k}}} \coth(\beta\epsilon_{\vec{k}}/2) \right] \tag{6.63}$$

$$<S_l^- S_m^+> = \frac{2}{N} <S^z> \sum_{\vec{k}} \frac{B_{\vec{k}}}{\epsilon_{\vec{k}}} \coth(\beta\epsilon_{\vec{k}}/2) e^{i\vec{k}\cdot(\vec{m}-\vec{l})} \tag{6.64}$$

For $S = 1/2$, we replace $< S_l^- S_l^+ >$ by $1/2 - < S^z >$ in (6.63). We finally arrive at

$$\frac{1}{2} = \frac{2}{N} < S^z > \sum_{\vec{k}} \frac{1}{\sqrt{1 - \gamma_{\vec{k}}^2}} \coth \left[\frac{\beta Z J < S^z > \sqrt{1 - \gamma_{\vec{k}}^2}}{2} \right] \qquad (6.65)$$

This implicit equation for $< S^z >$ should be solved by numerical iteration.

At low temperatures, one has

$$< S^z > \simeq \frac{1}{2} - \Delta S - a_2 T^2 - ... \qquad (6.66)$$

where ΔS is the zero-point spin contraction (at $T = 0$) (see chapter 5). As seen, the temperature dependence of $< S^z >$ is not the same as in the ferromagnetic case. This is a consequence of the linear dependence on k of $\epsilon_{\vec{k}}$ at small k.

The Néel temperature T_N is calculated by letting $< S^z > \rightarrow 0$ in (6.65). One has

$$\left(\frac{k_B T_N}{J} \right)^{-1} = \frac{4}{ZN} \sum_{\vec{k}} \left[\frac{1}{1 - \gamma_{\vec{k}}} + \frac{1}{1 + \gamma_{\vec{k}}} \right] \qquad (6.67)$$

6.5 Green's function method for non collinear magnets

For non collinear magnets, the Hamiltonian can be expressed in the local coordinates. This has been done in Eq. (5.136). We define the following Green's functions:

$$G_{k\ell}(t - t') = << S_k^+(t); S_\ell^-(t') >>$$
$$= -i\theta(t - t') < \left[S_k^+(t), S_\ell^-(t') \right] > \qquad (6.68)$$

$$F_{k\ell}(t - t') = << S_k^-(t); S_\ell^-(t') >>$$
$$= -i\theta(t - t') < \left[S_k^-(t), S_\ell^-(t') \right] > \qquad (6.69)$$

With the use of the Tyablikov decoupling, the equations of motion of these functions are given by

$$i\hbar \frac{dG_{k\ell}(t - t')}{dt} = 2 < S_k^z > \delta_{k\ell} \delta(t)$$

$$-\frac{1}{2} \sum_{k'} J_{k,k'} [< S_k^z > << S_{k'}^-; S_\ell^- >> (\cos \theta_{k,k'} - 1)$$

$$+ < S_k^z > << S_{k'}^+; S_\ell^- >> (\cos \theta_{k,k'} + 1)$$

$$-2 < S_{k'}^z > << S_k^+; S_\ell^- >> \cos \theta_{k,k'}]$$

$$= 2 < S_k^z > \delta_{k\ell} \delta(t) - \frac{1}{2} \sum_{k'} J_{k,k'} [< S_k^z > (\cos \theta_{k,k'} - 1)$$

$$\times F_{k'\ell}(t - t') + < S_k^z > (\cos \theta_{k,k'} + 1) G_{k'\ell}(t - t')$$

$$-2 < S_{k'}^z > \cos \theta_{k,k'} G_{k\ell}(t - t')] \qquad (6.70)$$

$$i\hbar \frac{dF_{k\ell}(t - t')}{dt} = \frac{1}{2} \sum_{k'} J_{k,k'} [< S_k^z > << S_{k'}^+; S_\ell^- >> (\cos \theta_{k,k'} - 1)$$

$$+ < S_k^z > << S_{k'}^-; S_\ell^- >> (\cos \theta_{k,k'} + 1)$$

$$-2 < S_{k'}^z > << S_k^-; S_\ell^- >> \cos \theta_{k,k'}]$$

$$= \frac{1}{2} \sum_{k'} J_{k,k'} [< S_k^z > (\cos \theta_{k,k'} - 1) G_{k'\ell}(t - t')$$

$$+ < S_k^z > (\cos \theta_{k,k'} + 1) F_{k'\ell}(t - t')$$

$$-2 < S_{k'}^z > \cos \theta_{k,k'} F_{k\ell}(t - t')] \qquad (6.71)$$

Note that the sinus terms in Eq. (5.136) are canceled out upon summing over symmetric neighbors at each lattice site (inversion symmetry) in the above equations. The next steps are similar to those used in the antiferromagnetic case in the previous section: using the Fourier transformations, the solution of the resulting coupled equations gives the dispersion relation which allows us to calculate the magnetization and the transition temperature. Some applications are given as problems in section 6.7.

6.6 Conclusion

We have studied in this chapter the Green's function method as applied to spin systems. We have considered the ferromagnetic, antiferromagnetic and non linear cases. The method gives the temperature-dependence of the magnon spectrum and a compact expression which allows us to calculate the magnetization up to the transition temperature. Of course, to keep the method simple and tractable, we have used the Tyablikov decoupling scheme to reduce higher-order Green's functions. There is room for improving it but we will loose the simplicity of the method. The dispersion relation can be explicitly obtained in simple cases. However, for complicated systems where we have to define several Green's functions, we often obtain a system of coupled linear equations which can be numerically solved to obtain the dispersion relation which is used to compute temperature-dependent physical quantities. Such applications are treated in chapter 9.

We note that the method is not accurate enough to allow us to calculate critical exponents of the phase transition. However, it can detect first-order transitions as well as multiple phase transitions of the system at different temperatures [113].

6.7   Problems

Problem 1. Give proofs of the formula (6.13).

Problem 2. Give the demonstration of Eq. (6.22).

Problem 3. Helimagnet by Green's function method:

Consider a crystal of simple cubic lattice with Heisenberg spins of amplitude $1/2$. The interaction J_1 between nearest neighbors is ferromagnetic. Suppose that along the y axis there exists an antiferromagnetic interaction J_2 between next nearest neighbors, in addition to J_1.

a) Follow the method in section 5.4, show that the ground state is helimagnetic in the y direction if $|J_2|/J_1$ is larger than a critical value α_c. Determine α_c.

b) Let θ be the helical angle between two nearest neighboring spins in the y direction. Express the Hamiltonian in terms of θ.

c) Define two Green's functions which allow to calculate the spin-wave spectrum using the RPA decoupling scheme. Calculate that spectrum.

d) Show that this spectrum is reduced to the ferromagnetic and antiferromagnetic spectra when $J_2 = 0$ and $J_1 \gtrless 0$.

Problem 4. Apply the Green's function method to a system of Ising spins $S = \pm 1$ in one dimension, supposing a ferromagnetic interaction between nearest neighbors under an applied magnetic field.

Problem 5. Apply the Green's function method to a system of Heisenberg spins on a simple cubic lattice, supposing ferromagnetic interactions between nearest neighbors and between next-nearest neighbors.

Problem 6. Magnon spectrum in Heisenberg triangular antiferromagnet: Green's function method

Calculate the magnon spectrum of a triangular lattice with spins 1/2 interacting with each other via an antiferromagnetic interaction between nearest neighbors. Estimate numerically the zero-point spin contraction. Guide: use the spin configuration obtained in Problem 6 of section 5.6.

Problem 7. Quantum gas by Green's function method:

Consider a gas of N particles, of volume Ω. Define the following Green's function:

$iG_{\alpha,\beta}(\vec{r}t, \vec{r}'t') = <\varphi_0|\hat{T}[\hat{\Psi}_{H\alpha}(\vec{r}t)\hat{\Psi}^+_{H\beta}(\vec{r}'t')]|\varphi_0>$

where $|\varphi_0>$ denotes the ground-state eigenfunction, $\hat{\Psi}_{H\alpha}(\vec{r}t)$ field operator of the particle at $(\vec{r}t)$ of spin α in the Heisenberg representation, and $\hat{T}$ time-ordering operator defined by $\hat{T}[A(t)B(t')] = A(t)B(t')$ if $t > t'$ and $\hat{T}[A(t)B(t')] = \pm B(t')A(t)$ if $t' > t$ (two signs denote the boson and fermion cases).

a) Show that for free electrons at $T = 0$, we have

$$iG_{\alpha,\beta}(\vec{r}t, \vec{r}'t') = iG^0_{\alpha,\beta}(\vec{r}t, \vec{r}'t')$$

$$= \delta_{\alpha\beta}(\frac{1}{2\pi})^3 \int d\vec{k} e^{i\vec{k}\cdot(\vec{r}-\vec{r}')} e^{-i\omega_k(t-t')}$$

$$\times [\Theta(t-t')\Theta(k-k_F) - \Theta(t'-t)\Theta(k_F-k)]$$

where $\Theta(y)$ is the Heavyside function, and $\omega_k = \frac{\hbar k^2}{2m}$.

b) Show that the time and spatial Fourier transform of $G^0_{\alpha,\beta}(\vec{r}t, \vec{r}'t')$ is written as

$G^0_{\alpha,\beta}(\vec{k}, \omega) = \delta_{\alpha\beta} \left[\frac{\Theta(k-k_F)}{\omega-\omega_k+i\eta} + \frac{\Theta(k_F-k)}{\omega-\omega_k-i\eta} \right]$

where η is an infinitesimal positive constant.

c) Show the following relations between the Green's function and observables in the general case (bosons and fermions, upper and lower signs):

The total number of particles:

$$N = \pm i \frac{\Omega}{(2\pi)^4} \lim_{\eta\to 0} \int d\vec{k} \int_{-\infty}^{\infty} d\omega e^{i\omega\eta} \text{Tr} G(\vec{k}, \omega)$$

The kinetic energy:

$$< \hat{T} >= \pm i \int d\vec{r} \lim_{\vec{r}' \to \vec{r}} [-\frac{\hbar^2 \nabla^2}{2m} \text{Tr} G(\vec{r}t, \vec{r}'t^+)]$$

The total energy:

$$E = \pm \frac{i}{2} \frac{\Omega}{(2\pi)^4} \lim_{\eta \to 0} \int d\vec{k} \int_{-\infty}^{\infty} d\omega e^{i\omega\eta} (\frac{\hbar^2 k^2}{2m} + \hbar\omega) \text{Tr} G(\vec{k}, \omega)$$

Chapter 7

Phase Transition

7.1 Introduction

Phase transitions and critical phenomena constitute one of the most important domains of statistical physics. This domain concerns not only numerous areas of physics but also other sciences such as biology and chemistry. We have introduced in chapter 4 examples of phase transitions occurring in magnetic materials studied by the mean-field theory. However, as discussed in that chapter, this theory is a first approximation to the problem, it cannot determine with precision the characteristics of a phase transition. Since these characteristics are intimately related to microscopic interactions between particles and the system symmetry, the more we know them the better we understand the mechanisms inside a material which govern its properties.

The study of phase transitions has been spectacularly developed since the 70's after the introduction of the renormalization group theory [139, 7, 24, 150] and Monte Carlo simulations. We have now a good understanding of second-order phase transitions in spite of the fact that there is still much to be done for more and more complex systems. The concept of the renormalization group has been formulated by K. G. Wilson in the early 70's [139, 140]. It has been applied with success in many fields of physics ranging from condensed matter to quantum field theory. We can mention a few well-known examples in condensed matter such as the Kondo problem [141] and the Kosterlitz-Thouless transition [80]. We will briefly present it together with other notions and methods in this chapter. Methods of Monte Carlo simulation are shown in chapter 8. It provides a precious complementary tool for the study of phase transitions.

We present hereafter some fundamental notions necessary to understand a phase transition. There exists a huge number of reviews and books specialized in this field [7, 24, 42, 150].

All systems do not have obligatorily a phase transition. In general, the existence of a phase transition depends on a few general parameters such as the space dimension, the nature of the interaction between particles and the system symmetry. For spin systems, we can give a brief summary here. In one dimension, in general there is no phase transition at a non zero temperature for systems of short-range interactions regardless of the spin model. The long-range ordering at $T = 0$ is destroyed as soon as $T \neq 0$. However, in two dimensions discrete spin models such as the Ising and Potts models have a phase transition at a finite temperature T_c, while continuous spin models such as the Heisenberg model do not have a transition at a finite temperature. The case of XY model is particular: in spite of the absence of a long-range order at finite temperatures, there is a phase transition of a special kind called the "Kosterlitz-Thouless" transition [80]. We will return to this point later in this chapter. In three dimensions, all known spin models have in general a phase transition at $T_c \neq 0$.

7.1.1   *Order parameter*

We consider a system of interacting particles. A transition from one phase to another may take place when an external parameter varies. Such a parameter can be the temperature or an external applied magnetic field. An often cited example is the transition from ice to water (solid to liquid phase). Another popular transition is the loss of magnetic attraction of a permanent magnet with increasing temperature: this is a transition from a magnetically ordered phase to a magnetically disordered (or paramagnetic) phase.

In order to measure the degree of ordering, we define an order parameter appropriate for each system, depending on its symmetry. A good order parameter should be non zero in one phase and zero elsewhere because changing from one phase to another breaks the phase symmetry. In some systems, the choice of the order parameter is natural. For example in a ferromagnet, the magnetization is the natural order parameter (see chapter 4), namely the thermal average of the spin component on its quantization axis. For the Ising model, the order parameter is given by

$$q = \frac{1}{N} |\sum_i \sigma_i| \tag{7.1}$$

where $\sigma_i = \pm 1$ is the spin at site i. In the ground state where all spins are parallel, we have $q = 1$, and in the paramagnetic phase where there is an equal mixing of $+1$ and -1 spins, we have $q = 0$. For the Heisenberg spins of amplitude 1 in a ferromagnet, the order parameter is also the magnetization which is defined by

$$M = \frac{1}{N} |\sum_i \vec{S}_i| \qquad (7.2)$$

We see that in the ground state all spins are parallel so that M is 1 whatever the orientation of spins with respect to the crystal axes. In the disordered state, each spin has a random spatial orientation so that $\sum_i \vec{S}_i = 0$.

7.1.2 *Order of the phase transition*

A phase transition takes place when physical quantities of the system undergo an anomaly. In order to define properly a phase transition, we examine various physical quantities at the transition point. If physical quantities such as the specific heat C_V and the susceptibility χ which are second derivatives of the free energy F (see Appendix A) diverge, we call the corresponding phase transition a "phase transition of second order". In this case, the first derivatives of F such as the average energy $\overline{E}$ and the average magnetization $\overline{M}$ are continuous functions at T_c. On the other hand, in a first-order phase transition, these first-derivative quantities undergo a discontinuity at the transition point.

7.1.3 *Correlation function — Correlation length*

An important function in the study of phase transitions is the correlation function defined by

$$G(\vec{r}) = < \vec{S}(0) \cdot \vec{S}(\vec{r}) > \qquad (7.3)$$

where $\vec{S}(0)$ is the spin at a site chosen as the origin and $\vec{S}(\vec{r})$ the spin at $\vec{r}$, $< ... >$ denotes the thermal average. In an isotropic system, $G(\vec{r})$ depends on r. In a phase where $\vec{S}(0)$ and $\vec{S}(\vec{r})$ are independent, namely their fluctuations are not correlated, $G(\vec{r})$ is zero. This is the case of the paramagnetic phase far from the transition temperature T_c and/or when the distance r is large.

When the temperature T_c is approached from the high temperature side, a correlation resulting from the interaction between spins sets in, $G(r)$

becomes non zero for spins at short distances. We can define the "correlation length" ξ as the distance beyond which $G(r)$ is no more significant. The fluctuations of two spins at a distance $r < \xi$ are said "correlated".

The correlation length ξ is written as

$$G(r) = < \vec{S}(0) \cdot \vec{S}(r) >= A\frac{\exp(-r/\xi)}{r^{(d-1)/2}} \qquad (7.4)$$

where d is the space dimension and A a constant.

In a second-order transition, the correlation length diverges at the transition, namely all spins are correlated at the transition regardless of their distance. On the contrary, at a first-order transition, the correlation length is finite and there is a coexistence of the two phases at the transition point.

7.1.4   *Critical exponents*

When the transition is of second order, we can define in the vicinity of T_c the following critical exponents

$$C_V = A\left|\frac{T - T_c}{T_c}\right|^{-\alpha} \qquad (7.5)$$

$$\overline{M} = B(\frac{T_c - T}{T_c})^\beta \qquad (7.6)$$

$$\chi = C\left|\frac{T - T_c}{T_c}\right|^{-\gamma} \qquad (7.7)$$

$$\xi \propto (\frac{T - T_c}{T_c})^{-\nu} \qquad (7.8)$$

$$\overline{M} = H^{1/\delta} \qquad (7.9)$$

The same α is defined for $T > T_c$ and $T < T_c$ but the coefficient A is different for each side of T_c. This is also the case for γ. However, β is defined only for $T < T_c$. The definition of δ is valid only at $T = T_c$ when the system is under an applied magnetic field of amplitude H. Finally, at $T = T_c$ we define exponent η of the correlation function by

$$G(r) \propto \frac{1}{r^{d-2+\eta}} \qquad (7.10)$$

There is another exponent called "dynamic exponent" z defined via the relaxation time τ of the spin system near T_c:

$$\tau \propto \xi^z \qquad (7.11)$$

Only the six exponents α, β, γ, δ, ν and η are critical exponents. We see below that there are four relations between them. Therefore, there are only two of them which are to be determined in an independent manner.

It is obvious that we cannot define such exponents in a first-order transition because there is no divergence of physical quantities. For this reason, we say that first-order transitions are not critical. We call "critical temperature" for second-order transitions but "transition temperature" for first-order transitions.

7.1.5 *Universality class*

Phase transitions of second order are distinguished by their "universality class". Phase transitions having the same values of critical exponents belong to the same university class. Renormalization group analysis shows that the universality class depends only on a few very general parameters such as the space dimension, the symmetry of the order parameter and the nature of the interaction. So, for example, Ising spin systems with short-range ferromagnetic interaction in two dimensions belong to the same universality class whatever the lattice structure is. Of course, the critical temperature T_c is not the same for square, hexagonal, rectangular, honeycomb, ...lattices, but T_c is not a universal quantity. It depends on the interaction value, the coordination number, ... but these quantities do not affect the values of the critical exponents.

We list in the following table the critical exponents of some known universality classes.

Class	Symmetry	α	β	γ	ν	η
2d Ising	Z_2	0	1/8	7/4	1	1/4
2d Potts (q=3)	Z_3	1/3	1/9	13/9	5/6	4/15
2d Potts (q=4)	Z_4	2/3	1/12	7/6	2/3	1/4
3d Ising	Z_2	0.11	0.325	1.241	0.63	0.031
3d XY	$O(2)$	-0.007	0.345	1.316	0.669	0.033
3d Heisenberg	$O(3)$	0.115	0.3645	1.386	0.705	0.033
Mean-field		0	1/2	1	1/2	0

Note that when a system is invariant by the following local transformation for a spin $\vec{S}_i$: $J \rightarrow -J$, $\vec{S}_i \rightarrow -\vec{S}_i$, the universality class of the new system does not change. This is understood immediately if we look at the partition function: such a local transformation does not change the argument of the exponential of the partition function. By consequence, nothing changes. However, the local transformation when operated on one spin every two in a square lattice for example (see Fig. 7.1), do change a ferromagnetic crystal into an antiferromagnetic one. We conclude that, a ferromagnetic crystal and its antiferromagnetic counterpart have the same critical temperature and the same critical exponents.

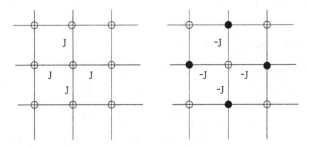

Fig. 7.1   Local transformation $J \to -J$, $\vec{S}_i \to -\vec{S}_i$ operated on one spin out of every two changes the square ferromagnetic lattice (left) into an antiferromagnetic lattice (right). White circles: ↑ spins, black circles: ↓ spins.

We emphasize here that there exist many systems in which it is impossible to operate local transformation without changing the argument of the partition function. One of these systems is the triangular lattice: it is impossible to find a spin configuration to satisfy all interactions if we change J into $-J$ everywhere. The energy of the system is not conserved, so the partition function changes. Such systems are called "frustrated systems" which are shown in section 7.5.

7.2   Landau-Ginzburg theory

We have shown in chapter 4 the mean-field theory which is a first approximation to study a phase transition. The validity of this theory has been briefly discussed. Essentially, several serious problems are encountered due to the fact that instantaneous spin fluctuations in the critical region have been neglected in this theory. We observe in particular an overestimation of T_c and the existence of a phase transition at finite temperatures in any space dimension which is not true. In addition, the critical exponents calculated with the mean-field theory are known to be incorrect (see the above table).

There exist several more efficient theories such as the high- and low-temperature series expansions [42], the Landau-Ginzburg theory and the renormalization group. In what follows, we present the Landau-Ginzburg theory and the concepts of the renormalization group.

The Landau-Ginzburg theory is an extension of the mean-field theory which includes a great part of fluctuations so far neglected near the transition. The main idea is to start from an expansion of the free energy per spin f in the vicinity of T_c when the magnetization m is sufficiently small:

$$f = -\frac{k_B T}{N} \ln Z \tag{7.12}$$

$$f_{MF} = -\frac{k_B T}{N} \ln Z_{MF}$$
$$\simeq C + A(T - T_c^{MF})m^2 + Bm^4 + Dhm + ... \tag{7.13}$$

where C, A, B and D are constants, h is an applied magnetic field and MF denotes quantities coming from the mean-field theory. The form of this expansion and the sign of B reflect the system symmetry. In the case where $h = 0$ and $B > 0$, f_{MF} presents a minimum at $m = 0$ for $T > T_c^{MF}$ and two symmetric minima at $\pm m_0$ for $T < T_c^{MF}$ (see Fig. 7.2), indicating two degenerate ordered states. This degeneracy is removed when $h \neq 0$ as shown in Fig. 7.3.

When $B < 0$, a first-order transition is possible. At $T = T_c$, there are three equivalent minima of f_{MF} at $0, \pm m_0$ (see Fig. 7.4) contrary to the case $B > 0$. This means that at the transition the three phases $m = 0$ (paramagnetic state) and $\pm m_0$ (ordered states) coexist. The energy distribution at $T = T_c$ is thus bimodal, the peak at low energy corresponds to the energy of the ordered phase while that at high energy corresponds to the energy of the disordered state. The distance between the two peaks is the latent heat which is observed at a first-order transition.

When an m^3 term is present in the expansion (7.13), the transition is always of first order.

7.2.1 *Mean-field critical exponents*

The critical exponents calculated by the mean-field theory are (see chapter 4) $\beta = 1/2$ [see (4.25)], $\gamma = 1$ [see (4.37)]. We can find them again here by using (7.13). Putting $t = (T - T_c^{MF})/T_c^{MF}$, we have

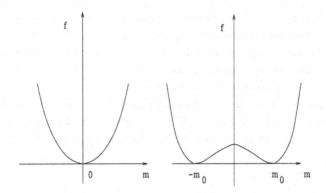

Fig. 7.2   Mean-field free energy at $T > T_c$ (left) and at $T < T_c$ (right).

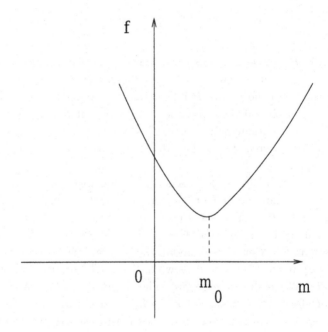

Fig. 7.3   Mean-field free energy at $T < T_c$ in an applied magnetic field.

- When $t < 0$ and $h = 0$, f_{MF} is minimum at $m_0 \propto (-t)^{\frac{1}{2}}$, so that $\beta = 1/2$.
- When $h \neq 0$ and $t \geq 0$, we have $m \propto h/t$ so that $\chi \propto t^{-1}$, hence $\gamma = 1$.
- At $t = 0$, f_{MF} is minimum at $m \propto (\frac{Dh}{B})^{1/3}$, hence $\delta = 3$.
- For $t > 0$, the minimum of f_{MF} is equal to C and for $t < 0$, it is equal

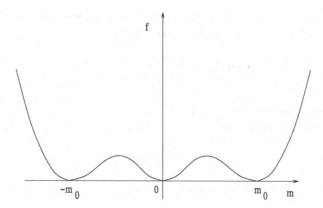

Fig. 7.4   Mean-field free energy at a first-order transition.

to $C + \mathcal{O}(\frac{A^2}{B}t^2)$. We have $C_V \propto \frac{\partial^2 f}{\partial T^2}$ which is 0 for $t > 0$ and equal to $\frac{A^2}{B}$ for $t < 0$. C_V is thus independent of t. The discontinuity of C_V at $t = 0$ is an artefact of the mean-field theory. When fluctuations are included, we find $\alpha = 0$ as will be seen below.

7.2.2   *Correlation function*

We consider a system of Ising spins for simplicity. We use in this paragraph a simplified notation: r instead of $\vec{r}$. We recover the correct notation when necessary. The Hamiltonian is given by

$$\mathcal{H} = -\sum_{(r',r'')} J(r' - r'')S(r')S(r'') \tag{7.14}$$

The correlation function is calculated as follows

$$
\begin{aligned}
G(r) &= <S(0)S(r)> \\
&= \frac{\mathrm{Tr}\,S(0)S(r)\exp(\frac{1}{2}\beta\sum_{r',r''}J(r'-r'')S(r')S(r''))}{\mathrm{Tr}\exp(\frac{1}{2}\beta\sum_{r',r''}J(r'-r'')S(r')S(r''))} \\
&= \frac{\mathrm{Tr}\,S(r)\exp(\frac{1}{2}\beta\sum_{r',r''}J(r'-r'')S(r')S(r''))}{\mathrm{Tr}\exp(\frac{1}{2}\beta\sum_{r',r''}J(r'-r'')S(r')S(r''))} \\
&= <S(r)> = \tanh[\beta\sum_{r'}J(r-r')S(r')] \tag{7.15}
\end{aligned}
$$

where we have taken $S(0) = 1$. The trace was taken over all configurations with $S(0) = 1$. In the mean-field spirit, we replace $S(r')$ in the right-hand side of (7.15) by its average value $< S(r') >$, then we make an expansion around T_c when $< S(r') >$ is small, we obtain

$$m(r) \simeq \beta \sum_{r'} J(r - r')m(r') \qquad (7.16)$$

with the notation $m(r) =< S(r) >$.

The Fourier transform of this equation gives

$$m(k) = \beta J(k)m(k) + C \qquad (7.17)$$

where C is a constant. For small k, we have

$$\begin{aligned}
J(k) &= \sum_{r=-\infty}^{\infty} J(r) \exp(ikr) \\
&\simeq \sum_{r=-\infty}^{\infty} J(r)[1 + ikr + (ikr)^2] \\
&\simeq \sum_{r=-\infty}^{\infty} J(r)[1 - k^2 r^2] \\
&\simeq \sum_{r=-\infty}^{\infty} J(r) - k^2 \sum_{r=-\infty}^{\infty} r^2 J(r) \\
&\simeq \sum_{r=-\infty}^{\infty} J(r) - k^2 \sum_{r'=-\infty}^{\infty} J(r') \sum_{r=-\infty}^{\infty} r^2 J(r) / \sum_{r'=-\infty}^{\infty} J(r') \\
&\simeq \sum_{r} \tilde{J}[1 - k^2 R^2] \qquad (7.18)
\end{aligned}$$

where the sum on the term $J(r)r$ of the second equality is zero because $J(r)r$ is an odd function of r [$J(r)$ is an even function: $J(r) = J(-r)$], $\tilde{J}$ is defined by

$$\tilde{J} = \sum_{r} J(r) \qquad (7.19)$$

and R^2 is defined by

$$R^2 = \frac{\sum_r r^2 J(r)}{\sum_r J(r)} \qquad (7.20)$$

R^2 is thus the order of the interaction range.

We obtain from (7.17) and (7.18)

$$m(k) = \frac{C}{1 - \beta \tilde{J}(1 - R^2 k^2)} \tag{7.21}$$

We recall here that in the mean-field theory $k_B T_c = \tilde{J}$. We write

$$
\begin{aligned}
m(k) &= \frac{C}{1 - \beta k_B T_c (1 - R^2 k^2)} \\
&= \frac{C}{1 - \frac{T_c}{T}(1 - R^2 k^2)} \\
&= \frac{C}{t + \frac{T_c}{T} R^2 k^2} \\
&\simeq \frac{C R^{-2}}{t R^{-2} + k^2}
\end{aligned} \tag{7.22}
$$

where we have taken $T \simeq T_c$. We put now

$$\xi^{-2} = t R^{-2} \tag{7.23}$$

and we use $m(r) = G(r)$ of (7.15), we finally arrive at

$$G(k) = \frac{C R^{-2}}{\xi^{-2} + k^2} \tag{7.24}$$

The inverse Fourier transform of (7.24) gives (see chapter 2)

$$G(r) \propto \frac{\exp(-r/\xi)}{r^{(d-1)/2}} \tag{7.25}$$

This form of $G(r)$ justifies the fact that we call ξ the "correlation length" in the mean-field theory. The expression (7.24) is called "Ornstein-Zernike correlation function".

From (7.23) we see that $\xi \propto t^{-1/2}$, therefore $\nu = 1/2$ in this mean-field theory. In addition, at $t = 0$, ξ tends to ∞ (see the following paragraph), $G(k)$ of (7.24) is then proportional to $1/k^2$. The inverse Fourier transform of this function is $\frac{1}{r^{d-2}}$ for large r, indicating therefore that exponent η of the mean-field theory is equal to 0 [see definition (7.10)].

7.2.3   *Corrections to mean-field theory*

We decompose the following term [see (5.6)]

$$
\begin{aligned}
J(r - r')\vec{S}_r \cdot \vec{S}_{r'} &= J(r - r')[\vec{S}_r^z + \delta \vec{S}_r] \cdot [\vec{S}_{r'}^z + \delta \vec{S}_{r'}] \\
&\simeq J(r - r') < \vec{S}_r^z > \cdot < \vec{S}_{r'}^z > \\
&\simeq J(r - r')m(r)m(r')
\end{aligned} \tag{7.26}
$$

where the average values of the linear terms in $\delta\vec{S}_r$ and $\delta\vec{S}_{r'}$ are zero by symmetry (rotating vectors in the xy plane). In the last equality, we have neglected, in the mean-field spirit, the following term

$$J(r - r')\delta\vec{S}_r \cdot \delta\vec{S}_{r'} \qquad (7.27)$$

Using the correlation function

$$G(r - r') = <\vec{S}_r \cdot \vec{S}_{r'}> = \text{constant} + <\delta\vec{S}_r \cdot \delta\vec{S}_{r'}>$$

we write

$$\sum_{r'} J(r - r') <\delta\vec{S}_r \cdot \delta\vec{S}_{r'}> = \sum_{r'} J(r - r')G(r - r') \qquad (7.28)$$

We are interested in a long-distance behavior. Using the Fourier transform of (7.24) for small k, we write

$$\sum_{r'} J(r - r') <\delta\vec{S}_r \cdot \delta\vec{S}_{r'}> \simeq \sum_{r'} J(r - r')\frac{C}{R^2} \int_{ZB} \frac{d^d k}{k^2 + \xi^{-2}}$$

$$= \tilde{J}\frac{C}{R^2} \int_{ZB} \frac{d^d k}{k^2 + \xi^{-2}} \qquad (7.29)$$

where the integral is performed in the first Brillouin zone. We decompose this integral as follows

$$\int_{ZB} \frac{d^d k}{k^2 + \xi^{-2}} = \int_{ZB} \frac{d^d k}{k^2} - \xi^{-2} \int_{ZB} \frac{d^d k}{k^2(k^2 + \xi^{-2})} \qquad (7.30)$$

The first integral does not depend on ξ, namely independent of T. It contributes to shift the value of T_c calculated by mean-field theory. The second integral, dependent on ξ thus of T, converges if $k = \pm\pi/a \to \infty$, namely $a \to 0$ (continuum limit), and if d (space dimension) < 4. By a simple dimension analysis, we see that this integral is proportional to ξ^{2-d} at the limit $\xi \to \infty$. The term which depends on T is thus proportional to $\frac{\tilde{J}}{R^2}\xi^{2-d}$. This term has been neglected before in the mean-field theory because we thought, wrongly of course, that it is always small with respect to the term coming from the mean field

$$\frac{\tilde{J}}{R^2}\xi^{2-d} \ll \tilde{J}t \qquad (7.31)$$

where $t = \frac{T-T_c}{T}$. Now we know that $\xi = Rt^{-1/2}$ [see (7.23)], expression (7.31) thus becomes

$$\xi^{4-d} \ll R^4 \qquad (7.32)$$

This condition is called "Ginzburg's criterion". We see that when $d < 4$ this criterion is not satisfied near T_c where $\xi \to \infty$. By consequence, the mean-field theory which neglects fluctuations characterized by ξ^{4-d} is not valid in the critical region for $d < 4$. The dimension $d = 4$ is called "upper critical dimension" for the Ising model with short-range interaction.

7.3 Renormalization group

7.3.1 *Transformation of renormalization group — Fixed point*

We present here the concepts of the renormalization group. The central idea of the approach is to replace the set of parameters which define the system by another set of parameters which are simpler to deal with but essential physical ingredients, in particular the system symmetry, are conserved. In the study of phase transitions, this approach consists in dividing the system into blocks of spins then replacing each block by a single spin. This "new" spin interacts with the others by the renormalized interactions calculated while decimating the block. The distances between the new spins are measured with a new lattice constant. We repeat this procedure with the new system of spins to obtain the next generation of spins, and repeat over and over again. At each iteration, also called decimation or transformation, we have a relation between the new interaction $K' = \beta J'$ and the previous one $K = \beta J$:

$$K' = f(K) \qquad (7.33)$$

The new correlation length ξ which is a function of K' is equal to the previous correlation length divided by the factor called "dilatation" b defined as the ratio between the new and old lattice constants. We have

$$\xi(K') = \frac{\xi(K)}{b} \qquad (7.34)$$

Near the phase transition, ξ becomes very large, the measuring distance unit is no more important. The interaction constants K' and K become identical $K' = K = K^*$. This point is called the "fixed point" in the renormalization group language. The fixed point is thus determined by

$$K^* = f(K^*) \tag{7.35}$$

The one-dimensional case is simple to proceed as seen in Problem 2 in section 7.7. However, in the case of a general dimension $d > 1$, relation (7.33) is rather complicated. It is often impossible to find a solution of (7.35). We then have to take into account physical considerations to find appropriate approximations.

We consider a point K in the proximity of the fixed point K^*. If the iteration process takes K away from the fixed point, K^* is an unstable fixed point. This is a "run away" case. On the other hand, in the case where K tends to K^* by iteration, K^* is a stable fixed point (see Problem 2 in section 7.7). The map of these trajectories near a fixed point with indicated moving directions is called a "flow diagram" . An example is shown in Fig. 7.5. This figure corresponds to the case of a square lattice of Ising spins with interactions $K_1 = \frac{J_1}{k_B T}$ in the x direction and $K_2 = \frac{J_2}{k_B T}$ in the y direction. P is the fixed point. For a given ratio K_2/K_1, we follow the discontinued line: intersection P_1 with the line separating regions of different flows is the critical point corresponding to that ratio K_2/K_1. The line of flow separation shown by the heavy solid line is the critical line. We see that P_1 runs toward P on the critical line, therefore P_1 and P belong to the same universality class: the universality class does not thus depend on the ratio K_2/K_1 (see subsection 7.3.3).

We can calculate the critical exponents using the renormalization group if we know how K' depends on K in the vicinity of a fixed point K^*. We make then an expansion of their relation (7.33) around K^*:

$$\begin{aligned} K' &\simeq f(K^*) + (K - K^*)f'(K^*) \\ &\simeq K^* + b^y(K - K^*) \end{aligned} \tag{7.36}$$

where we have replaced $f(K^*)$ by K^* using (7.35), and

$$y = \frac{\ln f'(K^*)}{\ln b}.$$

Now, near K^* we know that $\xi \propto (K - K^*)^{-\nu}$ (definition of ν). Therefore, by using (7.34) and (7.36) we obtain

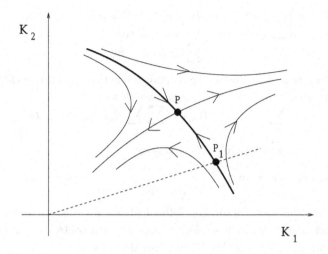

K_2

K_1

Fig. 7.5 Flow diagram for a square lattice of Ising spins with nearest-neighbor interactions K_1 along the x axis and K_2 along the y axis. See text for comments.

$$(K - K^*)^{-\nu} = b(K' - K^*)^{-\nu} = b[b^y(K - K^*)]^{-\nu} \qquad (7.37)$$

By identifying the two sides of the above equation, we get

$$\nu = 1/y \qquad (7.38)$$

This example shows that the critical exponent ν is a derivative of the renormalization group equation (7.33). The other critical exponents are calculated by the use of the free energy as seen below.

7.3.2 *Scaling variables*

We generalize (7.36) to the case where there are several interactions K_m ($m = 1, 2, ...$). An expansion of (7.33) around the fixed point yields

$$K'_m - K^*_m = \sum_n M_{mn}(K'_n - K^*_n) \qquad (7.39)$$

where

$$M_{mn} = \frac{\partial K'_m}{\partial K_n}\Big|_{K=K^*} \qquad (7.40)$$

Let λ^i be the eigenvalue corresponding to the eigenvector e^i_n of the matrix $\mathbf{M}$ defined by

$$\sum_m e^i_m M_{mn} = \lambda^i e^i_n \qquad (7.41)$$

We define now the scaling variables u_i by a linear combination of the deviations $(K'_m - K^*_m)$ from the fixed point in the e^i_n basis:

$$u_i = \sum_m e^i_m (K_m - K^*_m) \qquad (7.42)$$

We show that when we make a renormalization group transformation, u_i is transformed near the fixed point as follows:

$$u'_i = \sum_m e^i_m (K'_m - K^*_m) = \sum_m \sum_n e^i_m M_{mn} (K'_n - K^*_n)$$

$$= \sum_n \lambda^i e^i_n (K'_n - K^*_n) = \lambda^i u_i \qquad (7.43)$$

For commodity, we put

$$\lambda^i = b^{y_i} \qquad (7.44)$$

b being the dilatation factor. y_i defined above is called "renormalization group eigenvalue". We will see below that y_i is connected to critical exponents by simple relations. We distinguish three cases:

- If $y_i > 0$, u_i is a relevant variable. Repeated iterations of renormalization group take it more and more away from its value at the fixed point.

- If $y_i < 0$, u_i is an irrelevant variable. Repeated iterations make it go to zero.

- If $y_i = 0$, u_i is a marginal variable because we do not know if it goes away from its value at the fixed point or it tends to that value.

7.3.3   *Critical surfaces*

In the space of interactions K_m, the flow diagram around the fixed points gives information on the stability of each of these points. Critical surfaces are surfaces which separate regions of different flows. For example, in Fig. 7.5, the heavy solid line separates flows toward large values of K from flows toward small values. In the space of more than two interactions, this line generates a surface. Each point of the generated surface is a critical point of the system corresponding to the interaction parameters of that point. Properties at long distances of that critical point are controlled by the fixed point P because a flow line on the critical surface brings it to P with successive renormalization group iterations. In general, if the space around a fixed point has n' dimensions among which there are n relevant variables and $(n' - n)$ irrelevant ones, the critical surface has $(n' - n)$ dimensions. This shows the importance of irrelevant variables in the determination of critical properties of a system.

7.3.4 *Critical exponents*

The partition function is written as

$$Z = \text{Tr}_s \exp[-\mathcal{H}(x)] = \text{Tr}_{s'} \exp[-\mathcal{H}(x')] \tag{7.45}$$

where the traces are taken over the spin configurations s and s', before and after the renormalization group transformation. The temperature is included in the definition de $\mathcal{H}$. The free energy $\mathcal{F}$ per spin reads

$$\mathcal{F}(K) = -\frac{1}{N} \ln Z \tag{7.46}$$

where the temperature is absorbed in the definition of $\mathcal{F}$. After the transformation, the number of spins N becomes N' with $N' = N/b^d$. The free energy, being extensive, becomes

$$\mathcal{F}(K) = b^{-d}\mathcal{F}(K') + g(K) \simeq b^{-d}\mathcal{F}(K') \tag{7.47}$$

where we have neglected $g(K)$ because it is not a function which diverges at the transition.

We consider in the following only the variables t (reduced temperature) and h (magnetic field) for simplicity. The associated scaling variables are u_t and u_h. Using (7.43) and (7.44), we express $\mathcal{F}$ as a function of these variables

$$\begin{aligned}
\mathcal{F}(u_t, u_h) &= b^{-d}\mathcal{F}(u_t', u_h') \\
&= b^{-d}\mathcal{F}(b^{y_t}u_t, b^{y_h}u_h) \\
&= b^{-nd}\mathcal{F}(b^{ny_t}u_t, b^{ny_h}u_h)
\end{aligned} \tag{7.48}$$

where the last equality has been obtained after n iterations. The values of u_t and u_h increase with iterations, we should limit the number of iterations so that after n iterations we have $|b^{ny_t}u_t| = u_{t_0}$ where u_{t_0} is a chosen value sufficiently small to keep the validity of the linear expansion near the fixed point. We deduce $b = |\frac{u_{t_0}}{u_t}|^{\frac{1}{ny_t}}$ and $b^{ny_h}u_h = |\frac{u_t}{u_{t_0}}|^{-\frac{y_h}{y_t}}u_h$. Equation (7.48) becomes

$$\mathcal{F}(u_t, u_h) = |\frac{u_t}{u_{t_0}}|^{\frac{d}{y_t}} \mathcal{F}(\pm u_{t_0}, u_h|\frac{u_t}{u_{t_0}}|^{-\frac{y_h}{y_t}}) \tag{7.49}$$

$$\mathcal{F}(t, h) = |\frac{u_t}{u_{t_0}}|^{\frac{d}{y_t}} \theta(\frac{h/h_0}{|t/t_0|^{y_h/y_t}}) \tag{7.50}$$

where θ is a universal scaling function, while t_0 and h_0 depend on the characteristics of the system under consideration.

Using (7.50) we obtain

- the specific heat $C_V = \frac{\partial^2 F}{\partial t^2}|_{h=0} \propto |t|^{d/y_t - 2}$, whence

$$\alpha = 2 - \frac{d}{y_t} \tag{7.51}$$

- the magnetization $m = \frac{\partial F}{\partial h}|_{h=0} \propto |-t|^{(d-y_h)/y_t}$, whence

$$\beta = \frac{d - y_h}{y_t} \tag{7.52}$$

- the susceptibility $\chi = \frac{\partial^2 F}{\partial h^2}|_{h=0} \propto |t|^{(d-2y_h)/y_t}$, whence

$$\gamma = \frac{2y_h - d}{y_t} \tag{7.53}$$

- the critical exponent δ: we have

$$m = \frac{\partial F}{\partial h} = |t/t_0|^{(d-y_h)/y_t} \theta'\left(\frac{h/h_0}{|t/t_0|^{y_h/y_t}}\right) \tag{7.54}$$

If we want m to have a finite limit when $t \to 0$, the function $\theta'(x)$ should behave as $x^{d/y_h - 1}$ when $x \to \infty$. We then have $m \propto h^{d/y_h - 1}$ which leads to

$$\delta = \frac{y_h}{d - y_h} \tag{7.55}$$

All these exponents are functions of y_t, of y_h or of both. By eliminating these two variables from the above expressions, we obtain the following scaling relations

$$\alpha + 2\beta + \gamma = 2 \tag{7.56}$$
$$\alpha + \beta(1 + \delta) = 2 \tag{7.57}$$

These relations have been verified in all "normal" cases known so far. However, for complex cases such as multicritical phase transition, fractals etc. caution must be taken when using them.

We have seen that exponent ν is obtained simply from the correlation length [see (7.38)]. We can make a similar analysis for the correlation function in the general case as we did above for the other exponents. Without giving all details, we obtain the following results:

$$\nu = 1/y_t \tag{7.58}$$
$$\eta = d + 2 - 2y_h \tag{7.59}$$

These two exponents concerning the spin-spin correlation which are to be added to the four thermodynamic exponents given in (7.52)-(7.55). By a complicated calculation, we obtain two following "hyperscaling relations"

$$\alpha = 2 - d\nu \qquad (7.60)$$

$$\gamma = \nu(2 - \eta) \qquad (7.61)$$

In summary, there are six critical exponents and four relations between them. It suffices to determine two among six exponents. The other four can be then calculated using the four relations.

To close this section, we emphasize that, in addition to the results shown above, another important result of the renormalization group is the relations connecting the system size to the critical exponents: physical quantities calculated at a finite system size are shown to depend on powers of the linear system size. These powers are simple functions of critical exponents. The finite-size scaling relations are presented in section 8.4. They are very useful for the determination of critical exponents by Monte Carlo simulation.

7.4 Phase transition in some particular systems

We have seen so far various methods used to study phase transitions and critical phenomena. In general, the nature of a phase transition depends on the symmetry of the order parameter, the spatial dimension and the nature of the interaction (short or long range). Standard methods presented above can be used to determine it with satisfactory precision. However, in some particular systems we need special methods to deal with. Let us discuss some of these remarkable systems in the following.

7.4.1 *Exactly solved spin systems*

There are several families of systems in one or two dimensions with short-range non-crossing interactions which can be exactly solved. The spin models in those solvable systems are often Ising and Potts models. We need exact solutions in simple systems to test approximations conceived to be used in more complicated systems or systems in three dimensions. Methods for searching exact solutions are lengthy to present here. The reader is referred to the book by R. J. Baxter [16] for general methods and exactly solved models. For some exactly solved frustrated spin systems, the reader is referred to the review by Diep and Giacomini [41]. In a word, to find a solution, most frequently we transform the studied system into a vertex model where solutions for critical surfaces are known. Among the most popular models, we mention the 8-, 16- and 32-vertex models. Let us briefly present how these models work by showing the 16-vertex case.

Mapping between Ising models and vertex models: 16-vertex case

The 2D Ising model with non-crossing interactions is exactly soluble. The problem of finding the partition function can be transformed in a free-fermion model. If the lattice is a complicated one, the mathematical problem to solve is very cumbersome.

For numerous 2D Ising models with non-crossing interactions, there exists another easier method to find the exact partition function. This method consists in mapping the model on a 16-vertex model or a 32-vertex model. If the Ising model does not have crossing interactions, the resulting vertex model will be exactly soluble. We can apply this method for finding the exact solution of several Ising models in 2D lattices with non-crossing interactions.

Let us introduce the 16-vertex model. The 16-vertex model we consider is defined on a square lattice of N points, connected by edges between neighboring sites. These edges can assume two states, symbolized by right-and left- or up-and down-pointing arrows, respectively. The allowed configurations of the system are characterized by specifying the arrangement of arrows around each lattice point. In characterizing these so-called vertex configurations, we follow the enumeration of Baxter [16] (see Fig.7.6).

To each vertex we assign an energy $\epsilon_k (k = 1, 2, ..., 16)$ and a corresponding vertex weight (Boltzmann factor) $\omega_k = e^{\beta \epsilon_k}$, where $\beta = (1)/(k_B T)$, T being the temperature and k_B the Boltzmann constant. Then the partition function is

$$Z = \sum_C e^{-\beta(n_1 \epsilon_1 + ... + n_{16} \epsilon_{16})} \qquad (7.62)$$

where the sum runs over all allowed configurations C of arrows on the lattice, n_j is the number of vertex arrangements of type j in configuration C. It is clear from Eq. (7.62) that Z is a function of the eight Boltzmann

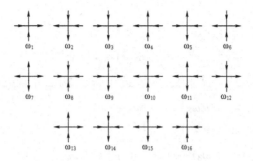

Fig. 7.6   Arrow configurations and vertex weights of the 16-vertex model.

weights $\omega_k(k = 1, 2, ..., 16)$:

$$Z = Z(\omega_1, ..., \omega_{16}) \tag{7.63}$$

So far, exact results have only been obtained for three subclasses of the general 16-vertex model, i.e. the 6-vertex (or ferroelectric) model, the symmetric 8-vertex model and the free-fermion model [16, 53]. Here we will consider only the case where the free-fermion condition is satisfied, because in these cases the 16-vertex model can be related to 2D Ising models without crossing interactions. Generally, a vertex model is soluble if the vertex weights satisfy certain conditions so that the partition function is reducible to the S matrix of a many-fermion system [53]. In the present problem these constraints are the following :

$$\omega_1 = \omega_2 \,,\; \omega_3 = \omega_4$$
$$\omega_5 = \omega_6 \,,\; \omega_7 = \omega_8$$
$$\omega_9 = \omega_{10} = \omega_{11} = \omega_{12}$$
$$\omega_{13} = \omega_{14} = \omega_{15} = \omega_{16}$$
$$\omega_1\omega_3 + \omega_5\omega_7 - \omega_9\omega_{11} - \omega_{13}\omega_{15} = 0 \tag{7.64}$$

If these conditions are satisfied, the free energy of the model can be expressed, in the thermodynamical limit, as follows :

$$f = -\frac{1}{4\pi\beta} \int_0^{2\pi} d\phi \log\{A(\phi) + [Q(\phi)]^{1/2}\} \tag{7.65}$$

where

$$A(\phi) = a + c\cos(\phi)$$
$$Q(\phi) = y^2 + z^2 - x^2 - 2yz\cos(\phi) + x^2\cos^2(\phi)$$
$$a = \frac{1}{2}(\omega_1^2 + \omega_3^2 + 2\omega_1\omega_3 + \omega_5^2 + \omega_7^2 + 2\omega_5\omega_7) + 2(\omega_9^2 + \omega_{13}^2)$$
$$c = 2[\omega_9(\omega_1 + \omega_3) - \omega_{13}(\omega_5 + \omega_7)]$$
$$y = 2[\omega_9(\omega_1 + \omega_3) + \omega_{13}(\omega_5 + \omega_7)]$$
$$z = \frac{1}{2}[(\omega_1 + \omega_3)^2 - (\omega_5 + \omega_7)^2] + 2(\omega_9^2 - \omega_{13}^2)$$
$$x^2 = z^2 - \frac{1}{4}[(\omega_1 - \omega_3)^2 - (\omega_5 - \omega_7)^2]^2 \tag{7.66}$$

Phase transitions occur when one or more pairs of zeros of the expression $Q(\phi)$ close in on the real ϕ axis and "pinch" the path of integration in the expression on the right-hand side of Eq. (7.65). This happens when $y^2 = z^2$, i.e. when

$$\omega_1 + \omega_3 + \omega_5 + \omega_7 + 2\omega_9 + 2\omega_{13} = 2\max\{\omega_1 + \omega_3, \omega_5 + \omega_7, 2\omega_9, 2\omega_{13}\} \tag{7.67}$$

The nature of the singularity in the specific heat depends on whether

$$(\omega_1 - \omega_3)^2 - (\omega_5 - \omega_7)^2 \neq 0 \quad \text{(logarithmic \ singularity)}$$

or

$$(\omega_1 - \omega_3)^2 - (\omega_5 - \omega_7)^2 = 0 \quad \text{(inverse square-root singularity)} \tag{7.68}$$

We will show an application of the above result in section 7.5.

7.4.2 *Kosterlitz-Thouless transition*

We consider the XY spins on a two-dimensional lattice with a ferromagnetic interaction between nearest neighbors. In the ground state, the spin configuration is a perfect ferromagnetic state, namely all spins are parallel. However, this system does not have a normal second-order transition at a finite temperature. There is no long-range ordering when the temperature is not zero, as indicated by the Mermin-Wagner theorem [95] for two-dimensional systems with spin continuous degrees of freedom (see the discussion in the conclusion of chapter 5). Kosterlitz and Thouless [80] have shown that the system has a special phase transition due to the unbinding of vortex-antivortex pairs at a finite temperature below (above) which the correlation function decays as a power law (exponential law) with increasing distance. This transition, called Kosterlitz-Thouless (KT) or Kosterlitz-Thouless-Berezinskii transition, is of infinite order. Let us outline in the following some important points which help us understand the mechanism behind the KT transition.

The Hamiltonian of the system is given by

$$H = -J \sum_{<i,j>} \vec{S}_i \cdot \vec{S}_j = -J \sum_{<i,j>} \cos(\theta_i - \theta_j) \tag{7.69}$$

where the sum runs over all nearest neighbor spin pairs $\vec{S}_i$ and $\vec{S}_j$ in the lattice, and θ_i denotes the angle of the spin $\vec{S}_i$ with respect to some (arbitrary) polar direction in the two dimensional vector space containing the spins. At low temperatures ($T << J$), the spins are nearly parallel on neighboring sites, we can expand the cosine to the second order to obtain the lattice gaussian model

$$H = E_0 + \frac{J}{2} \sum_{<i,j>} (\theta_i - \theta_j)^2 \tag{7.70}$$

where $E_0 = -CJN/2$ (C: coordination number) is the energy per spin of the parallel spin configuration. The continuum limit is written as

$$H = E_0 + \frac{J}{2} \int d\vec{r} [\vec{\nabla}\theta(\vec{r})]^2 \tag{7.71}$$

The partition function is given by

$$Z = e^{-\beta E_0} \int D[\theta] \exp\{-\beta \frac{J}{2} \int d\vec{r} [\vec{\nabla}\theta(\vec{r})]^2\} \tag{7.72}$$

where $D[\theta]$ is performed over all configurations of $\theta(\vec{r})$. This integration can be divided into two parts: a sum over the local minima θ_v of $H[\theta]$ (v: vortex) and a sum on the fluctuations θ_{sw} (sw: spin waves) around the minima, namely $\int D[\theta] = \sum_{\theta_v} \int D[\theta_{sw}]$.... We have

$$Z = e^{-\beta E_0} \sum_{\theta_v} \int D[\theta_{sw}] \exp\{-\beta [H(\theta_v) + \frac{J}{2} \int d\vec{r}_1 \int d\vec{r}_1$$

$$\times \theta_{sw}(\vec{r}_1) \frac{\partial^2 H}{\partial \theta(\vec{r}_1) \partial \theta(\vec{r}_2)} \theta_{sw}(\vec{r}_2)]\} \tag{7.73}$$

Note the absence of the term $\frac{\partial H}{\partial \theta(\vec{r})}$ because it is zero at the minimum. This corresponds to $|\vec{\nabla}\theta(\vec{r})| = 0$ which has two possible solutions

i) $\theta(\vec{r})$=constant, namely the ferromagnetic ground state,

ii) solutions of vortices of the director field around vortex centers. There are two cases

- For all closed curves encircling the position r_0 of the center of the vortex, we have

$$\oint \vec{\nabla}\theta(\vec{r}) \cdot d\vec{\ell} = 2\pi n \tag{7.74}$$

- For all paths that do not encircle the vortex position r_0

$$\oint \vec{\nabla}\theta(\vec{r}) \cdot d\vec{\ell} = 0 \tag{7.75}$$

At a given distance from r_0, we have $\theta(\vec{r}) = \theta(r)$, therefore the solution (7.74) yields $|\nabla\theta(r)|2\pi r = 2\pi n$, namely $|\nabla\theta(r)| = n/r$. Replacing this in Eq. (7.71), we obtain

$$
\begin{aligned}
E - E_0 &= \frac{J}{2} \int d\vec{r} [\vec{\nabla}\theta(\vec{r})]^2 \\
&= \frac{Jn^2}{2} \int_0^{2\pi} d\theta \int_a^L r dr \frac{1}{r^2} \\
&= \pi J n^2 \ln \frac{L}{a} \tag{7.76}
\end{aligned}
$$

where a is the lattice constant (distance between nearest neighbors) and L is the system linear size. Note that n is called "charge" of the vortex. When $n > 1$, it is called a multi-charge. The energy of a large system is large even with a single charge. When we encircle a vortex, the integral (7.74) gives 2π, and when we encircle an antivortex, the integral is equal to -2π. If we encircle a pair of vortex-antivortex at a large enough distance, the integral is zero. So, the vortex-antivortex energy is just the energy of the vortex and antivortex at their distance R given by the following expression

$$E_{2\ vort} = 2E_c + E_1 \ln \frac{R}{a} \tag{7.77}$$

where E_c is the energy of the vortex core and E_1 is proportional to J. We emphasize here that $E_{2\ vort}$ does not go to infinity when $L \to \infty$, unlike the energy of a single vortex E given by Eq. (7.76).

Now, we consider the free energy $F = E - TS$ where S is the entropy. We imagine a system of independent vortices excited at T. The entropy is the number of ways to place these vortex centers at centers of the square lattice cells: the number of cell centers is L^2/a^2. Therefore $S = k_B \ln(L^2/a^2)$. We have

$$F = E_0 + (\pi J - 2k_B T) \ln(\frac{L}{a}) \tag{7.78}$$

We see here that when $L \to \infty$, F tends to ∞ if $\pi J - 2k_B T > 0$, namely $T < \pi J/2k_B$. We conclude that single vortices cannot be excited at low temperatures for large systems. However, pairs of vortices can be excited at low temperatures because their energy, Eq. (7.77), does not diverge. So, the only thing which can happen at temperature $T < \pi J/2k_B$ is the excitation of pairs of bound vortices. Of course, the energy of bound vortices depends on T through $< R >$ which can be calculated using the partition function. For $T > \pi J/2k_B$ pairs of bound vortices are unbound to become independent single vortices which lower the free energy ($F \to -\infty$ for large systems). The unbinding of pairs of vortices occurs at $T = T_{KT} = \pi J/2k_B$ which is called the KT transition temperature.

Let us show the absence of long-range ordering in the present two-dimensional XY-spin system at finite temperature. We calculate $< S_x >$ as follows:

$$< S_x > = < \cos\theta(\vec{r}) > = \frac{\int D[\theta] \cos\theta(\vec{r}) e^{-\beta H}}{\int D[\theta] e^{-\beta H}} \qquad (7.79)$$

$$\theta(\vec{r}) = \sum_{\vec{k}} \theta(\vec{k}) e^{i\vec{k}\cdot\vec{r}}$$

$$|\vec{\nabla}\theta(\vec{r})| = \sum_{\vec{k}} k\theta(\vec{k}) e^{i\vec{k}\cdot\vec{r}}$$

$$|\vec{\nabla}\theta(\vec{r})|^2 = \sum_{\vec{k},\vec{k'}} kk'\theta(\vec{k})\theta(\vec{k'}) e^{i(\vec{k}+\vec{k'})\cdot\vec{r}}$$

$$\int d\vec{r}|\vec{\nabla}\theta(\vec{r})|^2 = \sum_{\vec{k},\vec{k'}} kk'\theta(\vec{k})\theta(\vec{k'}) \int d\vec{r} e^{i(\vec{k}+\vec{k'})\cdot\vec{r}}$$

$$= \sum_{\vec{k},\vec{k'}} kk'\theta(\vec{k})\theta(\vec{k'})\delta_{\vec{k},-\vec{k'}}$$

$$= \sum_{\vec{k}} k^2\theta(\vec{k})\theta(-\vec{k})$$

$$= \int \frac{d\vec{k}}{(2\pi)^d} k^2\theta(\vec{k})\theta(-\vec{k}) \qquad (7.80)$$

Replacing the last equality into the exponential argument of Eq. (7.79) [see the partition function Z given in Eq. (7.72)], we obtain after some calculations

$$< S_x > = \exp[-\frac{T}{2Ja^{2-d}}S_d \int_{\pi/L}^{\pi/a} dk\, k^{d-3}] \qquad (7.81)$$

where S_d is a constant. $< S_x >$ depends thus on the integral

$$I(L) = \int_{\pi/L}^{\pi/a} dk\, k^{d-3} \tag{7.82}$$

For $d < 2$, $I(L) \propto L^{2-d} \to \infty$ as $L \to \infty$. Thus $< S_x >= 0$: there is no ordering for $d < 2$. For $d > 2$, we have

$$I(L) = \frac{1}{d-2}(\frac{\pi}{a})^{d-2}$$

Therefore, $< S_x >$ is not zero at finite T: the ordering exists for this case. The case $d = 2$ is special: for $d = 2$ the integral $I(L)$ is logarithmically divergent $I(L) = \ln(L/a)$ which tends to ∞ for large L. $< S_x >$ thus goes to zero for any non-zero temperature. In spite of the absence of a long-range ordering at finite T, there is a special phase transition, called KT transition, of infinite order as shown above.

We examine now the behavior of the correlation function. We write

$$\begin{aligned}
< \vec{S}(\vec{r}) \cdot \vec{S}(0) > &= < \cos[\theta(\vec{r}) - \theta(0)] > \\
&= \mathrm{Re} < \exp\left[i[\theta(\vec{r}) - \theta(0)]\right] >
\end{aligned} \tag{7.83}$$

We expand in the Fourier modes

$$\theta(\vec{r}) - \theta(0) = \sum_{\vec{k}} \theta(\vec{k})[e^{i\vec{k}\cdot\vec{r}} - 1] = \int \frac{d\vec{k}}{(2\pi)^d}\theta(\vec{k})[e^{i\vec{k}\cdot\vec{r}} - 1] \tag{7.84}$$

Using Eqs. (7.72), (7.80) and (7.84), we rewrite Eq. (7.83) as

$$\begin{aligned}
G(\vec{r}) &= < \exp\left[i[\theta(\vec{r}) - \theta(0)]\right] > \\
&= \frac{\int D[\theta] \exp\left[i[\theta(\vec{r}) - \theta(0)]\right] \exp\{-\beta\frac{J}{2}\int d\vec{r}[\vec{\nabla}\theta(\vec{r})]^2\}}{\int D[\theta] \exp\{-\beta\frac{J}{2}\int d\vec{r}[\vec{\nabla}\theta(\vec{r})]^2\}} \\
&= \frac{\int D[\theta] \exp\{-\int \frac{d\vec{k}}{(2\pi)^d}\left[\beta\frac{J}{2}k^2\theta(\vec{k})\theta(-\vec{k}) - i\theta(\vec{k})[e^{i\vec{k}\cdot\vec{r}} - 1]\right]\}}{\int D[\theta] \exp\{-\beta\frac{J}{2}\int \frac{d\vec{k}}{(2\pi)^d}k^2\theta(\vec{k})\theta(-\vec{k})\}} \\
&= \exp[-g(\vec{r})] \tag{7.85}
\end{aligned}$$

where

$$g(\vec{r}) = \frac{k_B T}{J}\sum_{\vec{k}} \frac{1 - \cos(\vec{k}\cdot\vec{r})}{k^2} \tag{7.86}$$

The demonstration of Eq. (7.85) is shown in Problem 7 of section 7.7. In $d = 2$, we have, in changing the sum into an integral,

$$
\begin{aligned}
g(\vec{r}) &= \frac{k_B T}{J} \frac{1}{(2\pi)^2} \Big[\int_0^{\pi/r} 2\pi k dk \frac{1 - \cos(\vec{k} \cdot \vec{r})}{k^2} + \int_{\pi/r}^{\pi/a} 2\pi k dk \frac{1 - \cos(\vec{k} \cdot \vec{r})}{k^2} \Big] \\
&= \frac{k_B T}{J} \frac{1}{2\pi} \Big[\int_0^{\pi/r} 2\pi k dk \frac{1 - 1}{k^2} + \int_{\pi/r}^{\pi/a} 2\pi k dk \frac{1}{k^2} \Big] \\
&= \frac{k_B T}{J} \frac{1}{2\pi} \int_{\pi/r}^{\pi/a} dk \frac{1}{k} \\
&= \frac{k_B T}{2\pi J} \ln(r/a)
\end{aligned}
\tag{7.87}
$$

where we have replaced the cosine term by 1 in the region where $0 < k < \pi/r$ (very small k indeed) canceling thus the first integral, and for $kr > \pi$ we considered that the cosine term oscillates so rapidly that it makes a zero contribution to the second integral. Equation (7.85) then gives

$$
G(\vec{r}) \propto r^{-\eta} \quad \text{where } \eta = \frac{k_B T}{2\pi J}
\tag{7.88}
$$

The correlation function obeys thus a power law with the exponent η depending on T. Note that in the case of superfluid, J is replaced by the reduced superfluid density $\bar{\rho}_s = (\hbar/m)^2 \rho_s$. At the KT transition $T = T_{KT} = \pi J/2k_B$, we have $\eta = 1/4$. In the ordinary second-order transition, the power law happens only at the "critical point", but in the two-dimensional XY case here, the power law is found for the whole temperature range from T_{KT} down to zero. This is the reason why we call the line below the KT transition a "critical line". Above T_{KT}, the correlation function obeys an exponential law. It is noted that in the calculation of the correlation, we did not take into account the vortices. This is because, as shown above, independent vortices cannot occur at $T < T_{KT}$, otherwise the free energy will go to infinity for large systems. There exist of course pairs of vortex-antivortex at short distances below T_{KT} as seen above but they do not affect long-distance correlation. Note that due to the absence of vortices in Eq. (7.88), this result cannot be used at temperature above T_{KT}. The renormalization group treats in a much more elegant manner the whole temperature range, but this advanced calculation is so lengthy to present in this book.

7.5   Frustrated spin systems

We introduce in this section properties of some well-known frustrated systems. In particular, we show that the frustration causes many spectacular effects in their ground-state configurations and in their phase transition.

7.5.1   *Definition*

A system is said "frustrated" when the interaction bonds between a spin with its neighbors cannot be fully satisfied. An example is the triangular antiferromagnet: i) in the case of Ising spin model, the three spins cannot find orientations to satisfy the three bonds, ii) in the case of XY or Heisenberg spin models, the spins make a "compromise" to form a non collinear configuration in order to partially satisfy each bond as shown in Fig. 7.7. A few exercises to find the ground-state spin configuration using the method described in section 5.4 are given in Problem 6 in section 5.6 and in Problem 6 in section 7.7. Note that we can also use a numerical method described in Problem 4 in section 8.8.

Effects due to the frustration are numerous. We can mention a few of them: i) high ground-state degeneracy, ii) non collinear spin configuration, iii) multiple phase transitions, iv) reentrance phenomenon, v) disorder lines, vi) partial ordering at equilibrium, vii) difficulty in determining the nature of phase transitions in several systems, etc. In this section, we will show some of these spectacular effects observed in exactly solved two-dimensional systems. It is believed that these effects persist in three-dimensional systems and other more complicated unsolved models. We will also briefly mention one of the long-standing questions concerning the phase transition in XY and Heisenberg stacked triangular antiferromagnets.

7.5.2   *Phase transition*

7.5.2.1   *Kagomé lattice with nn and nnn interactions*

The Kagomé Ising lattice with nearest neighbor (nn) interaction J_1 shown in Fig. 7.8 has been solved a long time ago [73] showing no phase transition at finite T when J_1 is antiferromagnetic ($J_2 = 0$). Taking into account the next-nearest neighbor (nnn) interaction J_2 [see Fig.7.8], the Hamiltonian is

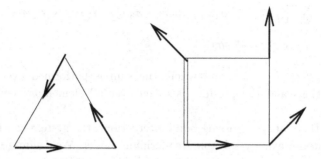

Fig. 7.7 Examples of frustrated spin systems. Left: antiferromagnetic triangular lattice with vector spins (XY or Heisenberg spins), Right: Villain's model with XY spins.

written as

$$H = -J_1 \sum_{(ij)} \sigma_i \sigma_j - J_2 \sum_{(ij)} \sigma_i \sigma_j \qquad (7.89)$$

where the first and second sums run over the spin pairs connected by single and double bonds, respectively.

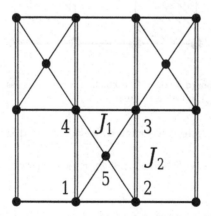

Fig. 7.8 Kagomé lattice. Interactions between nearest neighbors and between next-nearest neighbors, J_1 and J_2, are shown by single and double bonds, respectively. The lattice sites in a cell are numbered for decimation demonstration.

The partition function is written as

$$Z = \sum_{\sigma} \prod_{c} \exp[K_1(\sigma_1\sigma_5 + \sigma_2\sigma_5 + \sigma_3\sigma_5 + \sigma_4\sigma_5 + \sigma_1\sigma_2 + \sigma_3\sigma_4)$$
$$+ K_2(\sigma_1\sigma_4 + \sigma_3\sigma_2)] \tag{7.90}$$

where $K_{1,2} = J_{1,2}/k_B T$ and where the sum is performed over all spin configurations and the product is taken over all elementary cells of the lattice.

Since there are no crossing bond interactions, the system can be transformed into an exactly solvable free-fermion model. We decimate the central spin of each elementary cell of the lattice. In doing so, we obtain a checkerboard Ising model with multispin interactions (see Fig. 7.9).

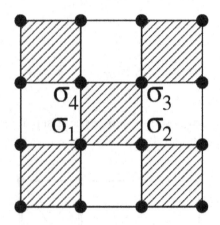

Fig. 7.9   The checkerboard lattice. At each shaded square is associated the Boltzmann weight $W(\sigma_1, \sigma_2, \sigma_3, \sigma_4)$, given in the text.

The Boltzmann weight associated to each shaded square is given by

$$W(\sigma_1, \sigma_2, \sigma_3, \sigma_4) = 2\cosh(K_1(\sigma_1 + \sigma_2 + \sigma_3 + \sigma_4))\exp[K_2(\sigma_1\sigma_4 + \sigma_2\sigma_3)$$
$$+ K_1(\sigma_1\sigma_2 + \sigma_3\sigma_4)] \tag{7.91}$$

The partition function of this checkerboard Ising model is given by

$$Z = \sum_{\sigma} \prod W(\sigma_1, \sigma_2, \sigma_3, \sigma_4) \tag{7.92}$$

where the sum is performed over all spin configurations and the product is taken over all the shaded squares of the lattice.

In order to map this model onto the 16-vertex model, let us introduce another square lattice where each site is placed at the center of each shaded square of the checkerboard lattice, as shown in Fig. 7.10.

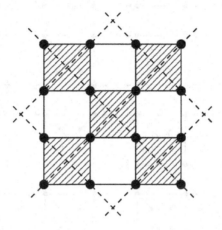

Fig. 7.10 The checkerboard lattice and the associated square lattice with their bonds indicated by dashed lines.

At each bond of this lattice we associate an arrow pointing out of the site if the Ising spin that is traversed by this bond is equal to $+1$, and pointing into the site if the Ising spin is equal to -1, as it is shown in Fig. 7.11.

In this way, we have a 16-vertex model on the associated square lattice. The Boltzmann weights of this vertex model are expressed in terms of the Boltzmann weights of the checkerboard Ising model, as follows

$$\omega_1 = W(-,-,+,+) \qquad \omega_5 = W(-,+,-,+)$$
$$\omega_2 = W(+,+,-,-) \qquad \omega_6 = W(+,-,+,-)$$
$$\omega_3 = W(-,+,+,-) \qquad \omega_7 = W(+,+,+,+)$$
$$\omega_4 = W(+,-,-,+) \qquad \omega_8 = W(-,-,-,-)$$
$$\omega_9 = W(-,+,+,+) \qquad \omega_{13} = W(+,-,+,+)$$
$$\omega_{10} = W(+,-,-,-) \qquad \omega_{14} = W(-,+,-,-)$$
$$\omega_{11} = W(+,+,-,+) \qquad \omega_{15} = W(+,+,+,-)$$
$$\omega_{12} = W(-,-,+,-) \qquad \omega_{16} = W(-,-,-,+)$$

$$(7.93)$$

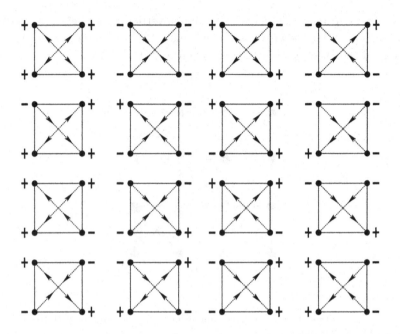

Fig. 7.11   The relation between spin configurations and arrow configurations of the associated vertex model.

Taking Eq. (7.91) into account, we obtain

$$\omega_1 = \omega_2 = 2e^{-2K_2+2K_1}$$
$$\omega_3 = \omega_4 = 2e^{2K_2-2K_1}$$
$$\omega_5 = \omega_6 = 2e^{-2K_2-2K_1}$$
$$\omega_7 = \omega_8 = 2e^{2K_2+2K_1}\cosh(4K_1)$$
$$\omega_9 = \omega_{10} = \omega_{11} = \omega_{12} = \omega_{13} = \omega_{14} = \omega_{15} = \omega_{16} = 2\cosh(2K_1)$$

$$(7.94)$$

As can be easily verified, the free-fermion conditions Eq. (7.64) are identically satisfied by the Boltzmann weights Eq. (7.94), for arbitrary values of K_1 and K_2. If we replace Eq. (7.94) in Eq. (7.65) and Eq. (7.66), we can obtain the explicit expression of the free energy of the model. Moreover, by replacing Eq. (7.94) in Eq. (7.67) we obtain the critical condition for this system :

$$\frac{1}{2} \left[\exp(2K_1 + 2K_2) \cosh(4K_1) + \exp(-2K_1 - 2K_2) \right]$$

$$+ \cosh(2K_1 - 2K_2) + 2\cosh(2K_1) = 2\max\{ \frac{1}{2} \left[\exp(2K_1 + 2K_2) \cosh(4K_1) \right.$$

$$+ \exp(-2K_1 - 2K_2)]; \ \cosh(2K_2 - 2K_1); \ \cosh(2K_1)\} \tag{7.95}$$

which is decomposed into four critical lines depending on the values of J_1 and J_2. The singularity of the free energy is everywhere logarithmic.

For the whole phase diagram, the reader is referred to the paper by Azaria *et al.* [11]. We show in Fig. 7.12 only the small region of J_2/J_1 in the phase diagram which has the reentrant paramagnetic phase and a disorder line. The reentrant phase is a disordered phase lying between two ordered phases . The disorder line is defined as a line separating two zones of pre-ordering fluctuations of different symmetries [123]. The disorder line in Fig. 7.12 has a dimension reduction.

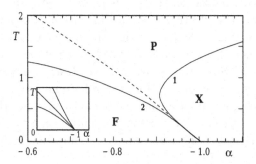

Fig. 7.12   Phase diagram of the Kagomé lattice with nnn interaction in the region $J_1 > 0$ of the space $(\alpha = J_2/J_1, T)$. T is measured in the unit of J_1/k_B. Solid lines are critical lines, dashed line is the disorder line. P, F and X stand for paramagnetic, ferromagnetic and partially disordered phases, respectively. The inset shows schematically enlarged region of the endpoint.

The phase X indicates a partially ordered phase where the central spins are free (the nature of ordering was determined by MC simulations) [11]. Here again, the reentrant phase takes place between a low-T ordered phase and a partially disordered phase. This suggests that a partial disorder at the high-T phase is necessary to ensure that the entropy is larger than that of the reentrant phase.

The general Kagomé lattice has been exactly solved with a rich phase space [29]. Other frustrated Ising models such as the centered honeycomb lattice [38] and dilute centered square lattices [30] have also been solved using vertex models.

7.5.2.2 *Antiferromagnetic stacked triangular lattices*

The phase transition in frustrated systems with non Ising spin models has been one of the most studied subjects in magnetism during the last 30 years, theoretically, numerically and experimentally [40]. The most studied system is no doubt the stacked triangular antiferromagnet (STA) with XY spins or Heisenberg spins. In Ref. [40], Delamotte *et al.* have given an exhaustive theoretical review on this question and Loison has provided principal numerical simulations published so far. The non perturbative renormalization group has already given evidence of a first-order transition in the STA with XY and Heisenberg spins around the year 2000. Numerical simulations also began to find some first-order evidence: using an improved Monte Carlo renormalization-group scheme to investigate the renormalization group flow of the effective Hamiltonian used in field-theoretical studies for the XY STA, Itakura [68] found that the XY STA exhibits a clear first-order behavior and there are no chiral fixed points of renormalization-group flow for XY spins. Since that year, there has been a number of numerical works which have made further steps toward the final conclusion. In 2004, Peles *et al.* [109] have used a continuous model to study the XY STA by Monte Carlo simulation. They found evidence of a first-order transition. In 2006, Kanki *et al.* [72], using a microcanonical method, have found a first-order signature of the XY STA. While these simulations have demonstrated evidence of first-order transition for the XY STA in agreement with the nonperturbative RG analysis, the definite answer come in 2008 with a work using a very high-performance technique for weak first-order transitions, the so-called Wang-Landau flat-histogram method [136] for the XY STA [101]. A first-order transition in that system has been clearly established, confirming results of other authors.

For the Heisenberg case, Itakura [68] found, as in the XY case mentioned above, the absence of chiral fixed points of renormalization-group flow. However, he could not find numerical evidence of the first-order transition. He predicted that if the transition is of first order for the Heisenberg spins, it should occur at much larger lattice sizes. Using the high performance of the Wang-Landau method, the Heisenberg case was studied with very large sizes [102]: the result shows clearly a first-order transition, putting an end to the 20-year long controversy.

The well-known simple cubic fully frustrated Villain's model shown in Fig. 7.13 has also been investigated with the Wang-Landau method: the phase transition was found to be of first order for the Ising [104], classical XY [105] and Heisenberg [106] spin models. The HCP antiferromagnet with Ising and XY spins was shown to have also a first-order transition [64]. To summarize, frustrated models with classical spins known so far undergo a first-order transition.

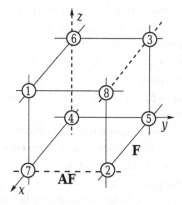

Fig. 7.13 Fully frustrated simple cubic lattice. Discontinued (continued) lines denote antiferromagnetic (ferromagnetic) bonds.

7.6 Conclusion

In this chapter, we have introduced the basic notions as well as some fundamental methods which are widely used in the field of phase transitions. We have also presented some complementary methods in section 7.7 to solve several problems related to the subject of this chapter. There are of course more advanced methods which are technically difficult for students of graduate level for whom this book is intended. We can mention a few of them: quantum phase transitions of low-dimensional systems [117, 82, 98], the Hubbard model [45] in particular in two dimensions, and the low- and high-temperature series expansions [42, 150]. Such advanced methods are not included here to keep the contents of the book suitable for lectures in a graduate course.

7.7   Problems

Problem 1. Solution for an Ising chain:
  Calculate the partition function of a chain of N Ising spins using the periodic boundary condition. Calculate the free energy, the averaged energy and the heat capacity as functions of the temperature. Show that there is no phase transition at finite temperature.

Problem 2. Renormalization group applied to an Ising chain:
  Study by the renormalization group a chain of Ising spins with a ferromagnetic interaction between nearest neighbors. Show that there is no phase transition at finite temperature.

Problem 3. Transfer matrix method applied to an Ising chain:
  Study by the transfer matrix method the chain of Ising spins in the previous exercise using the periodic boundary condition.

Problem 4. Low- and high-temperature expansions of the Ising model on the square lattice:
  The low- and high-temperature expansions are useful not only for studying physical properties of a spin system in these temperature regions, but also for introducing a new concept such called duality which allows to map a system of weak coupling into a system of strong coupling, as seen in the problem below.
  Consider N Ising spins on a square lattice with the Hamiltonian

$$\mathcal{H} = -J \sum_{<i,j>} \sigma_i \, \sigma_j$$

where the sum is performed over nearest neighbors and $\sigma_{i(j)} = \pm 1$. The periodic boundary conditions are used.

  a) Write the partition function Z. Calculate Z for the ground state (GS).
  b) Low-temperature expansion: Consider the GS in which one reverses one spin, two nearest neighboring spins, a block of three nearest spins, ... Count for each case the number of "broken" links, namely links between the reversed spins with the remaining spins. Calculate the degeneracy of each case. Write Z with the first excited states.
  c) Draw a path P crossing the broken links around each reversed spin cluster. Verify that each cluster has an even number of broken links and P is always a closed path. Let $\ell(P)$ the number of broken links

crossed by the path. Show that

$$Z = 2e^{N_b K} \sum_P e^{-2K\ell(P)} \qquad (7.96)$$

where N_b is the total number of links and $K = J/(k_B T)$.

d) High-temperature expansion: Using Eq. (11.225) or Eq. (11.234) show that the partition function is written as

$$Z = \sum_{\sigma_1 = \pm 1, \ldots} \prod_{<ij>} (\cosh K + \sigma_i \, \sigma_j \sinh K)$$

$$= (\cosh K)^{N_b} \sum_{\sigma_1 = \pm 1, \ldots} \prod_{<ij>} (1 + \sigma_i \, \sigma_j \tanh K) \qquad (7.97)$$

Expand the product in the last equation and show that

$$Z = 2^N (\cosh K)^{N_b} \sum_P (\tanh K)^{\ell(P)} \qquad (7.98)$$

e) Duality: The partition function Z in (7.96) and (7.98) has the same structure: the prefactors are non singular, the summations over the paths determine the singularity of Z. Show that the two Z have the same critical behavior if

$$K^* = -\frac{1}{2} \ln \tanh K \qquad (7.99)$$

where K^* corresponds to the low-T phase and K to the high-T phase. The relation (7.99) is called the "duality" condition which connects the low- and high-T phases.

f) Deduce the critical temperature of the Ising model on the square lattice.

Problem 5. Critical temperatures of the triangular lattice and the honeycomb lattice by duality:

Consider the triangular lattice with Ising spins with a ferromagnetic interaction between nearest neighbors. Construct its dual lattice. Calculate the partition functions of the two lattices. Deduce the critical temperature of each of them by following the method outlined in the previous problem.

Problem 6. Villain's model:

We study the ground state spin configuration of the 2D Villain's model with XY spins defined in Fig. 7.7. Write the energy of the elementary plaquette. By minimizing this energy, determine the ground state as a function of the antiferromagnetic interaction $J_{AF} = -\eta J_F$ where η is a positive coefficient. Determine the angle between two neighboring spins as a function of η. Show that the critical value of η beyond which the spin configuration is not collinear is $\eta_c = 1/3$.

Problem 7. Give the proofs of Eq. (7.85).

Problem 8. Critical line of an antiferromagnet in an applied magnetic field:

In chapter 4 we have seen that an antiferromagnet in a field can have a phase transition at a finite temperature T_c, in contrast to a ferromagnet. We calculate in this exercise T_c as a function of a weak field H.

Consider an Ising antiferromagnet on the square lattice in a uniform magnetic field $\vec{H}$ applied in the z direction, with the following Hamiltonian

$$-\frac{\mathcal{H}}{k_B T} = K \sum_{<i,j>} S_i S_j + |K| H \sum_i S_i \qquad (7.100)$$

where S_i is the Ising spin at the lattice site i pointing along $\pm z$ direction, and $< i, j >$ indicates the nearest neighbors pairs. Determine T_c as a function of H by using the zero-field magnetic susceptibility given by

$$\chi_s = Dt \ln(\frac{1}{|t|})$$

where $D = 0.1935951863$ and $t = (T - T_c^0)/T_c^0$ with $T_c^0 = T_c(H = 0)$ (this problem is based on Ref. [76]).

Chapter 8

Methods of Monte Carlo Simulation

To understand methods of Monte Carlo simulation for the study of phase transitions we need a background taught in statistical physics and a programming language. The first method of simulation has been introduced in the 50's by Metropolis [96]. But simulations become popular only from the early 90's when several new methods of simulation of high precision have been introduced and at the same time powerful computers have become accessible for a large number of researchers [17]. Today, numerical simulations are considered as an investigation method which is as important as theory and experiment in the study of properties of nature. Numerical simulations serve as a complementary tool, and in many cases they are the only way to study very complex systems that theory and experiment cannot investigate properly. Numerical simulations allow us to test the validity of theoretical approximations, to compare quantitatively simulated results with experimental data, and to propose interpretations. In addition, real systems have many parameters but often only a few of them govern their main properties. Theories and experiments cannot test all of them for a simple reason that experimental realizations and theoretical calculations take time and cost. Simulations can help identify relevant mechanisms before realizing experimental setups or constructing models.

It is obvious that numerical simulations also have particular difficulties. The first kind of difficulty is related to the capacity of computers: limited memory and speed. The second kind of difficulty concerns the efficiency of the simulation method to get good statistical averages, to treat correctly particular physical effects, to reduce errors, ... Of course, there are remedies to improve and to get rid of most of these difficulties as seen below. A good simulation requires a great care in every step from the choice of model to a deep analysis of results. We recall that simulation is a method to study a problem: a good knowledge on theoretical background and experimental data prior to the simulation is necessary. As said above, numerical simulations have been intensively used for investigation of phase transitions and critical phenomena. Basic notions on this domain have been given in chapters 4 and 7.

8.1   Principle

In this section, we present the principle of the Monte Carlo simulation and its implementation. Although Monte Carlo simulations can be made using the micro-canonical and grand-canonical statistical descriptions (see Appendix A), in this chapter we employ the canonical description to show it because this description is most used in simulations. The system we consider is kept at a constant temperature T. The internal variables such as energy and magnetization are free to fluctuate at T.

To illustrate the principle of the Monte Carlo simulation, we consider when it is necessary the Ising model with ferromagnetic interaction J between nearest neighbors. This simple example, however, does not cause the loss of the generality of the method.

In a simulation, we wish to calculate average values of physical quantities such as the average energy $< E >$, the heat capacity C_V, the average magnetization $< M >$ and the susceptibility χ. The average value of a physical quantity A is defined by par

$$< A >= \frac{1}{Z(T)} \sum_s A(s) \exp[-\beta E(s)] \qquad (8.1)$$

where $Z(T)$ is the partition function at the temperature T, $E(s)$ and $A(s)$ are the system energy and the value of A in the microscopic state s. For a spin system, s is a spin configuration.

In principle, we have to sum over all spin configurations s. The total number of spin configurations in a system of N Ising spins is 2^N. This is a huge number when N is large. We can overcome this difficulty by two following ways:

8.1.1   *Simple sampling*

We generate a number C of random spin configurations by giving randomly a value +1 or -1 to each spin. We then calculate $< A >$ using these C states

$$< A >= \frac{\sum_{s=1}^{C} A(s) \exp[-\beta E(s)]}{\sum_{s=1}^{C} \exp[-\beta E(s)]} \qquad (8.2)$$

It is obvious that the precision on the obtained average value depends on the number C of configurations: the larger C is, the more precise $< A >$ becomes. The simple sampling suffers from a more serious problem: the randomly generated spin configurations correspond to disordered states at high temperatures, unless we choose a large concentration of up spins with respect to down spins (or vice-versa) for low temperatures. But then, how do we know which concentration corresponds to which temperature? The average value so calculated contains certainly uncontrolled errors.

8.1.2 *Importance sampling*

The importance sampling is based on the following principle: we generate most probable spin configurations s at T, namely we generate states using the canonical probability. Once these states are generated, the average value $< A >$ is calculated by a simple addition

$$< A >= \frac{1}{C} \sum_{s=1}^{C} A(s) \qquad (8.3)$$

Now, the question is "how to generate these states according to the canonical probability at T in a convenient way for simulation"?

An answer to this question is the following algorithm. Instead of generating these states in an independent manner, we can generate a series of states called "Markov chain" where the (i+1)-th state, s_{i+1}, is obtained from the precedent state s_i with an appropriate transition probability $w(s_i \to s_{i+1})$ between these two states. The choice of $w(s_i \to s_{i+1})$ should obey the following probability at equilibrium (see Appendix A)

$$P(s_i) = \frac{1}{Z(T)} \exp[-\beta E(s_i)] \qquad (8.4)$$

This is possible if we impose on $w(s_i \to s_{i+1})$ the following principle of detailed balance

$$P(s_i)w(s_i \to s_k) = P(s_k)w(s_k \to s_i) \qquad (8.5)$$

We obtain

$$\frac{w(s_i \to s_k)}{w(s_k \to s_i)} = \exp(-\beta \Delta E) \qquad (8.6)$$

where $\Delta E = E(s_k) - E(s_i)$.

If the above relation for equilibrium is obeyed, there is no problem for the system to reach equilibrium. The choices frequently used to respect the detailed balance are

$$w(s_i \to s_k) = \frac{\exp(-\beta \Delta E)}{1 + \exp(-\beta \Delta E)} \tag{8.7}$$

and

$$w(s_i \to s_k) = \exp(-\beta \Delta E) \qquad \text{if } \Delta E \geq 0$$
$$= 1 \qquad \text{if } \Delta E < 0 \tag{8.8}$$

The second choice will be used in the following to write a Monte Carlo program for the Ising spin model.

Remarks:

- The state s_{i+1} can be different from the state s_i by just the state of a single spin. This facilitates the simulation task: the difference of energy ΔE of the two states is equal to the difference of energy of the single spin under consideration.
- The power of the Monte Carlo method resides in the facility to make the system transit from the state s_i to the state s_{i+1}: we just reverse one single spin or a block of spins. The first choice is called "single-spin flip algorithm" or "Metropolis algorithm", and the second choice is called "cluster-flip algorithm" which is shown in section 8.6.

8.2   Construction of computer program

We implement now the principle of the Monte Carlo simulation presented in the previous section. To simplify the presentation, we use the Metropolis algorithm hereafter. Let us consider the Ising model on a square lattice with a ferromagnetic interaction J between nearest neighbors. The Hamiltonian is given by

$$\mathcal{H} = -J \sum_{(k,m)} S(\vec{R}_k) S(\vec{R}_m) \tag{8.9}$$

where the sum is made over pairs of nearest neighbors. $S(\vec{R}_k)$ indicates the spin at the position $\vec{R}_k$. We suppose the periodic boundary conditions. Next, we proceed to the following steps:

- 1. Initial spin configuration:

In the case of the Ising model, we create a square lattice of dimension $N \times N$ where at each site we attribute a spin. For the square lattice, each site is defined by two cartesian indices (i, j) corresponding to the position of the site (see Fig. 8.1). The spin occupying this site is $S(i, j)$ with an attributed initial value (1 or -1).

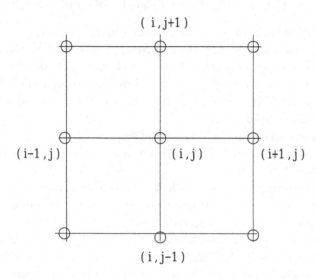

Fig. 8.1   Coding of the square lattice. Each site is defined by two cartesian indices.

- 2. Choice of physical parameters:

Here the parameters of the model are J and T. We take $J = 1$ as energy unit.
- 3. Calculation of the energy of spin $S(i, j)$, using the periodic boundary conditions if it is at an edge:

The best and fast way to identify the neighbors of $S(i, j)$ is to call the left neighbor of (i, j) by (im, j), the right one by (ip, j), the top one by (i, jp) and the below by (i, jm) with

$$im = i - 1 + (1/i) * N \qquad (8.10)$$

$$ip = i + 1 - (i/N) * N \qquad (8.11)$$

$$jm = j - 1 + (1/j) * N \qquad (8.12)$$

$$jp = j + 1 - (j/N) * N \qquad (8.13)$$

where we see that $im = i - 1$ as long as $i \neq 1$ and $ip = i + 1$ as long as $i \neq N$ because the integer divisions $(1/i)$ and (i/N) give zero (with Fortran). When $i = 1$, we have $im = N$ and when $i = N$ we have $ip = 1$. This automatically respects the periodic condition in the x direction: the left neighbor of the spin $S(1, j)$ is $S(N, j)$, and the right neighbor of $S(N, j)$ is $S(1, j)$. The same explanation is for the y direction. We write the energy of the spin $S(i, j)$ as

$$E1 = -J * S(i, j) * (S(im, j) + S(ip, j) + S(i, jm) + S(i, jp)) \quad (8.14)$$

- 4. Updating the spin $S(i, j)$ and calculating its new energy:
  For an Ising spin, the new state of $S(i, j)$ is obtained by changing its sign. Its new energy is thus $E2 = -E1$. If $E2 < E1$, the new orientation of $S(i, j)$ is accepted. If $E2 > E1$, the new orientation is accepted with a probability $e^{-(E2-E1)/T}$ (we take $k_B = 1$ for simplicity). This step is called "spin update".
- 5. Taking another spin and repeating steps 3 and 4:
  We continue until all spins have been updated: we say we have made one Monte Carlo step.
- 6. Equilibrating the system with N_1 Monte Carlo steps:
  We have to make many Monte Carlo steps to equilibrate the system at T since the initial configuration does not in general correspond to the temperature at which we make simulation. The repetition of spin updates is necessary to bring the system to equilibrium at T.
- 7. Averaging physical quantities with N_2 Monte Carlo steps:
  During N_2 Monte Carlo steps, we calculate average values of physical quantities of interest such as the energy

$$< E > = \frac{1}{N_2} \sum_{t=N_1+1}^{N_1+N_2} E(t) \quad (8.15)$$

where t is the Monte Carlo "time" and $E(t)$ the instantaneous energy of the system at t.

Remarks:

- During the Monte Carlo steps for equilibrating, we can record the instantaneous energy and magnetization at each Monte Carlo step in order to observe their time evolution. We can then see the time necessary to get equilibrium. An example is shown in Fig. 8.2.

- At the end of step 7, we record all average values in a file, and then restart the simulation at another temperature. We have at the end average values for several temperatures so that we can plot average values of physical quantities versus T. From those curves, we can recognize the phase transition as well as other properties of the system. We show some examples in the next section.

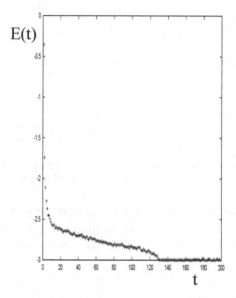

Fig. 8.2 Example of time evolution (in unit of Monte Carlo step), during equilibrating, of the energy per spin $E(t)$ (in unit of J), at $k_B T/J = 1$ for the ferromagnetic Ising model on the triangular lattice with a random initial spin configuration.

8.3 Phase transition as seen in Monte Carlo simulations

We perform a Monte Carlo simulation for a system of linear size L at various temperatures as described in the previous section. We examine the curves of average values of different physical quantities versus T. A phase transition is recognized by the anomalies of these quantities at some temperature.

In a second-order transition, the internal energy changes its curvature at the transition temperature T_c, the heat capacity and the susceptibility have a peak at T_c, the magnetization falls to zero but with a small tail above T_c due to a finite-size effect. These are schematically shown in Fig. 8.3.

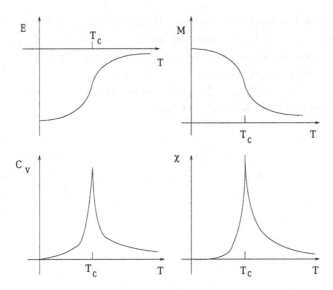

Fig. 8.3   Physical quantities of a second-order phase transition as seen in a simulation with a linear size L: energy E, heat capacity C_V, magnetization M, susceptibility χ.

In a first-order transition, for a sufficiently large L, the energy and the magnetization are discontinued at the transition, the heat capacity and the susceptibility therefore cannot be defined at the transition. We schematically show E and M in Fig. 8.4.

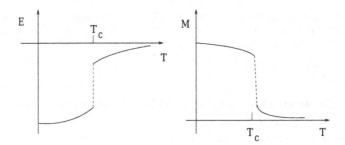

Fig. 8.4   First-order transition as seen in a simulation for a large size L: discontinuities of E and M are observed at the transition (discontinued vertical lines).

Several remarks are in order:

- We have to pay attention to size effects while analyzing the results. A first-order transition can look like a second-order transition if L is not larger than the correlation length at the transition (see paragraph 7.1.3). If we cannot take large-enough L because of the computer memory limit and/or of a too long CPU time, then we have to use a finite-size scaling analysis shown below to determine the order of the transition.

- Some systems show a maximum of the heat capacity at a temperature but do not have a phase transition. One of these systems is the case of a chain of Ising spins with a ferromagnetic interaction between nearest neighbors. We have to verify the existence of a phase transition by several methods. For example, if it is a real transition, the heights of the maxima of C_V and χ as well as of other quantities (see below) should depend on the system size.

8.4 Size effects and scaling laws

Finite-size effects on thermodynamic quantities and on spin-spin correlation have been shown by a renormalization group analysis [13]. The calculations are too complicated to be reproduced here. We just recall some results which are very helpful while analyzing data obtained from Monte Carlo simulations.

Let L be the linear dimension of the system under consideration. We distinguish two cases:

8.4.1 *Second-order phase transition*

- The height of the maximum of the heat capacity depends on L as follows

$$C_V^{max}(L) \propto L^{\alpha/\nu} \tag{8.16}$$

- The height of the maximum of the susceptibility behaves as

$$\chi^{max}(L) \propto L^{\gamma/\nu} \tag{8.17}$$

- The magnetization at the transition depends on L through

$$M_{T_c}(L) \propto L^{-\beta/\nu} \tag{8.18}$$

- The moment of n-th order is defined as

$$V_n = <(\ln M^n)'> = \frac{<\partial \ln M^n>}{\partial \frac{1}{k_B T}} \tag{8.19}$$

We can show that

$$V_1 = \frac{<ME> - <E>}{<M>} \tag{8.20}$$

$$V_2 = \frac{<M^2E> - <M^2><E>}{<M^2>} \tag{8.21}$$

The finite-size effects on the maxima of these moments are given by [13]

$$V_1^{max}, \quad V_2^{max} \propto L^{1/\nu} \tag{8.22}$$

- The Binder cumulant [17] is defined by

$$U_E = 1 - \frac{<E^4>}{3<E^2>^2} \tag{8.23}$$

In a second-order phase transition, we have

$$U_E(L) - U_E(\infty) \propto L^{-\theta} \tag{8.24}$$

where $U_E(\infty) = 2/3$ and $\theta < d$.

- The critical temperature depends on L through the relation

$$T_c(L) - T_c(\infty) \propto L^{-1/\nu} \tag{8.25}$$

We use the above relations to determine the critical exponents by realizing simulations with many sizes L. Note that L has to be chosen so as the system sizes are different by at least two orders.

8.4.2   *First-order phase transition*

When the system size is not sufficiently large, a first-order phase transition can have aspects of a second-order one, namely E and M are continuous at the transition with a maximum of C_V and χ. In the case of a first-order transition, we should have

- The height of the maximum of C_V and χ are proportional to the system volume:

$$C_V^{max}, \chi^{max} \propto L^d \tag{8.26}$$

- The Binder cumulant depends on L as

$$U_E(L) - U_E^*(\infty) \propto L^{-d} \tag{8.27}$$

where $U_E^*(\infty) \neq 2/3$.

Note that finite-size effects are seen in the transition region around T_c. This is the reason why they depend on the critical exponents. They act therefore in different manners on different physical quantities. At a finite size, the "pseudo" transition temperature where C_V is maximum is not that of χ. Only at the infinite size that these maxima occur at the same temperature which is the "real" critical temperature. This is schematically shown in Fig. 8.5.

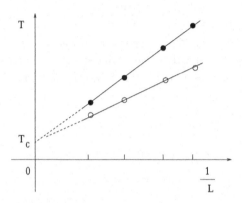

Fig. 8.5 Finite-size effects on the temperatures corresponding to the maxima of C_V (black circles) and χ (white circles). These temperatures coincide only when $L \to \infty$ (extrapolated by discontinued lines).

8.5 Error estimations

One of the problems encountered in Monte Carlo simulations is the estimation of errors due to a number of causes: simulation time, artificial procedures used to accelerate the convergence toward equilibrium, etc. There are two types of principal errors (i) statistical errors due to autocorrelation and finite-size effects on these errors (ii) errors due to the fitting of raw data with some laws. Errors of the second type are directly given by the computer according to the chosen fitting procedure (mean least squares, for example). Errors of the first type become less and less important because computers are more and more powerful, rapid in execution with huge available memories. Simulation time nowadays is much longer than autocorrelation time and the relaxation time in most systems. Therefore, errors are extremely small. Most of Monte Carlo works at the present time do not show anymore errors since they are often smaller than the size of the presented data points. We give anyway in the following some notions on error sources and how to calculate errors if we wish to do so.

8.5.1 *Autocorrelation*

The autocorrelation function of a quantity A at the time t with itself at $t = 0$ is defined by

$$\phi(t) = \frac{< A(0)A(t) > - < A >^2}{< A^2 > - < A >^2} \tag{8.28}$$

where $< ... >$ is the thermal average taken between $t = 0$ and t, $A(t)$ the instantaneous value of A. By definition, we have $\phi(0) = 1$ and $\phi(\infty) = 0$. In a simulation, we can calculate $\phi(t)$ and obtain the integrated autocorrelation time $\bar{\tau}$ by

$$\bar{\tau} = \int_0^\infty \phi(t)dt \tag{8.29}$$

The autocorrelation time τ is defined by

$$\phi(t) = C \exp(-t/\tau) \tag{8.30}$$

We then have

$$\int_0^\infty \phi(t)dt = C \int_0^\infty \exp(-t/\tau)dt = C\tau \tag{8.31}$$

Comparing to (8.29), we get

$$\tau = \bar{\tau}/C \tag{8.32}$$

The error on A is given by

$$< (\delta A)^2 > = < \frac{1}{N} \sum_{n=1}^{N} (A(t_n) - < A >)^2 >$$

$$= \frac{< A^2 > - < A >^2}{N}[1 + \frac{2}{\Delta t} \int_0^{t_N} (1 - \frac{t}{t_N})\phi(t)dt] \tag{8.33}$$

where N is the total number of measures and $A(t_n)$ the instantaneous value of A measured at t_n. The second equality has been obtained by expanding the square of the first equality and then replacing the sum by an integral using (8.28). Δt is the interval between two measures. To de-correlate successive values $A(t_n)$ we should not measure A at each Monte Carlo step. We can for example take a measure once every ten steps, namely $\Delta t = 10$. Since $\phi(t) \simeq$ constant for $t \gg \tau$, we replace the upper limit t_N by ∞. In addition, we can neglect t/t_N with respect to 1. We then obtain

$$< (\delta A)^2 > \simeq (< A^2 > - < A >^2) \frac{2\tau + \Delta t}{N \Delta t} \qquad (8.34)$$

where $N \Delta t = N_{MC}$ is the total number of Monte Carlo steps used in the simulation.

The number of independent measures among N measures is

$$n = \frac{N \Delta t}{2\tau + \Delta t} \qquad (8.35)$$

The relative error on A can be estimated by

$$\rho = \frac{< (\delta A)^2 >^{1/2}}{< A >} \simeq \left[\frac{< A^2 > - < A >^2}{< A >^2} \frac{2\tau}{N_{MC}} \right]^{1/2} \qquad (8.36)$$

where we neglected Δt while comparing to 2τ. Since τ is given by (8.32), $< A^2 >$ and $< A >$ are known from the simulation, we obtain ρ by the above formula.

8.5.2 Size effects on errors

In the vicinity of the transition, the relaxation time τ is very long. Spins are correlated even at very long distances. This phenomenon is called "critical slowing-down". The first error due to the finite system size comes from error on the relaxation time. We have [65]

$$\tau \propto \xi^z \propto (\frac{T_c}{T - T_c})^{z\nu} \qquad (8.37)$$

where ξ is the correlation length, z the dynamic exponent (see chapter 7). It has been shown that with the Metropolis algorithm (single-spin flip) we have $z \simeq 2$ for many systems. In addition, theoretically ξ diverges at the transition of second order, but in simulation the transition takes place when $\xi \simeq L$ which is the infinite limit due to the periodic boundary conditions. For this reason, in a second-order transition, τ depends on the size L through

$$\tau(L) \propto L^z \qquad (8.38)$$

In a first-order transition, we have [13, 17]

$$\tau(L) \simeq L^a \exp(2\sigma L^{d-1}) \qquad (8.39)$$

where a depends on the algorithm and $(2\sigma L^{d-1})$ is the height of the barrier of the free energy at the transition. We have $a \simeq 2.1$ obtained by Metropolis algorithm for the 10-state Potts model, but $a \simeq 1.5$ obtained by Swendsen-Wang algorithm (see description in 8.6.1) for the same model.

Replacing $\tau(L)$ of (8.38) and (8.39) in (8.36), we obtain for a second-order transition

$$\rho(L) \simeq \frac{1}{N_{MC}^{1/2}} L^{(z-x)/2} \tag{8.40}$$

and for a first-order transition

$$\rho(L) \simeq \frac{1}{N_{MC}^{1/2}} L^{(a-x)/2} \exp(\sigma L^{d-1}) \tag{8.41}$$

where x is a fitting parameter.

The systematic error is defined by

$$\sigma^2(n) \equiv (<A^2> - <A>^2)(1 - \frac{1}{n}) \tag{8.42}$$

The relative error on the variance $<(\delta A)^2>$ is

$$\epsilon = \frac{<A^2> - <A>^2 - \sigma^2(n)}{<A^2> - <A>^2} = \frac{1}{n} = \frac{2\tau + \Delta t}{N\Delta t} \tag{8.43}$$

For an error of $\epsilon = 1\%$, we see that the simulation time should be 200 times τ ($\Delta t \ll \tau$). As said at the beginning of this section, with the power of today's computers, such a simulation time is not a problem. Also, the necessity for large-enough system sizes is no longer a real problem in Monte Carlo simulations with ever-increasing computer capacity.

8.6 Advanced techniques in Monte Carlo simulations

In simulations, we have to solve two practical problems: (i) to shorten the waiting time (ii) to find better algorithms to study physical phenomena which cannot be treated with precision by the standard Monte Carlo algorithm. To solve the first kind of problems, mainly we have to find ways to accelerate convergence to equilibrium so that equilibrating time is shortened and the quality of physical quantities during a given averaging time is better. One of the best methods proposed so far is the cluster-flip method due to Wolff [142] and Swendsen-Wang [125] described below. For the second kind of problems, we can mention difficulties encountered by the Metropolis algorithm in the calculation of critical exponents and in the detection of extremely weak first-order transition. To deal with such difficulties, histogram and multiple histogram techniques as well as flat-histogram methods have been proposed in the literature [46, 136]. We will describe them below.

8.6.1 *Cluster-flip algorithm*

The central idea comes from the following observation. In the vicinity of the transition, the correlation length is very long (see chapter 7), there are large clusters of parallel spins just above T_c (we are in the case of a ferromagnetic crystal). The Metropolis algorithm which updates spin after spin will take a long time to consider unnecessarily all spins in a cluster because, except for a small fraction at spins at the boundary, the spins in a cluster do not need to change orientation (they are ready for the ferromagnetic state at T_c). Therefore, to perform single-spin flips for all spins in a cluster near T_c is a loss of time. In addition, due to the existence of large clusters near T_c, the relaxation time is very long [see Eq. (8.37)]. This phenomenon is called "critical slowing-down". Wolff [142] and Swendsen and Wang [125] have proposed to update simultaneously all spins of a cluster by flipping the entire cluster (we have in mind the case of the Ising model). The method is simple to implement. We can write the updating part and insert it into a standard program such as the one shown in Appendix B. To simplify the description, we consider a system of Ising spins with a ferromagnetic interaction J between nearest neighbors. The principle of the algorithm is the following:

- For a given spin configuration, we consider a spin S_i and we "construct" a "cluster" around S_i as follows. We examine the neighboring spins: if a neighbor in one direction is parallel to S_i, then it belongs to the cluster with a probability $p = 1 - \exp(-2\beta J)$ where $\beta = (k_B T)^{-1}$. We consider the spin next to that spin in that direction, and we continue the cluster construction. The limit of the cluster in the considered direction is where the cluster encounters an antiparallel spin or if a random number taken between 0 and 1 is larger than p. We have to go to all directions to determine the boundary of the cluster. Note that there is a very efficient algorithm for the cluster construction proposed by Hoshen and Kopelman [66].
- We flip the cluster and we calculate ΔE the difference in energy with the previous state: $\Delta E = 2J[C(++) - C(-+)]$ where $C(++)$ the number of broken parallel links along the boundary of the cluster, and $C(-+)$ the number of antiparallel links along the boundary. An example is shown in Fig. 8.6.

  The new orientation of the cluster is accepted or not following the Metropolis criterion as for a single spin.

- We take another spin outside the above cluster and we begin again another cluster construction etc.

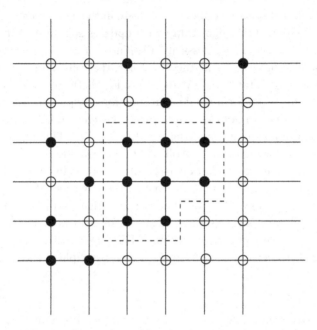

Fig. 8.6 Example of a cluster, limited by the discontinued contour, constructed by cluster-construction algorithm. Black circles = ↑ spins, white circles = ↓ spins. The number of broken parallel links $C(++)$ is 1, that of antiparallel spin links $C(-+)$ is 11.

The difference between the method by Wolff and that by Swendsen-Wang is the following. In the first method, we flip only large clusters while in the second one we flip all clusters. It is true that near T_c, the two methods are equivalent because most of clusters are large. However, a little bit further from T_c where there are many small clusters, we spend much time to flip small clusters in the Swendsen-Wang method, which does not significantly improve the result.

In practice, we can combine the Metropolis algorithm and the cluster-flip algorithm: we use the latter from time to time because the cluster construction takes time. We need it only very near T_c to get rid of the critical slowing-down. The result is striking: the dynamic exponent $z \simeq 2$ as obtained by the Metropolis algorithm becomes $z \simeq 0.5$ using the cluster-flip method.

8.6.2 *Histogram method*

In standard Metropolis Monte Carlo simulations, we calculate average physical quantities at discrete temperatures. We extrapolate results between discrete temperatures. However, near the transition temperature, extrapolation is impossible because physical quantities diverge at a precise single temperature value. Practically, we cannot find the exact location of the transition temperature. If the chosen temperature is not T_c, we can wonder about the precision of the critical exponents calculated with the heights of C_V and χ which are not at T_c.

To avoid this difficulty, Ferrenberg and Swendsen [46] have proposed the histogram method which consists of making a simulation at a temperature T_0 as close as possible to T_c and recording the instantaneous system energy E during the simulation as long as possible to establish a histogram $H(E)$. Using this histogram we can calculate the canonical probabilities $P(T, E)$ at neighboring temperatures T around T_0 at as many points as we wish. Using these probabilities $P(T, E)$, we can calculate average values of physical quantities as continuous functions of T around T_0, by the formulas of the canonical description (see Appendix A), not by simulations. It suffices to choose T_0 close to T_c (not necessarily at T_c), we can find the exact location of the maximum of C_V and χ, for example.

The method is described in the following. We consider the partition function at T_0:

$$Z(T_0) = \sum_s \exp[-\beta_0 E(s)] \tag{8.44}$$

$$= \sum_E W(E) \exp(-\beta_0 E) \tag{8.45}$$

where the sum is taken over all energies of microscopic states, $W(E)$ denotes the degeneracy of the energy level E independent of T_0. The probability of the level E at T_0 is then

$$P(T_0, E) = \frac{W(E) \exp[-\beta_0 E(s)]}{Z(T_0)} \tag{8.46}$$

Now, by simulation we obtain the energy histogram $H(E)$ at T_0. The probability $P(T_0, E)$ is nothing but

$$P(T_0, E) = \frac{H(E)}{N_{MC}} \tag{8.47}$$

where N_{MC} is the number of Monte Carlo steps used to establish $H(E)$. By comparison of (8.47) to (8.46) we have

$$H(E) = N_{MC} P(T_0, E) = N_{MC} \frac{W(E) \exp[-\beta_0 E]}{Z(T_0)} \tag{8.48}$$

We consider now a temperature T near T_0. The probability $P(T, E)$ is written by

$$
\begin{aligned}
P(T, E) &= \frac{W(E) \exp[-\beta E]}{Z(T)} \\
&= \frac{Z(T_0) H(E) \exp[(\beta_0 - \beta)E]}{N_{MC} Z(T)} \\
&= \frac{H(E) \exp[(\beta_0 - \beta)E]}{\sum_E H(E) \exp[(\beta_0 - \beta)E]}
\end{aligned}
\tag{8.49}
$$

where we have used (8.48) to replace $W(E)$ and the following relation to replace $Z(T)$

$$
Z(T) = \sum_E W(E) \exp[-\beta E] = \frac{Z(T_0)}{N_{MC}} \sum_E H(E) \exp[(\beta_0 - \beta)E]
\tag{8.50}
$$

The histogram $H(E)$ established for T_0 is thus used to calculate the probability at another temperature T by (8.49). Using this probability we can calculate without simulation the average value of a quantity A by the formula

$$
< A > = \sum_E A P(T, E)
\tag{8.51}
$$

Several remarks are in order:

- To get a precise $H(E)$, we have to use a very large number of Monte Carlo steps N_{NC} in order to include as many as possible microscopic states in the sum.
- Since $H(E)$ is obtained with an importance sampling at T_0, if T is rather far from T_0, the probability $P(T, E)$ calculated by (8.49) using $H(E)$ is not precise. It is therefore very important to verify the form of each $P(T, E)$ by looking at its plot before using it. In general, we have a gaussian form as shown in Fig. 8.7. If T is far from T_0, $P(T, E)$ calculated has an irregular, asymmetric form. We should not use it [46].
- To determine with precision the transition temperature $T_c(L)$, we have to choose T_0 in the critical region, as close as possible to the presumed $T_c(L)$. To have a good choice, we have to make several simulations with Metropolis algorithm to detect a good value of T_0.
- When the transition is of first order, $P(T_0, E)$ presents a double peak if T_0 coincides or very close to $T_c(L)$. This is shown in Fig. 8.8 where the energies at the peaks E_1 and E_2 correspond to energies of the ordered

and disordered phases, which coexist at T_0. The histogram between two peaks is almost zero, indicating an energy discontinuity, namely a latent heat, of the system.

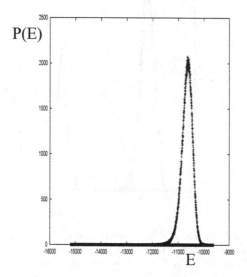

Fig. 8.7 Probability $P(E)$ obtained by simulation in the case of a face-centered cubic antiferromagnet with interaction J between nearest neighbors at temperature $T_0 = 1.76J$ ($k_B = 1$), just above the transition temperature. E is the total energy of the system.

The determination of the critical exponents with the histogram method is very precise (see for example the original papers of Ferrenberg and Swendsen [46]).

8.6.3 *Multiple histogram technique*

The multiple histogram technique is known to reproduce with very high accuracy the critical exponents of second-order phase transitions [47, 23]. It is more complicated to be implemented because we have to realize many histograms in independent simulations, but it gives much better results in difficult cases.

The principle consists of the following steps [47]:

i) First, to realize independent simulations at n temperatures T_i ($i = 1, ..., n$). For each temperature T_i, the number of Monte Carlo steps is N_i. The histogram taken during the simulation at that temperature is $H(E, T_i)$: one has $\sum_E H(E, T_i) = N_i$,

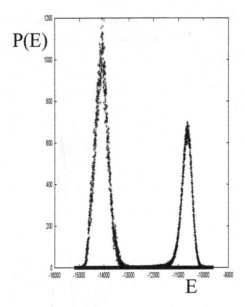

Fig. 8.8   Probability $P(E)$ obtained by simulation in the case of a face-centered cubic antiferromagnet with interaction J between nearest neighbors at the transition temperature $T_0 = T_c = 1.755J$ ($k_B = 1$). E is the total energy of the system.

ii) Second, to calculate the density of states $\Omega(E)$ by

$$\Omega(E) = \frac{\sum_{i=1}^{n} H(E, T_i)}{\sum_{i=1}^{n} N_i Z(T_i)^{-1} e^{-E/k_B T_i}} \tag{8.52}$$

where the partition function $Z(T_i)$ is

$$Z(T_i) = \sum_E \Omega(E) e^{-E/k_B T_i} \tag{8.53}$$

We see here that $\Omega(E)$ and $Z(T_i)$ should be calculated self-consistently. The choice of neighboring temperatures $T_1, T_2, ..., T_n$ should be guided as the choice of T_0 discussed in the single histogram technique shown above.

iii) Once $\Omega(E)$ and $Z(T)$ are obtained, we can calculate the thermal average of any physical quantity A at T by

$$\langle A(T) \rangle = \frac{\sum_E A(E)\, \Omega(E) e^{-E/k_B T}}{Z(T)} \tag{8.54}$$

Thermal averages of physical quantities are thus calculated as continuous functions of T. The results are valid over a much wider range of temperature than for any single histogram. The calculation of the critical exponents is much more precise than with a single histogram technique.

8.6.4 *Wang-Landau flat-histogram method*

Recently, Wang and Landau [136] proposed a Monte Carlo algorithm for classical statistical models which allowed us to study systems with difficultly accessed microscopic states. In particular, it permits to detect with efficiency weak first-order transitions [101, 102]. The algorithm uses a random walk in energy space in order to obtain an accurate estimate for the density of states $g(E)$ which is defined as the number of spin configurations for any given E. This method is based on the fact that a flat energy histogram $H(E)$ is produced if the probability for the transition to a state of energy E is proportional to $g(E)^{-1}$.

We summarize how this algorithm is implemented here. At the beginning of the simulation, the density of states is set equal to one for all possible energies, $g(E) = 1$. We begin a random walk in energy space (E) by choosing a site randomly and flipping its spin with a probability proportional to the inverse of the temporary density of states. In general, if E and E' are the energies before and after a spin is flipped, the transition probability from E to E' is

$$p(E \to E') = \min\left[g(E)/g(E'), 1\right]. \qquad (8.55)$$

Each time an energy level E is visited, the density of states is modified by a modification factor $f > 0$ whether the spin flipped or not, i.e. $g(E) \to g(E)f$. At the beginning of the random walk, the modification factor f can be as large as $e^1 \simeq 2.7182818$. A histogram $H(E)$ records the number of times a state of energy E is visited. Each time the energy histogram satisfies a certain "flatness" criterion, f is reduced according to $f \to \sqrt{f}$ and $H(E)$ is reset to zero for all energies. The reduction process of the modification factor f is repeated several times until a final value f_{final} which is close enough to one. The histogram is considered as flat if

$$H(E) \geq x\%.\langle H(E)\rangle \qquad (8.56)$$

for all energies, where $x\%$ is chosen between 70% and 95% and $\langle H(E)\rangle$ is the average histogram.

The thermodynamic quantities [136, 22] can be evaluated by

$$\langle E^n \rangle = \frac{1}{Z} \sum_E E^n g(E) \exp(-E/k_B T) \tag{8.57}$$

$$C_v = \frac{\langle E^2 \rangle - \langle E \rangle^2}{k_B T^2} \tag{8.58}$$

$$\langle M^n \rangle = \frac{1}{Z} \sum_E M^n g(E) \exp(-E/k_B T) \tag{8.59}$$

$$\chi = \frac{\langle M^2 \rangle - \langle M \rangle^2}{k_B T} \tag{8.60}$$

where Z is the partition function defined by

$$Z = \sum_E g(E) \exp(-E/k_B T) \tag{8.61}$$

The canonical distribution at any temperature can be calculated simply by

$$P(E, T) = \frac{1}{Z} g(E) \exp(-E/k_B T) \tag{8.62}$$

We have to choose an energy range of interest [120, 92] (E_{min}, E_{max}). We divide this energy range into R subintervals, the minimum energy of each subinterval is E^i_{min} for $i = 1, 2, ..., R$, and the maximum of the subinterval i is $E^i_{max} = E^{i+1}_{min} + 2\Delta E$, where ΔE can be chosen large enough for a smooth boundary between two subintervals. The Wang-Landau algorithm is used to calculate the relative density of states of each subinterval (E^i_{min}, E^i_{max}) with the modification factor $f_{final} = \exp(10^{-9})$ and flatness criterion $x\% = 95\%$. We reject the suggested spin flip and do not update $g(E)$ and the energy histogram $H(E)$ of the current energy level E if the spin-flip trial would result in an energy outside the energy segment. The density of states of the whole range is obtained by joining the density of states of each subinterval $(E^i_{min} + \Delta E, E^i_{max} - \Delta E)$. Numerous examples of applications of this method can be found in the literature.

8.7   Conclusion

In this chapter, we have presented the principle of Monte Carlo simulation and we have shown its implementation to study properties of spin systems, in particular at the phase transition. Several advanced techniques have also been described to help solve a number of problems encountered in Monte Carlo simulations for phase transitions. Numerical simulations are considered today as important as theory and experiment in the study of materials.

To make a good simulation, we have to know results from previous simulations, theories and experiments. This knowledge allows us to decide ingredients to include and ingredients not to include in the model. We should bear in mind that simulation is just a method of investigation. Simulation results come from the model and should be interpretable. A deep understanding of difficulties encountered during simulations helps us modify technical details or introduce new techniques to circumvent those obstacles. For example, we have shown above that to get rid of the critical slowing-down, we can use cluster-flips, to detect weak first-order transitions we have to use new methods such as the Wang-Landau technique, and to calculate critical exponents with precision we should use techniques such as the multiple histogram method. To succeed in simulations, we should have, beyond a numerical skill, a good background in theories in the field we study. This helps us understand and interpret what comes out from the computer after a long waiting time.

8.8 Problems

Problem 1. Program for Ising model:
- **a)** Include in the program shown in Appendix B the calculation of the heat capacity and the magnetic susceptibility.
- **b)** Modify that program to study the case of a simple cubic lattice
- **c)** Modify that program to study the case of a centered-cubic lattice

Problem 2. Write a simple program for the classical Heisenberg spin model.

Problem 3. Write the instruction which realizes the energy histogram $H(E)$ in the program for the Ising model shown in Appendix B.

Problem 4. Program to search for the ground state:

We can determine in most cases the ground state of a spin system with Ising, XY, Heisenberg or Potts model by the steepest-descent method: at each spin, we minimize its energy by aligning it along its local field.

Describe the necessary steps to make a program to this end.

Write a program which realizes the above steps. Apply it to the

Ising model on a square lattice with nearest-neighbor interaction J_1 and next-nearest neighbor interaction J_2. Determine the phase diagram at temperature $T = 0$ in the space (J_1, J_2).

PART 2
Application to Surface Physics

Chapter 9

Magnetic Properties of Thin Films

9.1 Introduction

We have introduced in the preceding chapters basic methods to study general bulk properties of systems of interacting spins. We have taken advantage of the periodic crystalline structure to use the Fourier transforms and the periodic boundary conditions which allowed us to simplify calculations via the sum rules and the crystal symmetry.

When the invariance by translation is broken because of the presence of impurities, defects or surfaces, the calculation becomes more complicated. Often we have to modify methods established for the bulk and to introduce new techniques. Physically, the loss of the spatial periodicity causes a change in bulk properties of materials. The physics of disorder is a major research domain since more than 40 years. We can mention some well-studied disordered systems such as spin glasses, amorphous compounds and doped semiconductors. There is another domain which has been spectacularly developed in the past decades: the physics of nanomaterials. Nanomaterials such as ultrafine particles, ultrathin films and nanoribbons do not have periodic structures due to their nanometric dimension. These tiny objects have been used in many industrial applications which rapidly change our daily life: tiny computers, smart phones, high-speed internet, ... In this chapter, we study a family of simple magnetic systems where the loss of translational invariance is caused by the existence of surfaces or interfaces such as semi-infinite crystals, thin films, or multilayers. Effects of surface are important because they modify drastically properties of a bulk material when the so-called aspect ratio, i.e. the ratio of the number of surface atoms to the total number of atoms, becomes important as it is the case in small particles or in films of a few atomic layers.. Surface physics has been intensively developed during the last 30 years. Among the main reasons for that rapid and successful development we can mention the interest in understanding the physics of low-dimensional systems and an immense potential of industrial applications of thin films [147, 19, 32]. In particular, theoretically it has been shown that systems of continuous spins (XY and Heisenberg) in two dimensions (2D) with a short-range interaction cannot have a long-range order at finite temperatures [95]. In the case of thin films, it has been shown that low-lying localized spin waves can be found at the film surface [112] and effects of these localized modes on the surface magnetization at finite temperatures (T) and on the critical temperature have been investigated by the Green's function technique [35, 36]. Experimentally, objects of nanometric size such as ultrathin films and nanoparticles have also been intensively studied because of numerous and important applications in industry. An example is the so-called giant magnetoresistance used in data storage devices, magnetic sensors, etc. [12, 56, 14, 131]. Recently, much interest has been attracted towards practical problems such as spin transport, spin valves and spin-transfer torques, due to numerous applications in spintronics.

9.2   Surface physics

All crystals terminate in space by a surface. It can have various geometries and situations. The simplest one is a clean surface which is a perfect crystalline atomic plane such as (100) and (111) planes. It can be spherical as in the case of a spherical aggregates. In general, the surface can include impurities, dislocations, vacancies, islands, steps, ... A surface can also chemisorb or physisorb alien atoms. Chemisorbed atoms form a strong chemical binding with surface atoms while physisorbed atoms are physically bound to surface atoms by weak potentials such as the long-range van der Waals interaction. They are sometimes called "adatoms". It is obvious that the more the surface is disordered, the more it is difficult to study. Spectacular effects are often observed with well controlled and well characterized surfaces. Today, sophisticated techniques allow to create a surface with desired characteristics.

At the surface, atoms do not have the same environment as those inside the crystal. Due to the lack of neighbors and various neighboring defects and geometries, electronic states of surface atoms are modified in one way or in another, giving rise to changes in their effective interactions with neighboring atoms. The density of states shows often surface states which modify the filling of electronic bands, the position of the Fermi level and the magnetic moment of surface atoms. In addition, the surface anisotropy and surface exchange interaction can be very different from those of the interior atoms. We mention here some remarkable observations:

- For a thin film, the dipolar interaction favors an in-plane spin configuration. However, when the film thickness becomes very small a perpendicular anisotropy comes into play to favor a spin configuration perpendicular to the film surface at low temperatures [19]. In addition, the sign and amplitude of the surface exchange may also change with decreasing thickness. These modifications can cause interesting surface behaviors such as magnetic ordering reconstructions near the surface and localized surface spin waves. Also, the competition between the dipolar interaction and the perpendicular surface anisotropy can give rise to the so-called "re-orientation transition" which turns the perpendicular configuration into the in-plane one at a finite temperature.
- Perturbations in the cohesive interaction which binds surface atoms can yield a modification of the lattice constant at the surface (contraction or dilatation) and even a reconstruction of surface geometry.
- Perturbations in electronic energy bands can give rise to anomalies in electronic (charge and/or spin) transport observed in multilayers and near the surface.

It is not the purpose of this chapter to give experimental data on problems raised above. There exists a great number of handbooks and reviews [147, 19, 18, 32, 33]. In this chapter, we present some fundamental aspects which are well understood at present. The purpose is to provide a theoretical framework to understand microscopic mechanisms which lead to macroscopic surface effects such as low surface magnetization, low transition temperature, surface phase transition and surface instability. We will concentrate our attention to simple methods which allow us to study properties of magnetic semi-infinite crystals and thin films.

9.3 Surface magnetism: Generalities

We consider a ferromagnetic thin film of N_T atomic layers. The surface is denoted by index $n = 1$ and the last layer by $n = N_T$. We suppose the $\vec{Oz}$ axis is perpendicular to the film surface.

9.3.1 *Surface magnon modes*

We have studied the bulk spin waves in chapter 5. The amplitude of a bulk spin-wave mode does not vary in space. In general, near magnetic perturbation sources such as magnetic impurities and surfaces, spin-waves can be spatially localized. Such modes are called "localized modes". The amplitude of a surface localized mode decays when it propagates from the surface into the bulk. We show in Fig. 9.1 a surface mode and a bulk mode, for comparison.

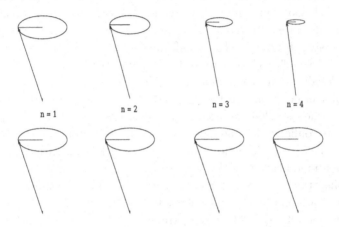

Fig. 9.1   A surface spin-wave mode (top) and a bulk mode (bottom).

We express the amplitude of a surface mode by $U_n(\vec{k}) = Ae^{i\vec{k}\cdot\vec{r}_n}$ where n denotes the index of the layer and $\vec{r}_n = \vec{r}_\| + \vec{z}$ is the position of a lattice site of the n-th layer at the z position $z = na$ on the $\vec{Oz}$ axis, a being the distance between two successive layers in the z direction. In such a notation, a surface mode corresponds to a complex wave-vector $\vec{k} = k_1 + ik_2$ where k_2 is nonzero. Its amplitude $U_n(\vec{k}) \propto e^{-k_2 na}$ therefore diminishes while propagating into the crystal interior (increasing n). We will show some examples in the following.

9.3.2 *Reconstruction of surface magnetic ordering*

As we said above, surface exchange interaction and surface anisotropy may suffer from modifications not only in their magnitudes but also in their signs. These modifications may result in a rearrangement of the magnetic ordering at the surface to minimize the system energy. In the case of

Heisenberg spin model, we can have a non collinear spin configuration in the vicinity of the surface as will be seen below.

9.3.3 *Surface phase transition*

Basic methods leading to fundamental properties of phase transitions in the bulk have been presented in chapter 7. Phase transition is a collective phenomenon which takes place when the system changes its symmetry. At the transition, the system particles become strongly correlated at a macroscopic scale. In a system with a finite size such as fine particles, the theoretical definition of the phase transition is not rigorously obeyed. For example, the correlation length in a second-order phase transition cannot go to infinity in a finite system. Nevertheless, anomalies in physical quantities can be observed in small systems. Finite-size scaling relations have been established [42, 7, 150] to allow us to obtain properties of the phase transition such as universality class at the infinite-size limit. In thin films, the infinite dimension of the film planes makes transitions possible. The characteristics of the phase transition in films depend on the surface conditions: if the film thickness is important then the influence of surface parameters is small since the aspect ratio is small. However, for ultrathin films, surface parameters become dominant making the surface phase transition very different from the bulk one.

There exists a large number of books on the surface phase transition. The reader is referred to, for example, reviews given in the references [18, 24, 32] for more details. One of the remarkable results is the existence of surface critical exponents and surface scaling laws which are different from the bulk ones [24].

9.4 Semi-infinite solids

One examines the case of a semi-infinite crystal. One calculates the magnon spectrum and shows the existence of surface modes in this section. The simplest way to do is to use the method of equation of motion described in chapter 5. To give a simple example, one considers the same system illustrated there with the inclusion of a surface: a semi-infinite ferrimagnetic crystal of body-centered cubic lattice. One uses the same Hamiltonian (5.110) and the same equations of motion (5.111) with the same hypothesis $J_2^A = J_2^B \equiv J_2$. Note that this system is a body-centered cubic antiferromagnet if $S_A = S_B$. Now, for a semi-infinite crystal, one uses the Fourier transformation only in the xy plane which is still periodic

$$S_l^- = \frac{1}{(2\pi)^3} \int_{-\infty}^{-\infty} d\omega e^{-i\omega t} \int_{BZ} d\vec{k}_\| e^{i\vec{k}_\| \cdot \vec{l}} U_n(\vec{k}_\|, \omega) \qquad (9.1)$$

$$S_m^+ = \frac{1}{(2\pi)^3} \int_{-\infty}^{-\infty} d\omega e^{-i\omega t} \int_{BZ} d\vec{k}_\| e^{i\vec{k}_\| \cdot \vec{l}} U_{n'}(\vec{k}_\|, \omega) \qquad (9.2)$$

where $\vec{k}_\|$ is a wave vector parallel to the surface, n and n' denote respectively the indices of the layers to which the spins l and m belong, BZ stands for the first Brillouin zone in the xy plane.

To simplify the presentation, we take the body-centered cubic lattice with the surface plane (001). We suppose that the sublattice of A spins ($\uparrow$) occupies the planes of even indices and the sublattice of B spins ($\downarrow$) takes the planes of odd indices as indicated in Fig. 9.2.

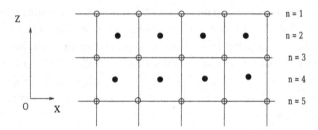

Fig. 9.2 Semi-infinite crystal of body-centered cubic lattice with a (001) surface (side view). The sublattices $\uparrow$ and $\downarrow$ are denoted by black and white circles, respectively. The surface has index $n = 1$.

The equations of motion for S_l^- and S_m^+ give two following coupled equations for $n \geq 2$

$$(E + E_A)U_{2n} = -(1 - \alpha)[4\gamma_1(\vec{k}_\|)(U_{2n-1} + U_{2n+1}) +$$
$$\epsilon(U_{2n-2} + U_{2n+2}) \qquad (9.3)$$

$$(E - E_B)U_{2n+1} = (1 + \alpha)[4\gamma_1(\vec{k}_\|)(U_{2n} + U_{2n+2}) +$$
$$\epsilon(U_{2n-1} + U_{2n+3}) \qquad (9.4)$$

For the first two layers, we have

$$(E - E_1)U_1 = (1 + \alpha)\left[4\gamma_1(\vec{k}_\|)U_2 + \epsilon U_3\right] \qquad (9.5)$$

$$(E + E_2)U_2 = -(1 - \alpha)\left[4\gamma_1(\vec{k}_\|)(U_1 + U_3) + \epsilon U_4\right] \qquad (9.6)$$

where we have replaced, at $T = 0$ neglecting zero-point spin contractions, $< S_A^z > \simeq S_A = (1 - \alpha)S$, $< S_B^z > \simeq S_B = -(1 + \alpha)S$ with $|\alpha| < 1$ and S a constant, $E = \frac{\hbar\omega}{J_1 S}$, $\epsilon = \frac{J_2}{J_1}$, and

$$E_A = 8(1 + \alpha) - 6\epsilon(1 - \alpha)[1 - \gamma_2(\vec{k}_\parallel)] \tag{9.7}$$

$$E_B = 8(1 - \alpha) - 6\epsilon(1 + \alpha)[1 - \gamma_2(\vec{k}_\parallel)] \tag{9.8}$$

$$E_2 = 8(1 + \alpha) - \epsilon(1 - \alpha)[5 - 4\gamma_2(\vec{k}_\parallel)] \tag{9.9}$$

$$E_1 = 4(1 - \alpha) - \epsilon(1 + \alpha)[5 - 4\gamma_2(\vec{k}_\parallel)] \tag{9.10}$$

$$\gamma_1(\vec{k}_\parallel) = \cos(\frac{k_x a}{2}) \cos(\frac{k_y a}{2}) \tag{9.11}$$

$$\gamma_2(\vec{k}_\parallel) = \frac{1}{2}[\cos(k_x a) + \cos(k_y a)] \tag{9.12}$$

a being the lattice constant. The factor $\gamma_1(\vec{k}_\parallel)$ couples the nearest neighbors belonging to the adjacent planes while $\gamma_2(\vec{k}_\parallel)$ connects the next-nearest neighbors belonging to the same sublattice.

We take the following forms for the bulk spin-wave amplitudes U_{2n} and U_{2n+1}

$$U_{2n} = U_e \exp(ik_z na) \tag{9.13}$$

$$U_{2n+1} = U_o \exp(ik_z(1 + 1/2)na) \tag{9.14}$$

where k_z is the real wave vector in the z direction. Replacing these in (9.3) and (9.4), we obtain the following secular equation for a non trivial solution:

$$(E + E_A)(E - E_B) = -64(1 - \alpha^2)\gamma_1(\vec{k}_\parallel)^2 \tag{9.15}$$

We deduce the dispersion relation for bulk modes

$$E_\pm = \frac{1}{2}\left[E_B - E_A \pm \sqrt{(E_A - E_B)^2 + [4E_A E_B - 64(1 - \alpha^2)\gamma_1(\vec{k}_\parallel)^2]} \right] \tag{9.16}$$

We are now interested in finding surface modes. We look for solutions of the form

$$U_{2n} = U_2 \phi^{n-1} \tag{9.17}$$

$$U_{2n+1} = U_1 \phi^n \tag{9.18}$$

where ϕ is a real factor defined by $\phi^n = e^{-k_2 na}$ where k_2 is the imaginary part of k_z. For a decaying wave, $\phi < 1$. ϕ is called "decay factor". Replacing these amplitudes in (9.3)-(9.6) we obtain a system of coupled equations. Surface modes correspond to solutions $\phi < 1$. We examine a particular case where $k_x = k_y = 0$. In this case, the following solution for a surface mode is found

$$E_s = -8\alpha + \frac{4\alpha\epsilon(1 + \alpha)}{2(1 - \alpha) - \epsilon} \tag{9.19}$$

$$\phi = 1 + \frac{4\alpha}{2(1 - \alpha) - \epsilon} \tag{9.20}$$

We see that in order to have $\phi < 1$, we should have $\alpha < 0$ for the case $k_x = k_y = 0$. For $k_x, k_y \neq 0$, there exists for $\alpha < 0$ and $\alpha > 0$ a surface spin-wave branch in the spectrum as shown in Fig. 9.3.

Note: In the case of an antiferromagnet, we just put $\alpha = 0$ in the above equations. There is no surface mode for $k_x = k_y = 0$.

The gap at $k_x = k_y = 0$ in the ferrimagnet is proportional to α. We show in Fig. 9.3 the magnon spectrum versus $k_x = k_y$.

We show in Fig. 9.4 the spatial variation of the amplitude of the surface mode at $k_x = k_y = 0$ with $\alpha = -1/3$. The decay factor calculated by (9.20) is equal to 0.5.

We discuss now the effect of the interaction between next nearest neighbors. The interaction between nearest neighbors J_1 is antiferromagnetic [$J_1 > 0$ as seen in the definition of the Hamiltonian (5.110)]. If J_2 is ferromagnetic, i.e. $\epsilon < 0$, the magnetic order is antiferromagnetic between the two sublattices. On the other hand, if J_2 is antiferromagnetic (> 0), there is a competition between J_1 and J_2. When $|J_2|$ is large enough, the collinear antiferromagnetic configuration is no more stable for $\epsilon = \frac{J_2}{J_1} > \epsilon_c$. The determination of ϵ_c with a surface is more complicated than in the bulk case because there may exist a non uniform spin configuration near the surface. In the present case, the critical value ϵ_c depends also on α. We show in Fig. 9.5 the spin configuration near the surface for a value of $\epsilon > \epsilon_c$.

9.5   Ferromagnetic films with different lattices

We show in this section that localized surface spin-wave modes affect strongly thermodynamic behaviors of ferromagnetic thin films. In particular, we will show that low-lying localized modes diminish the surface magnetization and the Curie temperature with respect to the bulk ones. These quantities depend of course on the surface interaction parameters

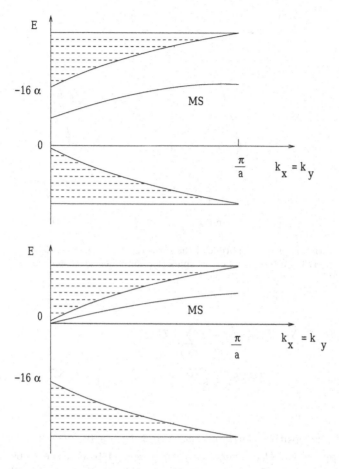

Fig. 9.3 Magnon spectrum of a semi-infinite ferrimagnet versus $k_x = k_y$ in the case where $\epsilon = 0$, $\alpha = -1/3$ (top) and $\alpha = 1/3$ (bottom). Surface spin-wave branches are indicated by MS. The hachured bands are the bulk continuum. The upper limit of each band corresponds to $k_z = \pi/a$ and the lower limit to $k_z = 0$.

and the film thickness.

The method which can cover correctly a large region of temperature is no doubt the Green's function method (cf. chapter 6). We shall use here that method to study properties of thin films from $T = 0$ up to the phase transition.

We consider a thin film of N_T layers with the Heisenberg quantum spin model. The Hamiltonian is written as

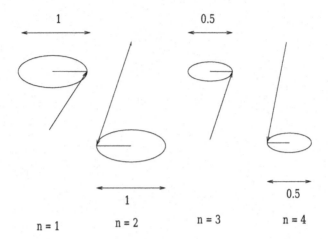

Fig. 9.4 Variation of the amplitude of the surface mode near the surface. $k_x = k_y = 0$, $\epsilon = 0$, $\alpha = -1/3$. Surface spins are B spins (for convenience, B spins are drawn as up spins).

$$\mathcal{H} = -2 \sum_{<i,j>} J_{ij} \vec{S}_i \cdot \vec{S}_j - 2 \sum_{<i,j>} D_{ij} S_i^z S_j^z$$

$$= -2 \sum_{\langle i,j \rangle} J_{ij} \left(S_i^z S_j^z + \frac{1}{2}(S_i^+ S_j^- + S_i^- S_j^+) \right) - 2 \sum_{<i,j>} D_{ij} S_i^z S_j^z$$

$$(9.21)$$

where J_{ij} is positive (ferromagnetic) and $D_{ij} > 0$ denotes an exchange anisotropy. When D_{ij} is very large with respect to J_{ij}, the spins have an Ising-like behavior. The factor 2 in front of the terms is used for historical reasons [see (3.51)].

9.5.1 *Method*

The Green's function method has been formulated in detail in chapter 6. It is useful to summarize here the main steps in its application to thin films.

We define one Green's function for each layer, numbering the surface as the first layer. We write next the equation of motion for each of the Green's functions. We obtain a system of coupled equations. We linearize these equations to reduce higher-order Green's functions by using the Tyablikov decoupling scheme. We are then ready to make the Fourier transforms for all Green's functions in the xy planes. We obtain a system of equations in

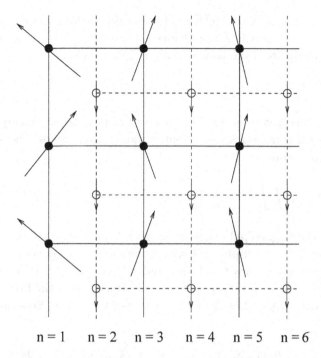

n = 1   n = 2   n = 3   n = 4   n = 5   n = 6

Fig. 9.5 Non collinear spin configuration near the surface (side view) for $\epsilon > \epsilon_c$, with $\alpha > 0$. Surface spins are B spins (for convenience, B spins are drawn as up spins).

the space $(\vec{k}_{xy}, \omega)$ where $\vec{k}_{xy}$ is the wave vector parallel to the xy plane and ω is the spin-wave frequency (pulsation). Solving this system we obtain the Green's functions and ω as functions of $\vec{k}_{xy}$. Using the spectral theorem, Eq. (6.39), we calculate the layer magnetization.

We define the following Green's function for two spins $\vec{S}_i$ and $\vec{S}_j$

$$G_{i,j}(t,t') = \langle\langle S_i^+(t); S_j^-(t') \rangle\rangle \tag{9.22}$$

The equation of motion of $G_{i,j}(t,t')$ is written as

$$i\hbar\frac{dG_{i,j}(t,t')}{dt} = (2\pi)^{-1}\langle [S_i^+(t), S_j^-(t')] \rangle + \langle\langle [S_i^+; \mathcal{H}](t); S_j^-(t') \rangle\rangle \tag{9.23}$$

where [...] is the boson commutator and $\langle...\rangle$ the thermal average in the canonical ensemble defined as

$$\langle F \rangle = \mathrm{Tre}^{-\beta\mathcal{H}}F/\mathrm{Tre}^{-\beta\mathcal{H}} \tag{9.24}$$

with $\beta = 1/k_B T$.

The commutator of the right-hand side of Eq. (9.23) generates functions of higher orders. In a first approximation, we can reduce these functions with the help of the Tyablikov decoupling [20] as follows

$$\langle\langle S_m^z S_i^+; S_j^-\rangle\rangle \simeq \langle S_m^z\rangle\langle\langle S_i^+; S_j^-\rangle\rangle, \tag{9.25}$$

We obtain then the same kind of Green's function defined in Eq. (9.22). As the system is translation invariant in the xy plane, we use the following Fourier transforms

$$G_{i,j}(t,t') = \frac{1}{\Delta}\int\int d\vec{k}_{xy}\frac{1}{2\pi}\int_{-\infty}^{+\infty}d\omega\, e^{-i\omega(t-t')}\, g_{n,n'}(\omega,\vec{k}_{xy})\, e^{i\vec{k}_{xy}\cdot(\vec{R}_i-\vec{R}_j)} \tag{9.26}$$

where ω is the magnon pulsation (frequency), $\vec{k}_{xy}$ the wave vector parallel to the surface, $\vec{R}_i$ the position of the spin at the site i, n and n' are respectively the indices of the planes to which i and j belong ($n = 1$ is the index of the surface). The integration on $\vec{k}_{xy}$ is performed within the first Brillouin zone in the xy plane. Let Δ be the surface of that zone. Equation (9.23) becomes

$$(\hbar\omega - A_n)g_{n,n'} + B_n(1-\delta_{n,1})g_{n-1,n'} + C_n(1-\delta_{n,N_T})g_{n+1,n'} = 2\delta_{n,n'} < S_n^z > \tag{9.27}$$

where the factors $(1 - \delta_{n,1})$ and $(1 - \delta_{n,N_T})$ are added to ensure that there are no C_n and B_n terms for the first and the last layer. The coefficients A_n, B_n and C_n depend on the crystalline lattice of the film. We give here some examples:

9.5.1.1 *Film of stacked triangular lattices*

$$\begin{aligned} A_n &= -2J_n < S_n^z > C\gamma_k + 2C(J_n + D_n) < S_n^z > \\ &\quad +2(J_{n,n+1} + D_{n,n+1}) < S_{n+1}^z > \\ &\quad +2(J_{n,n-1} + D_{n,n-1}) < S_{n-1}^z > \end{aligned} \tag{9.28}$$

$$B_n = 2J_{n,n-1} < S_n^z > \tag{9.29}$$

$$C_n = 2J_{n,n+1} < S_n^z > \tag{9.30}$$

where the following notations have been used:

i) J_n, D_n are the interactions in the layer n,

ii) $J_{n,n\pm1}$ and $D_{n,n\pm1}$ are the interactions between a spin in the layer n and a spin in the layer $(n \pm 1)$. Of course, $J_{n,n-1}$, $D_{n,n-1} = 0$ if $n = 1$, and $J_{n,n+1}$, $D_{n,n+1} = 0$ if $n = N_T$,

iii)$\gamma_k = [2\cos(k_x a) + 4\cos(k_x a/2)\cos(k_y a\sqrt{3}/2)]/C$

iv) $C = 6$ is the coordination number in the xy plane.

9.5.1.2 *Film of simple cubic lattice*

$$A_n = -2J_n < S_n^z > C\gamma_k + 2C(J_n + D_n) < S_n^z >$$
$$+2(J_{n,n+1} + D_{n,n+1}) < S_{n+1}^z >$$
$$+2(J_{n,n-1} + D_{n,n-1}) < S_{n-1}^z > \tag{9.31}$$
$$B_n = 2J_{n,n-1} < S_n^z > \tag{9.32}$$
$$C_n = 2J_{n,n+1} < S_n^z > \tag{9.33}$$

where $C = 4$ and $\gamma_k = \frac{1}{2}[\cos(k_x a) + \cos(k_y a)]$.

9.5.1.3 *Film of body-centered cubic lattice*

$$A_n = 8(J_{n,n+1} + D_{n,n+1}) < S_{n+1}^z >$$
$$+8(J_{n,n-1} + D_{n,n-1}) < S_{n-1}^z > \tag{9.34}$$
$$B_n = 8J_{n,n-1} < S_n^z > \gamma_k \tag{9.35}$$
$$C_n = 8J_{n,n+1} < S_n^z > \gamma_k \tag{9.36}$$

where $\gamma_k = \cos(k_x a/2)\cos(k_y a/2)$

Writing Eq. (9.27) for $n = 1, 2, ..., N_T$, we obtain a system of N_T equations which can be rewritten in a matrix form

$$\mathbf{M}(\omega)\mathbf{g} = \mathbf{u} \tag{9.37}$$

where $\mathbf{u}$ is a column matrix whose n-th element is $2\delta_{n,n'} < S_n^z >$.

For a given $\vec{k}_{xy}$ the magnon dispersion relation $\hbar\omega(\vec{k}_{xy})$ can be obtained by solving the secular equation $det|\mathbf{M}| = 0$. There are N_T eigenvalues $\hbar\omega_i$ ($i = 1, ..., N_T$) for each $\vec{k}_{xy}$. It is obvious that ω_i depends on all $\langle S_n^z \rangle$ contained in the coefficients A_n, B_n and C_n.

To calculate the thermal average of the magnetization of the layer n in the case where $S = \frac{1}{2}$, we use the following relation (see chapter 6)

$$\langle S_n^z \rangle = \frac{1}{2} - \langle S_n^- S_n^+ \rangle \tag{9.38}$$

where $\langle S_n^- S_n^+ \rangle$ is given by the following spectral theorem [see (6.39)]

$$\langle S_i^- S_j^+ \rangle = \lim_{\epsilon \to 0} \frac{1}{\Delta} \int\int d\vec{k}_{xy} \int\limits_{-\infty}^{+\infty} \frac{i}{2\pi} \left[g_{n,n'}(\omega + i\epsilon) - g_{n,n'}(\omega - i\epsilon) \right]$$

$$\times \frac{d\omega}{e^{\beta\omega} - 1} e^{i\vec{k}_{xy} \cdot (\vec{R}_i - \vec{R}_j)}. \tag{9.39}$$

ϵ being an infinitesimal positive constant. Equation (9.38) becomes

$$\langle S_n^z \rangle = \frac{1}{2} - \lim_{\epsilon \to 0} \frac{1}{\Delta} \int\int d\vec{k}_{xy} \int\limits_{-\infty}^{+\infty} \frac{i}{2\pi} \left[g_{n,n}(\omega + i\epsilon) - g_{n,n}(\omega - i\epsilon) \right] \frac{d\omega}{e^{\beta\hbar\omega} - 1} \tag{9.40}$$

where the Green's function $g_{n,n}$ is obtained by the solution of Eq. (9.37)

$$g_{n,n} = \frac{|\mathbf{M}|_n}{|\mathbf{M}|} \tag{9.41}$$

$|\mathbf{M}|_n$ is the determinant obtained by replacing the n-th column of $|\mathbf{M}|$ by **u**.

To simplify the notations we put $\hbar\omega_i = E_i$ and $\hbar\omega = E$ in the following. By expressing

$$|\mathbf{M}| = \prod_i (E - E_i) \tag{9.42}$$

we see that E_i ($i = 1, ..., N_T$) are the poles of the Green's function. We can therefore rewrite $g_{n,n}$ as

$$g_{n,n} = \sum_i \frac{f_n(E_i)}{E - E_i} \tag{9.43}$$

where $f_n(E_i)$ is given by

$$f_n(E_i) = \frac{|\mathbf{M}|_n(E_i)}{\prod_{j \neq i}(E_i - E_j)} \tag{9.44}$$

Replacing Eq. (9.43) in Eq. (9.40) and making use of the following identity

$$\frac{1}{x - i\eta} - \frac{1}{x + i\eta} = 2\pi i \delta(x) \tag{9.45}$$

we obtain

$$\langle S_n^z \rangle = \frac{1}{2} - \frac{1}{\Delta} \int \int dk_x dk_y \sum_{i=1}^{N_T} \frac{f_n(E_i)}{e^{\beta E_i} - 1} \tag{9.46}$$

where $n = 1, ..., N_T$.

As $< S_n^z >$ depends on the magnetizations of the neighboring layers via $E_i(i = 1, ..., N_T)$, we should solve by iteration the equations (9.46) written for all layers, namely for $n = 1, ..., N_T$, to obtain the layer magnetizations at a given temperature T.

The critical temperature can be calculated in a self-consistent manner by iteration, letting all $< S_n^z >$ tend to zero.

In the case where the film is homogeneous, namely in the case where the surface parameters are not different from the bulk ones, it is not exaggerated to calculate the critical temperature by supposing that all layer magnetizations are equal to a unique average value M which is to be determined self-consistently. The value of M is defined from $< S_n^z >$ of Eq. (9.46) as follows

$$M = \frac{1}{N_T} \sum_{n=1}^{N_T} \langle S_n^z \rangle = \frac{1}{2} - \frac{1}{N_T} \frac{1}{\Delta} \int \int dk_x dk_y \sum_{n=1}^{N_T} \sum_{i=1}^{N_T} \frac{f_n(E_i)}{e^{\beta E_i} - 1} \tag{9.47}$$

Replacing all $< S_n^z >$ in the matrix elements by M, we see that $\sum_n f_n(E_i) = 2M$ by using Eq. (9.44). We deduce

$$M = \frac{1}{2} - \frac{1}{N_T} \frac{1}{\Delta} \int \int dk_x dk_y \sum_{i=1}^{N_T} \frac{2M}{e^{\beta E_i} - 1} \tag{9.48}$$

When $T \to T_c$, $M \to 0$. We can then make an expansion of the exponential of the denominator of Eq. (9.48). We obtain

$$\left[\frac{J}{k_B T_c} \right]^{-1} = \frac{2}{\Delta} \int \int dk_x dk_y \frac{1}{N_T} \sum_{i=1}^{N_T} \frac{2M}{E_i} \tag{9.49}$$

We notice that all matrix elements A_n, B_n and C_n are proportional to M in the above hypothesis of uniform layer magnetization. We see that E_i is proportional to M. The right-hand side of Eq. (9.49) therefore does not depend on M.

9.5.2 *Results*

To calculate numerically the above equations, one must first determine the first Brillouin zone according to the lattice structure. For the iteration process, one starts in general at a very low T with input values of $< S_n^z >$ close to the spin amplitude. Next, one uses the solutions of $< S_n^z >$ as inputs for a temperature not far from the previous T in order to facilitate the convergence. Of course, if we use the hypothesis of uniform layer magnetizations, i.e. $< S_n^z > = M$ for all n, then there is only one solution M to find at a given T.

At each T, with input values for $< S_n^z >$, one calculates the eigenvalues of the magnon energy E_i by solving $det|\mathbf{M}| = 0$ for each value of $\vec{k}_{xy}$. Using the values so obtained of E_i one calculates the output values of $< S_n^z >$ $(n = 1, ..., N_T)$. If the output values are equal to the input values within a given precision, one stops the iteration. In general, a few iterations suffice at low T for a solution with a precision of 1%. For a temperature close to T_c, one needs a few dozens to a few hundreds of iterations. We show here the results of a few cases for comparison.

9.5.2.1 *Magnon spectrum*

We show in Figs. 9.6 and 9.7 the magnon spectra of ferromagnetic films of simple cubic and body-centered cubic lattices with a thickness $N_T = 8$. For simplicity, we suppose all exchange interactions are identical and equal to J. Also, all anisotropies D are equal to $0.01J$. We see in Figs. 9.6 and 9.6 that there is no surface mode in the simple cubic case while there are two branches of surface modes in the body-centered lattice. Two branches result from the interaction of the surface modes coming from the two surfaces of the film. If the thickness is thick enough (longer than the penetration length of the surface mode), then two branches are degenerate due to symmetrical surface conditions. This is not the case for $N_T = 8$.

9.5.2.2 *Layer magnetizations*

Figure 9.8 shows the results of the layer magnetizations for the first two layers in the cases considered above with $N_T = 4$. One observes that the layer magnetization at the surface is smaller than that of the second layer. This difference is larger in the case where there exists a surface mode as in the body-centered cubic lattice because surface modes are localized at the surface, making a larger deviation for surface spins.

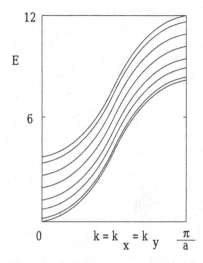

Fig. 9.6 Magnon spectrum $E = \hbar\omega$ of a ferromagnetic film with a simple cubic lattice versus $k \equiv k_x = k_y$ for $N_T = 8$ and $D/J = 0.01$. No surface mode is observed for this case.

One also sees that the critical temperature is strongly decreased in the case where surface modes of low-lying energy exist (acoustic surface modes).

9.6 Antiferromagnetic films

We can adapt the Green's function method presented in chapter 6 for bulk antiferromagnets to the case of antiferromagnetic thin films. We consider the following Hamiltonian

$$
\begin{aligned}
\mathcal{H} &= 2 \sum_{<i,j>} J_{ij} \vec{S}_i \cdot \vec{S}_j + 2 \sum_{<i,j>} D_{ij} S_i^z S_j^z \\
&= 2 \sum_{\langle i,j \rangle} J_{ij} \left(S_i^z S_j^z + \frac{1}{2}(S_i^+ S_j^- + S_i^- S_j^+) \right) + 2 \sum_{<i,j>} D_{ij} S_i^z S_j^z
\end{aligned} \quad (9.50)
$$

where $J_{ij} > 0$ and $D_{ij} > 0$. We divide the lattice into two sublattices: sublattice A contains ↑ spins and sublattice B ↓ spins. We define the following Green's functions

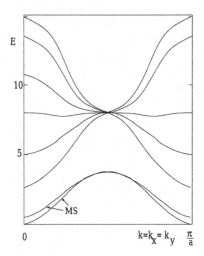

Fig. 9.7   Magnon spectrum $E = \hbar\omega$ of a ferromagnetic film with a body-centered cubic lattice versus $k \equiv k_x = k_y$ for $N_T = 8$ and $D/J = 0.01$. The branches of surface modes are indicated by MS.

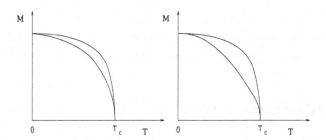

Fig. 9.8   Ferromagnetic films of simple cubic lattice (left) and body-centered cubic lattice (right): magnetizations of the surface layer (lower curve) and the second layer (upper curve), with $N_T = 4$, $D = 0.01J$, $J = 1$.

$$G_{j,j'}(t - t') = \,<< S_j^+(t); S_{j'}^-(t') >>$$
$$= -i\theta(t - t') < \left[S_j^+(t), S_{j'}^-(t') \right] > \qquad (9.51)$$
$$F_{i,j'}(t - t') = \,<< S_i^+(t); S_{j'}^-(t') >>$$
$$= -i\theta(t - t') < \left[S_i^+(t), S_{j'}^-(t') \right] > \qquad (9.52)$$

where j, j' belong to sublattice A and i to sublattice B.

9.6.1 *Films of simple cubic lattice*

We write the equations of motion for $G_{j,j'}(t)$ and $F_{i,j'}(t)$, then we use the Tyablikov decoupling [20] and the following Fourier transforms

$$G_{j,j'}(t,t') = \frac{1}{\Delta} \int \int d\vec{k}_{xy} \frac{1}{2\pi} \int_{-\infty}^{+\infty} d\omega \, e^{-i\omega(t-t')} g_{n,n'}(\omega, \vec{k}_{xy})$$
$$\times e^{i\vec{k}_{xy}\cdot(\vec{R}_j - \vec{R}_{j'})} \tag{9.53}$$

$$F_{i,j'}(t,t') = \frac{1}{\Delta} \int \int d\vec{k}_{xy} \frac{1}{2\pi} \int_{-\infty}^{+\infty} d\omega \, e^{-i\omega(t-t')} f_{n,n'}(\omega, \vec{k}_{xy})$$
$$\times e^{i\vec{k}_{xy}\cdot(\vec{R}_i - \vec{R}_{j'})} \tag{9.54}$$

where we use the same notations as in the above ferromagnetic case. We obtain

$$
\begin{aligned}
E g_{n,n'}(\omega) = {} & 2 < S_n^z > \delta_{n,n'} \\
& + 8 \left[\gamma_k < S_n^z > f_{n,n'}(\omega) + (1 + d_e) < S_n^z > g_{n,n'}(\omega) \right] \\
& + 2 \left[< S_n^z > f_{n+1,n'}(\omega) + (1 + d) < S_{n+1}^z > g_{n,n'}(\omega) \right] \\
& \times (1 - \delta_{N_T,n}) \\
& + 2 \left[< S_n^z > f_{n-1,n'}(\omega) + (1 + d) < S_{n-1}^z > g_{n,n'}(\omega) \right] \\
& \times (1 - \delta_{1,n}) \tag{9.55}
\end{aligned}
$$

$$
\begin{aligned}
E f_{n,n'}(\omega) = {} & -8 \left[\gamma_k < S_n^z > g_{n,n'}(\omega) + (1 + d_e) < S_n^z > f_{n,n'}(\omega) \right] \\
& - 2 \left[< S_n^z > g_{n+1,n'}(\omega) + (1 + d) < S_{n+1}^z > f_{n,n'}(\omega) \right] \\
& \times (1 - \delta_{N_T,n}) \\
& - 2 \left[< S_n^z > g_{n-1,n'}(\omega) + (1 + d) < S_{n-1}^z > f_{n,n'}(\omega) \right] \\
& \times (1 - \delta_{1,n}) \tag{9.56}
\end{aligned}
$$

where $n, n' = 1, ..., N_T$ and the following notations have been used

- $J_{i,j} = J$ for all pairs (i, j),
- $d = D_{i,j}/J$ for all pairs (i, j) except when (i, j) are both on the surface ($n = 1$ and $n = N_T$) where $d = d_s$,
- $E = \hbar\omega/J$,
- $d_e = d(1 - \delta_{1,n})(1 - \delta_{N_T,n}) + d_s(\delta_{1,n} + \delta_{N_T,n})$ (this complicated notation was used in order to include all cases in the same formula),
- $\gamma_k = \cos(k_x a) \cos(k_y a)$ (for convenience, we used the distance between the two neighbors of the same sublattice in the xy plane equal to $2a$ and k_x and k_y oriented along the axes of one sublattice).

It is noted that in Eqs. (9.55)-(9.56) we have changed the sign of the average values of B spins ($< S_n^z > \rightarrow - < S_n^z >$) so that all $< S_n^z >$ are positive in Eqs. (9.55)-(9.56).

9.6.2 *Films of body-centered cubic lattice*

In the case of a film of body-centered cubic lattice with a (001) surface, we suppose that $\uparrow$ sublattice and $\downarrow$ sublattice occupy even- and odd-index layers, respectively. Using the following Fourier transforms

$$G_{j,j'}(t,t') = \frac{1}{\Delta} \int\int d\vec{k}_{xy} \frac{1}{2\pi} \int_{-\infty}^{+\infty} d\omega\, e^{-i\omega(t-t')} g_{2n,2n'}(\omega, \vec{k}_{xy})$$
$$\times e^{i\vec{k}_{xy}\cdot(\vec{R}_j - \vec{R}_{j'})}$$

$$F_{i,j'}(t,t') = \frac{1}{\Delta} \int\int d\vec{k}_{xy} \frac{1}{2\pi} \int_{-\infty}^{+\infty} d\omega\, e^{-i\omega(t-t')} f_{2n+1,2n'}(\omega, \vec{k}_{xy})$$
$$\times e^{i\vec{k}_{xy}\cdot(\vec{R}_i - \vec{R}_{j'})}$$

we obtain

$$\begin{aligned}
E g_{2n,2n'}(\omega) = {}& 2 < S_{2n}^z > \delta_{n,n'} + 8[\gamma_k < S_{2n}^z > f_{2n+1,2n'}(\omega) \\
&+ (1+d) < S_{2n+1}^z > g_{2n,2n'}(\omega)](1 - \delta_{n,N}) \\
&+ 8[\gamma_k < S_{2n}^z > f_{2n-1,2n'}(\omega) \\
&+ (1+d_f) < S_{2n-1}^z > g_{2n,2n'}(\omega)]
\end{aligned} \tag{9.57}$$

$$\begin{aligned}
E f_{2n-1,2n'}(\omega) = {}& -8[\gamma_k < S_{2n-1}^z > g_{2n,2n'}(\omega) \\
&+ (1+d_f) < S_{2n}^z > f_{2n-1,2n'}(\omega)] \\
&- 8[\gamma_k < S_{2n-1}^z > g_{2n-2,2n'}(\omega) \\
&+ (1+d) < S_{2n-2}^z > f_{2n-1,2n'}(\omega)](1 - \delta_{1,n}) \quad (9.58)
\end{aligned}$$

where $n, n' = 1, ..., N$ with $N = N_T/2$ (N=number of layers in each sublattice), $d_f = d(1-\delta_{1,n})(1-\delta_{N,n})+d_s(\delta_{1,n}+\delta_{N,n})$, $\gamma_k = \cos(k_x a/2)\cos(k_y a/2)$. As before, we have redefined $< S_{2n\pm1}^z > \rightarrow - < S_{2n\pm1}^z >$ of B spins.

The next steps of the calculation are the same as in the ferromagnetic case: we calculate magnon energy eigenvalues E_i by solving $det|\mathbf{M}| = 0$. The layer magnetization is calculated by Eq. (9.46). The value of the spin in the layer n at $T = 0$ is calculated by (see chapter 5)

$$\langle S_n^z \rangle (T=0) = \frac{1}{2} + \frac{1}{\Delta} \int\int dk_x dk_y \sum_{i=1}^{N_T} f_n(E_i) \tag{9.59}$$

Table 9.1   Values of Néel temperature T_N and Curie temperature T_c for thicknesses $N_T = 2$ and 20, calculated with several values of anisotropy d.

	$N_T = 2$	$N_T = 2$	$N_T = 2$	$N_T = 20$	$N_T = 20$	$N_T = 20$
d_s	0.01	0.1	0.2	0.01	0.1	0.2
$k_B T_N / J$	1.37	1.91	2.25	1.86	2.02	2.14
$k_B T_c / J$	1.40	1.93	2.27	2.03	2.10	2.14

where the sum is performed over negative values of E_i (for positive values the Bose-Einstein factor is equal to 0 at $T = 0$).

The numerical results for $S = 1/2$, $N_T = 4$ and $d = 0.01$ are $S_1^z = 0.44075$, $S_2^z = 0.41885$. We conclude that the zero-point spin contraction (cf. chapter 5) is smaller for surface spins. This is not surprising because zero-point fluctuations are due to antiferromagnetic interaction: the weaker the antiferromagnetic interaction is, the smaller the zero-point spin contraction becomes. Due to the lack of neighbors, the antiferromagnetic local field acting on a surface spin is weaker than that on an interior spin [37].

The Néel temperature T_N is obtained in the same manner as for T_c. We have the following N_T coupled equations to solve by iteration

$$\left[\frac{J}{k_B T_N} \right]^{-1} = \frac{1}{\Delta} \int \int dk_x dk_y \sum_{i=1}^{N_T} \frac{f_n(E_i)}{E_i} \qquad (9.60)$$

where $n = 1, ..., N_T$. The uniform layer-magnetization approximation presented above for ferromagnetic films [see (9.47)-(9.49)] yields for the antiferromagnetic case the following expression

$$\left[\frac{J}{k_B T_N} \right]^{-1} = \frac{1}{\Delta} \int \int dk_x dk_y \frac{1}{N_T} \sum_{i=1}^{N_T} \frac{2M}{E_i} \qquad (9.61)$$

We show in Table 9.1 the values of T_N and T_c calculated for films of simple cubic lattice with two thicknesses $N_T = 2$ and $N_T = 20$. We see that the antiferromagnetic film has the critical temperature lightly but systematically smaller than that of the ferromagnetic film, except when there is a strong anisotropy which suppresses more or less antiferromagnetic fluctuations (last column). The critical temperature increases with increasing thickness as expected. It reaches the bulk value at a few dozens of layers.

9.7   Frustrated films

So far, we have studied thin films with collinear spin ordering. In this section we consider the effect of the frustration in thin films. We have studied in chapter 5 some frustrated bulk spin systems where the low-temperature spin configuration is non collinear. The situation in frustrated thin films is more complicated because the non collinear spin configuration may not be uniform near the film surface.

This paragraph deals with the effect of the frustration in magnetic thin films. The frustration is known to cause spectacular effects in various bulk spin systems. Its effects have been extensively studied during the last decade theoretically, experimentally and numerically. Frustrated model systems serve not only as testing ground for theories and approximations, but also to compare with experiments [40].

We consider an example in this section: a ferromagnetic film with frustrated surfaces [103]. We study, by the analytical Green's function method and extensive Monte Carlo simulations, effects of frustrated surfaces on the properties of thin films made of stacked triangular layers of atoms bearing Heisenberg spins with an Ising-like interaction anisotropy. We suppose that the in-plane surface interaction J_s can be antiferromagnetic or ferromagnetic while all other interactions are ferromagnetic. We show that the ground-state spin configuration is non linear when J_s is lower than a critical value J_s^c. The film surfaces are then frustrated. In the frustrated case, there are two phase transitions related to disordering of surface and interior layers. There is a good agreement between Monte Carlo and Green's function results.

9.7.1   *Model*

We consider a thin film made up by stacking N_z planes of triangular lattice of $L \times L$ lattice sites.

The Hamiltonian is given by

$$\mathcal{H} = - \sum_{\langle i,j \rangle} J_{i,j} \mathbf{S}_i \cdot \mathbf{S}_j - \sum_{<i,j>} I_{i,j} S_i^z S_j^z \qquad (9.62)$$

where $\mathbf{S}_i$ is the Heisenberg spin at the lattice site i, $\sum_{\langle i,j \rangle}$ indicates the sum over the nearest neighbor spin pairs $\mathbf{S}_i$ and $\mathbf{S}_j$. The last term, which will be taken to be very small, is needed to ensure that there is a phase transition at a finite temperature for the film with a finite thickness when all exchange interactions $J_{i,j}$ are ferromagnetic. Otherwise, it is known that a strictly two-dimensional system with an isotropic non-Ising spin model (XY or Heisenberg model) does not have a long-range ordering at finite temperatures [95].

We suppose that the interaction between two nearest neighbor surface spins is equal to J_s, and all other interactions are ferromagnetic and equal to $J = 1$ for simplicity. The two surfaces of the film are frustrated if J_s is antiferromagnetic ($J_s < 0$).

9.7.2 Ground state

In this paragraph, we suppose that the spins are classical. The classical ground state can be easily determined as shown below. Note that for antiferromagnetic systems, even for bulk materials, the quantum ground state cannot be exactly determined.

For $J_s > 0$ (ferromagnetic interaction), the magnetic ground state is ferromagnetic. However, when J_s is negative the surface spins are frustrated. There is a competition between the non collinear surface ordering and the ferromagnetic ordering of the spins of the beneath layer.

We first determine the ground state configuration for $I = I_s = 0.1$ by using the steepest descent method : starting from a random spin configuration, we calculate the magnetic local field at each site and align the spin of the site in its local field. In doing so for all spins and repeating until the convergence is reached, we obtain in general the ground state configuration, without metastable states in the present model. The result shows that when J_s is smaller than a critical value J_s^c the magnetic ground state is obtained from the planar 120° spin structure in the XY plane, by pulling them out of the XY plane by an angle β. The three spins on a triangle on the surface form thus an "umbrella" with an angle α between them and an angle β between a surface spin and its beneath neighbor (see Fig. 9.9). This non planar structure is due to the interaction of the spins on the beneath layer, just like an external applied field in the z direction. Of course, when $|J_s|$ is smaller than $|J_s^c|$ one has the collinear ferromagnetic ground state as expected: the frustration is not strong enough to resist the ferromagnetic interaction from the beneath layer.

We show in Fig. 9.10 $\cos(\alpha)$ and $\cos(\beta)$ as functions of J_s. The critical value J_s^c is found between -0.18 and -0.19. This value can be calculated analytically by assuming the "umbrella structure". For ground state analysis, it suffices to consider just a cell shown in Fig. 9.9. This is justified by the numerical determination discussed above. Furthermore, we consider as

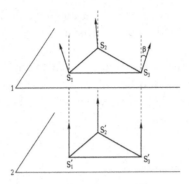

Fig. 9.9   Non collinear surface spin configuration. Angles between spins on layer 1 are all equal (noted by α), while angles between vertical spins are β.

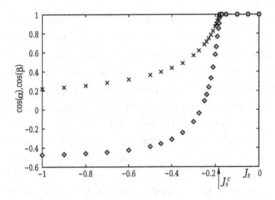

Fig. 9.10 $\cos(\alpha)$ (diamonds) and $\cos(\beta)$ (crosses) as functions of J_s. Critical value of J_s^c is shown by the arrow.

a single solution all configurations obtained from each other by any global spin rotation.

Let us consider the full Hamiltonian (9.62). For simplicity, the interaction inside the surface layer is set equal J_s $(-1 \leq J_s \leq 1)$ and all others are set equal to $J > 0$. Also, we suppose that $I_{i,j} = I_s$ for spins on the surfaces with the same sign as J_s and all other $I_{i,j}$ are equal to $I > 0$ for the inside spins including interaction between a surface spin and the spin on the beneath layer.

The spins are numbered as in Fig. 9.9: S_1, S_2 and S_3 are the spins in the surface layer (first layer), S_1', S_2' and S_3' are the spins in the internal layer (second layer). The Hamiltonian for the cell is written as

$$
\begin{aligned}
H_p = & -6 \left[J_s \left(\mathbf{S}_1 \cdot \mathbf{S}_2 + \mathbf{S}_2 \cdot \mathbf{S}_3 + \mathbf{S}_3 \cdot \mathbf{S}_1 \right) \right. \\
& + I_s \left(S_1^z S_2^z + S_2^z S_3^z + S_3^z S_1^z \right) \\
& + J \left(\mathbf{S}_1' \cdot \mathbf{S}_2' + \mathbf{S}_2' \cdot \mathbf{S}_3' + \mathbf{S}_3' \cdot \mathbf{S}_1' \right) \\
& \left. + I \left(S_1'^z S_2'^z + S_2'^z S_3'^z + S_3'^z S_1'^z \right) \right] \\
& - 2J \left(\mathbf{S}_1 \cdot \mathbf{S}_1' + \mathbf{S}_2 \cdot \mathbf{S}_2' + \mathbf{S}_3 \cdot \mathbf{S}_3' \right) \\
& - 2I \left(S_1^z S_1'^z + S_2^z S_2'^z + S_3^z S_3'^z \right),
\end{aligned} \tag{9.63}
$$

Let us decompose each spin into two components: an xy component, which is a vector, and a z component $\mathbf{S}_i = (\mathbf{S}_i^{\parallel}, S_i^z)$. Only surface spins have xy vector components. The angle between these xy components of nearest neighbor surface spins is $\gamma_{i,j}$ which is chosen by ($\gamma_{i,j}$ is in fact the projection of α defined above on the xy plane)

$$
\gamma_{1,2} = 0, \quad \gamma_{2,3} = \frac{2\pi}{3}, \quad \gamma_{3,1} = \frac{4\pi}{3}. \tag{9.64}
$$

The angles β_i and β_i' of the spin $\mathbf{S}_i$ and $\mathbf{S}_i'$ with the z axis are by symmetry

$$
\begin{cases}
\beta_1 = \beta_2 = \beta_3 = \beta, \\
\beta_1' = \beta_2' = \beta_3' = 0,
\end{cases}
$$

The total energy of the cell (9.63), with $S_i = S_i' = \frac{1}{2}$, can be rewritten as

$$
\begin{aligned}
H_p = & -\frac{9(J+I)}{2} - \frac{3(J+I)}{2} \cos\beta - \frac{9(J_s + I_s)}{2} \cos^2\beta \\
& + \frac{9J_s}{4} \sin^2\beta.
\end{aligned} \tag{9.65}
$$

By a variational method, the minimum of the cell energy corresponds to

$$
\frac{\partial H_p}{\partial \beta} = \left(\frac{27}{2} J_s + 9 I_s \right) \cos\beta \sin\beta + \frac{3}{2}(J+I) \sin\beta = 0 \tag{9.66}
$$

We have

$$
\cos\beta = -\frac{J+I}{9J_s + 6I_s}. \tag{9.67}
$$

For given values of I_s and I, we see that the solution (9.67) exists for $J_s \leq J_s^c$ where the critical value J_s^c is determined by $-1 \leq \cos\beta \leq 1$. For $I = -I_s = 0.1$, $J_s^c \approx -0.1889J$ in excellent agreement with the numerical result.

The classical ground state determined here will be used as input ground state configuration for quantum spins considered in the next section.

9.7.3   Green's function method

Let us consider the quantum spin case. For a given value of J_s, we shall use the Green's function method to calculate the layer magnetizations as functions of temperature. The details of the method in the case of non collinear spin configuration have been given in Ref. [113]. We briefly recall it here and show the application to the present model.

We can rewrite the full Hamiltonian (9.62) in the local framework of the classical ground state configuration as

$$
\mathcal{H} = - \sum_{<i,j>} J_{i,j} \left\{ \frac{1}{4} \left(\cos\theta_{ij} - 1 \right) \left(S_i^+ S_j^+ + S_i^- S_j^- \right) \right.
$$
$$
+ \frac{1}{4} \left(\cos\theta_{ij} + 1 \right) \left(S_i^+ S_j^- + S_i^- S_j^+ \right)
$$
$$
+ \frac{1}{2} \sin\theta_{ij} \left(S_i^+ + S_i^- \right) S_j^z - \frac{1}{2} \sin\theta_{ij} S_i^z \left(S_j^+ + S_j^- \right)
$$
$$
\left. + \cos\theta_{ij} S_i^z S_j^z \right\} - \sum_{<i,j>} I_{i,j} S_i^z S_j^z \qquad (9.68)
$$

where $\cos\left(\theta_{ij}\right)$ is the angle between two nearest-neighbor spins determined classically in the previous section. We define two double-time Green's functions by

$$
G_{ij}(t,t') = \ll S_i^+(t); S_j^-(t') \gg, \qquad (9.69)
$$

$$
F_{ij}(t,t') = \ll S_i^-(t); S_j^-(t') \gg. \qquad (9.70)
$$

The equations of motion for $G_{ij}(t,t')$ and $F_{ij}(t,t')$ read

$$
i\frac{d}{dt} G_{i,j}(t,t') = \left\langle \left[S_i^+(t), S_j^-(t') \right] \right\rangle \delta(t-t')
$$
$$
- \left\langle \left\langle \left[\mathcal{H}, S_i^+(t) \right]; S_j^-(t') \right\rangle \right\rangle, \qquad (9.71)
$$

$$
i\frac{d}{dt} F_{i,j}(t,t') = \left\langle \left[S_i^-(t), S_j^-(t') \right] \right\rangle \delta(t-t')
$$
$$
- \left\langle \left\langle \left[\mathcal{H}, S_i^-(t) \right]; S_j^-(t') \right\rangle \right\rangle, \qquad (9.72)
$$

We neglect higher order correlations by using the Tyablikov decoupling scheme [20] which is known to be valid for exchange terms. Then, we introduce the Fourier transforms

$$G_{i,j}(t,t') = \frac{1}{\Delta} \int\int d\mathbf{k}_{xy} \frac{1}{2\pi} \int_{-\infty}^{+\infty} d\omega e^{-i\omega(t-t')}.$$
$$g_{n,n'}(\omega, \mathbf{k}_{xy}) e^{i\mathbf{k}_{xy}\cdot(\mathbf{R}_i-\mathbf{R}_j)}, \tag{9.73}$$

$$F_{i,j}(t,t') = \frac{1}{\Delta} \int\int d\mathbf{k}_{xy} \frac{1}{2\pi} \int_{-\infty}^{+\infty} d\omega e^{-i\omega(t-t')}.$$
$$f_{n,n'}(\omega, \mathbf{k}_{xy}) e^{i\mathbf{k}_{xy}\cdot(\mathbf{R}_i-\mathbf{R}_j)}, \tag{9.74}$$

where ω is the spin-wave frequency, $\mathbf{k}_{xy}$ denotes the wave-vector parallel to xy planes, $\mathbf{R}_i$ is the position of the spin at the site i, n and n' are respectively the indices of the layers where the sites i and j belong to. The integral over $\mathbf{k}_{xy}$ is performed in the first Brillouin zone whose surface is Δ in the xy reciprocal plane.

The Fourier transforms of the retarded Green's functions satisfy a set of equations rewritten under a matrix form

$$\mathbf{M}(\omega)\mathbf{g} = \mathbf{u}, \tag{9.75}$$

where $\mathbf{M}(\omega)$ is a square matrix $(2N_z \times 2N_z)$, $\mathbf{g}$ and $\mathbf{u}$ are the column matrices which are defined as follows

$$\mathbf{g} = \begin{pmatrix} g_{1,n'} \\ f_{1,n'} \\ \vdots \\ g_{N_z,n'} \\ f_{N_z,n'} \end{pmatrix}, \qquad \mathbf{u} = \begin{pmatrix} 2\langle S_1^z \rangle \delta_{1,n'} \\ 0 \\ \vdots \\ 2\langle S_{N_z}^z \rangle \delta_{N_z,n'} \\ 0 \end{pmatrix}, \tag{9.76}$$

and

$$\mathbf{M}(\omega) = \begin{pmatrix} A_1^+ & B_1 & D_1^+ & D_1^- & \cdots \\ -B_1 & A_1^- & -D_1^- & -D_1^+ & \vdots \\ \vdots & \cdots & \cdots & \cdots & \vdots \\ \vdots & C_{N_z}^+ & C_{N_z}^- & A_{N_z}^+ & B_{N_z} \\ \cdots & -C_{N_z}^- & -C_{N_z}^+ & -B_{N_z} & A_{N_z}^- \end{pmatrix}, \tag{9.77}$$

where

$$
\begin{aligned}
A_n^\pm = \omega \pm \Big[&\frac{1}{2} J_n \langle S_n^z \rangle \, (Z\gamma) \, (\cos \theta_n + 1) \\
&- J_n \langle S_n^z \rangle \, Z \cos \theta_n - J_{n,n+1} \langle S_{n+1}^z \rangle \cos \theta_{n,n+1} \\
&- J_{n,n-1} \langle S_{n-1}^z \rangle \cos \theta_{n,n-1} - Z I_n \langle S_n^z \rangle \\
&- I_{n,n+1} \langle S_{n+1}^z \rangle - I_{n,n-1} \langle S_{n-1}^z \rangle \Big],
\end{aligned}
\tag{9.78}
$$

$$
B_n = \frac{1}{2} J_n \langle S_n^z \rangle \, (\cos \theta_n - 1) \, (Z\gamma),
\tag{9.79}
$$

$$
C_n^\pm = \frac{1}{2} J_{n,n-1} \langle S_n^z \rangle \, (\cos \theta_{n,n-1} \pm 1),
\tag{9.80}
$$

$$
D_n^\pm = \frac{1}{2} J_{n,n+1} \langle S_n^z \rangle \, (\cos \theta_{n,n+1} \pm 1),
\tag{9.81}
$$

in which, $Z = 6$ is the number of in-plane nearest neighbors, $\theta_{n,n\pm1}$ the angle between two nearest neighbor spins belonging to the layers n and $n\pm1$, θ_n the angle between two in-plane nearest neighbors in the layer n, and $\gamma = \left[2\cos(k_x a) + 4\cos(k_x a/2) \cos(k_y a\sqrt{3}/2) \right] /Z$. Here, for compactness we have used the following notations:

i) J_n and I_n are the in-plane interactions. In the present model J_n is equal to J_s for the two surface layers and equal to J for the interior layers. All I_n are set to be I.

ii) $J_{n,n\pm1}$ and $I_{n,n\pm1}$ are the interactions between a spin in the n-th layer and its neighbor in the $(n\pm1)$-th layer. Of course, $J_{n,n-1} = I_{n,n-1}=0$ if $n = 1$, $J_{n,n+1} = I_{n,n+1}=0$ if $n = N_z$.

Solving $\det|\mathbf{M}| = 0$, we obtain the spin-wave spectrum ω of the present system. The solution for the Green's function $g_{n,n}$ is given by

$$
g_{n,n} = \frac{|\mathbf{M}|_n}{|\mathbf{M}|},
\tag{9.82}
$$

where $|\mathbf{M}|_n$ is the determinant made by replacing the n-th column of $|\mathbf{M}|$ by $\mathbf{u}$ in (9.76). Writing now

$$
|\mathbf{M}| = \prod_i \left(\omega - \omega_i \left(\mathbf{k}_{xy} \right) \right),
\tag{9.83}
$$

one sees that $\omega_i \left(\mathbf{k}_{xy} \right)$, $i = 1, \cdots , N_z$, are poles of the Green's function $g_{n,n}$. $\omega_i \left(\mathbf{k}_{xy} \right)$ can be obtained by solving $|\mathbf{M}| = 0$. In this case, $g_{n,n}$ can be expressed as

$$
g_{n,n} = \sum_i \frac{h_n \left(\omega_i \left(\mathbf{k}_{xy} \right) \right)}{\left(\omega - \omega_i \left(\mathbf{k}_{xy} \right) \right)},
\tag{9.84}
$$

where $h_n\left(\omega_i\left(\mathbf{k}_{xy}\right)\right)$ is

$$h_n\left(\omega_i\left(\mathbf{k}_{xy}\right)\right) = \frac{|\mathbf{M}|_n\left(\omega_i\left(\mathbf{k}_{xy}\right)\right)}{\prod_{j \neq i}\left(\omega_j\left(\mathbf{k}_{xy}\right) - \omega_i\left(\mathbf{k}_{xy}\right)\right)}. \tag{9.85}$$

Next, using the spectral theorem which relates the correlation function $\langle S_i^- S_j^+ \rangle$ to the Green's functions [151], one has

$$\langle S_i^- S_j^+ \rangle = \lim_{\varepsilon \to 0} \frac{1}{\Delta} \int\int d\mathbf{k}_{xy} \int_{-\infty}^{+\infty} \frac{i}{2\pi} [g_{n,n'}\left(\omega + i\varepsilon\right)$$
$$- g_{n,n'}\left(\omega - i\varepsilon\right)] \frac{d\omega}{e^{\beta\omega} - 1} e^{i\mathbf{k}_{xy}\cdot(\mathbf{R}_i - \mathbf{R}_j)}, \tag{9.86}$$

where ϵ is an infinitesimal positive constant and $\beta = 1/k_B T$, k_B being the Boltzmann constant. For spin $S = 1/2$, the thermal average of the z component of the i-th spin belonging to the n-th layer is given by

$$\langle S_i^z \rangle = \frac{1}{2} - \langle S_i^- S_i^+ \rangle \tag{9.87}$$

In the following we shall use the case of spin one-half. Note that for the case of general S, the expression for $\langle S_i^z \rangle$ is more complicated since it involves higher quantities such as $\langle (S_i^z)^2 \rangle$.

Using the Green's function presented above, we can calculate self-consistently various physical quantities as functions of temperature T. The first important quantity is the temperature dependence of the angle between each spin pair. This can be calculated in a self-consistent manner at any temperature by minimizing the free energy at each temperature to get the correct value of the angle as it has been done for a frustrated bulk spin systems [118]. Here, we limit ourselves to the self-consistent calculation of the layer magnetizations which allows us to establish the phase diagram as seen in the following.

For numerical calculation, we used $I = 0.1J$ with $J = 1$. For positive J_s, we take $I_s = 0.1$ and for negative J_s, we use $I_s = -0.1$. A size of 80^2 points in the first Brillouin zone is used for numerical integration. We start the self-consistent calculation from $T = 0$ with a small step for temperature 5×10^{-3} or 10^{-1} (in units of J/k_B). The convergence precision has been fixed at the fourth figure of the values obtained for the layer magnetizations.

9.7.4 *Phase transition and phase diagram of the quantum case*

We first show an example where $J_s = -0.5$ in Fig. 9.11. As seen, the surface-layer magnetization is much smaller than the second-layer one. In addition there is a strong spin contraction at $T = 0$ for the surface layer. This is due to the antiferromagnetic nature of the in-plane surface interaction J_s. One sees that the surface becomes disordered at a temperature $T_1 \simeq 0.2557$ while the second layer remains ordered up to $T_2 \simeq 1.522$. Therefore, the system is partially disordered for temperatures between T_1 and T_2. This result is very interesting because it confirms again the existence of the partial disorder in quantum spin systems observed earlier in the bulk [113, 118]. Note that between T_1 and T_2, the ordering of the second layer acts as an external field on the first layer, inducing therefore a small value of its magnetization. A further evidence of the existence of the surface transition will be provided with the surface susceptibility in the Monte Carlo results shown below.

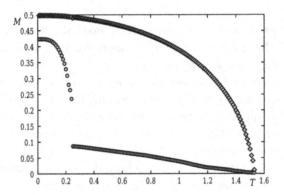

Fig. 9.11   First two layer-magnetizations obtained by the Green's function technique vs. T for $J_s = -0.5$ with $I = -I_s = 0.1$. The surface-layer magnetization (lower curve) is much smaller than the second-layer one. See text for comments.

Figure 9.12 shows the non frustrated case where $J_s = 0.5$, with $I = I_s = 0.1$. As seen, the first-layer magnetization is smaller than the second-layer one. There is only one transition temperature. Note the difficulty for numerical convergency when the magnetizations come close to zero.

We show in Fig. 9.13 the phase diagram in the space (J_s, T). Phase I denotes the ordered phase with surface non collinear spin configuration, phase II indicates the collinear ordered state, and phase III is the paramagnetic phase. Note that the surface transition does not exist for $J_s \geq J_s^c$.

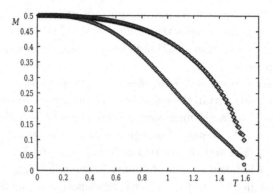

Fig. 9.12   First two layer-magnetizations obtained by the Green's function technique vs. T for $J_s = 0.5$ with $I = I_s = 0.1$.

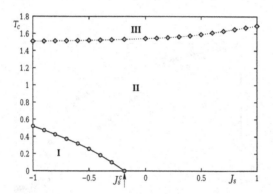

Fig. 9.13   Phase diagram in the space (J_s, T) for the quantum Heisenberg model with $N_z = 4$, $I = |I_s| = 0.1$. See text for the description of phases I to III.

9.7.5   *Monte Carlo results*

It is known that methods for quantum spins, such as the spin-wave theory or the Green's function method presented above, suffer in one way or in another at high temperatures. The spin-wave theory, even with magnon-magnon interactions taken into account, cannot go to temperatures close to T_c. The Green's function method on the other hand can go up to T_c but due to the decoupling scheme, it cannot give a correct critical behavior at T_c. An alternative method for high temperatures is to consider the counterpart classical spins and to use Monte Carlo simulations to obtain the phase diagram for comparison. This is somewhat justified because the quantum nature of spins is no more important at high T.

In this paragraph, we show the results obtained by Monte Carlo simulations with the Hamiltonian (9.62) but the spins are the classical Heisenberg model of magnitude $S = 1$.

The film sizes are $L \times L \times N_z$ where $N_z = 4$ is the number of layers (film thickness) taken as in the quantum case presented above. We use here $L = 24, 36, 48, 60$ to study finite-size effects. Periodic boundary conditions are used in the XY planes. The equilibrating time is about 10^6 Monte Carlo steps per spin and the averaging time is 2×10^6 Monte Carlo steps per spin. $J = 1$ is taken as unit of energy in the following.

Let us show in Fig. 9.14 the layer magnetization of the first two layers as a function of T, in the case $J_s = 0.5$ with $N_z = 4$ (the third and fourth layers are symmetric). Though we observe a smaller magnetization for the surface layer, there is clearly no surface transition just as in the quantum case.

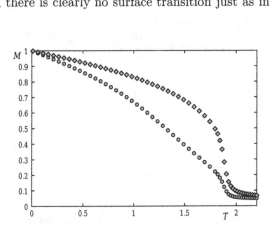

Fig. 9.14 Magnetizations of layer 1 (circles) and layer 2 (diamonds) versus temperature T in unit of J/k_B for $J_s = 0.5$ with $I = I_s = 0.1$, $L = 36$.

In Fig. 9.15 we show a frustrated case where $J_s = -0.5$. The surface layer in this case becomes disordered at a temperature much lower than that for the second layer. Note that the surface magnetization is not saturated to 1 at $T = 0$. This is because the surface spins make an angle with the z axis so their z component is less than 1 in the ground state.

To establish the phase diagram, the transition temperatures are taken at the change of curvature of the layer magnetizations, i. e. at the maxima of layer susceptibilities shown before. Figure 9.16 shows the phase diagram obtained in the space (J_s, T). Interesting enough, this phase diagram re-

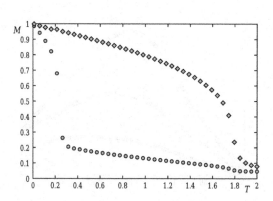

Fig. 9.15 Magnetizations of layer 1 (circles) and layer 2 (diamonds) versus temperature T in unit of J/k_B for $J_s = -0.5$ with $I = -I_s = 0.1$, $L = 36$.

sembles remarkably to that obtained for the quantum counterpart model shown in Fig. 9.13.

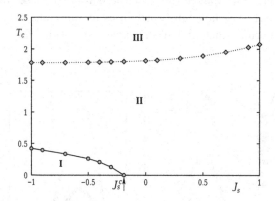

Fig. 9.16 Phase diagram in the space (J_s, T) for the classical Heisenberg model with $N_z = 4$, $I = |I_s| = 0.1$. Phases I to III have the same meanings as those in Fig. 9.13 .

Let us study the finite size effect of the phase transitions shown in Fig. 9.16. To this end we use the histogram technique which has been proved so far to be excellent for the calculation of critical exponents [46, 48]. The principle has been described in chapter 8. Figure 9.17 shows the suscepti-bility versus T for $L = 36, 48, 60$ in the case of $J_s = 0.5$. For presentation convenience, the size $L = 24$ has been removed since the peak for this case is rather flat in the scale of the figure. As seen, the maximum χ^{max} of the susceptibility increases with increasing L.

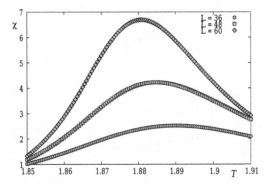

Fig. 9.17   Susceptibility versus T for $L = 36, 48, 60$ with $J_s = 0.5$ and $I = I_s = 0.1$.

In the frustrated case, i.e. $J_s < J_s^c$, we perform the same calculation for finite-size effect. Note that in this case there are two phase transitions. We show in Fig. 9.18 the layer susceptibilities as functions of T for different L. As seen, both surface and second-layer transitions have a strong size dependence.

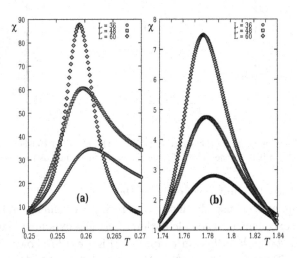

Fig. 9.18   Layer susceptibilities versus T for $L = 36, 48, 60$ with $J_s = -0.5$ and $I = -I_s = 0.1$. Left (right) figure corresponds to the first (second) layer susceptibility.

9.8   Conclusion

We have shown in this chapter how to calculate the magnon spectrum and properties due to magnon excitations in thin ferromagnetic and antiferromagnetic films with several lattice structures. The aim of this chapter is to provide basic theoretical methods to deal with simple models of clean surfaces. However, physical results obtained here such as origin of low surface magnetization and low critical temperature are explained in terms of localized surface magnons. The existence of surface magnons depends on the lattice structure, the surface orientation, the surface exchange interaction, etc. Many aspects of these results remain in real magnetic films where surface conditions are much more complicated. We did not show here all types of surface modes. In general, we can have acoustic surface modes which lie in the low energy region below the bulk magnon band and optical surface modes which lie in the high energy region above the bulk band. The low-lying acoustic modes lower the surface magnetization and the critical temperature as can be seen by examining the formulas shown above. The low surface magnetization corresponds what was called in the 80's "magnetically dead surface". On the other hand, high-energy optical surface modes make the surface magnetization larger than the bulk one. This situation occurs when the surface exchange interactions are much larger than the bulk values. The surface in this case was called "hard surface" [147].

9.9   Problems

Problem 1. Surface magnon:
  Calculate the surface magnon modes in the case of a semi-infinite ferromagnetic crystal of body-centered cubic lattice for $k_x = k_y = 0, \pi/a$ in using the method presented in section 9.4.
Problem 2. Critical next-nearest-neighbor interaction:
  Calculate the critical value of ϵ defined in section 9.4 for an infinite crystal.
Problem 3. Uniform magnetization approximation:
  Show that with the hypothesis of uniform layer-magnetization [Eq. (9.49)], the energy eigenvalue E_i is proportional to M.
Problem 4. Multilayers: critical magnetic field
  One considers a system composed of three films A, B and C, of Ising spins with respective thicknesses N_1, N_2 and N_3. The lattice

sites are occupied by Ising spins pointing in the $\pm z$ direction perpendicular to the films. The interaction between two spins of the same film is ferromagnetic. Let J_1, J_2 and J_3 the magnitudes of these interactions in the three films. One supposes that the interactions at the interfaces $A - B$ and $B - C$ are antiferromagnetic and both equal to J_s. One applies a magnetic field along the z direction. Determine the critical field above which all spins are turned into the field direction. For simplicity, consider the case $J_1 = J_2 = J_3$.

Problem 5. Mean-field theory of thin films:

Calculate the layer magnetizations of a 3-layer film by the mean-field theory (cf. chapter 4). One supposes the Ising spin model with values $\pm 1/2$ and a ferromagnetic interaction J for all pairs of nearest neighbors.

Problem 6. Holstein-Primakoff method:

Using the Holstein-Primakoff method of chapter 5 for a semi-infinite crystal with the Heisenberg spin model, write the expression which allows us to calculate the surface magnetization as a function of temperature. Show that a surface mode of low energy (acoustic surface mode) diminishes the surface magnetization.

Problem 7. Frustrated surface: surface spin rearrangement

Consider a semi-infinite system of Heisenberg spins composed of stacked triangular lattices. Suppose that the interaction between nearest neighbors J is everywhere ferromagnetic except for the spins on the surface: they interact with each other via an antiferromagnetic interaction J_s. Determine the ground state of the system as a function of J_s/J.

Problem 8. Ferrimagnetic film:

Write the equations of motion for a five-layer ferrimagnetic film of body-centered cubic lattice, using the model and the method presented in section 9.4. Consider the cases $k_x = k_y = 0, \pi/a$. Solve numerically these equations to find surface and bulk magnons.

Chapter 10

Monte Carlo Simulation of Spin Transport

The magnetic phase transition is experimentally known to give rise to an anomalous temperature-dependence of the electron resistivity in ferromagnetic crystals. Phenomenological theories based on the interaction between itinerant electron spins and lattice spins have been suggested to explain these observations. In this chapter, we recall these theories and introduce a method of Monte Carlo simulation to study the behavior of the resistivity of the spin current as a function of temperature (T) from low-T ordered phase to high-T paramagnetic phase in ferromagnets and antiferromagnets. Frustrated systems and thin films are also shown. As a demonstration of its efficiency, we show the results obtained for MnTe.

We shall introduce the Boltzmann's equation in the relaxation-time approximation. This part allows us to understand the basic mechanism of the transport phenomenon. The Boltzmann's equation for the spin transport can be solved using numerical data obtained from simulations.

10.1 Boltzmann's equation

Moving particles carry with them physical quantities such as electric current resulting from charge displacements under an electric field, heat current resulting from a transfer of particles carrying thermal energies from high- to low-temperature region etc. It is obvious that a transport phenomenon takes place only when the system is out of equilibrium. Statistical physics at equilibrium (Appendix A) is not applicable. For systems out of equilibrium, the most popular method is the Boltzmann's equation. Monte Carlo simulations are more and more employed in the transport problem. The methods shown in chapter 8 need several adaptations to be operational in the study of the transport.

We consider an electric current in a metal. The current density is given by

$$\vec{j} = -e \int \vec{v} n(\vec{v}) d^3 v \tag{10.1}$$

where $n(\vec{v})$ is the density of electrons moving at the velocity $\vec{v}$ and $-e$ the electron charge. The integral is taken over all velocities. Let n_0 be the electron density at equilibrium. We have

$$n(\vec{v}) = n_0 + \Delta n \tag{10.2}$$

where Δn is the contribution induced by an external cause, for example an applied electric field. It is obvious that in the absence of Δn the integral (10.1) is zero because the integrand is an odd function of $\vec{v}$. We thus have

$$\vec{j} = -e \int \vec{v} \Delta n d^3 v \tag{10.3}$$

To obtain $\vec{j}$, we have to know Δn. The objective for any transport theory is to estimate this quantity. We study here this question with the Boltzmann's equation. We consider in the following only systems not far from equilibrium. Systems far from equilibrium which are very difficult to deal with, are not treated here.

10.1.1 *Classical formulation*

Let $n(x, y, z, p_x, p_y, p_z) dx dy dz dp_x dp_y dp_z$ be the number of particles in the volume $d\vec{r} d\vec{p}$ around the point $(\vec{r}, \vec{p})$ in the classical phase space. The variation of the density of particles in the volume results from two causes: collisions between particles and interaction between particles with an external field. We write

$$\left(\frac{\partial n}{\partial t}\right)_C + \left(\frac{\partial n}{\partial t}\right)_F = \frac{dn}{dt} \tag{10.4}$$

where the indices C and F indicate "collision" and "field". In the regime out of equilibrium but stationary we have $\frac{dn}{dt} = 0$ which leads to

$$\left(\frac{\partial n}{\partial t}\right)_C + \left(\frac{\partial n}{\partial t}\right)_F = 0 \tag{10.5}$$

We explicit now the terms of this equation:

(i) Expression of $\left(\frac{\partial n}{\partial t}\right)_F$:

The particles at (x, y, z, p_x, p_y, p_z) at the time $t + \Delta t$ were at $(x - \dot{x}\Delta t, y - \dot{y}\Delta t, z - \dot{z}\Delta t, p_x - \dot{p}_x\Delta t, p_y - \dot{p}_y\Delta t, p_z - \dot{p}_z\Delta t)$ at the time t. So, we have

$$
\begin{aligned}
(\frac{\partial n}{\partial t})_F &= \lim_{\Delta t \to 0} \frac{n(x - \dot{x}\Delta t, ..., p_x - \dot{p}_x\Delta t, ...) - n(x, ..., p_x, ...)}{\Delta t} \\
&= -(\frac{\partial n}{\partial x})\dot{x} - ... - (\frac{\partial n}{\partial p_x})\dot{p}_x - ... \\
&= -\left[\dot{\vec{r}} \cdot \vec{\nabla}_{\vec{r}} n + \dot{\vec{p}} \cdot \vec{\nabla}_{\vec{p}} n\right] \\
&= -\left[\vec{v} \cdot \vec{\nabla}_{\vec{r}} n + \vec{F} \cdot \vec{\nabla}_{\vec{p}} n\right]
\end{aligned}
\tag{10.6}
$$

where $\vec{F}$ is the force acting on the particles. The first term of the right-hand side is called "term of diffusion" due to the presence of $\vec{v}$.

(ii) Expression of $(\frac{\partial n}{\partial t})_C$:

Let $P_{\vec{p}', \vec{p}}$ be the probability of the transition $\vec{p} \to \vec{p}'$ during a collision. $(\frac{\partial n}{\partial t})_C$ is equal to the difference between the number of particles entering the volume $d\vec{r}d\vec{p}$ and the number of out-going particles. We write

$$
(\frac{\partial n}{\partial t})_C = \int [n(\vec{r}, \vec{p}')P_{\vec{p}, \vec{p}'} - n(\vec{r}, \vec{p})P_{\vec{p}', \vec{p}}] \, d\vec{p}'
\tag{10.7}
$$

The Boltzmann's equation (10.5) becomes

$$
\vec{v} \cdot \vec{\nabla}_{\vec{r}} n + \vec{F} \cdot \vec{\nabla}_{\vec{p}} n = \int [n(\vec{r}, \vec{p}')P_{\vec{p}, \vec{p}'} - n(\vec{r}, \vec{p})P_{\vec{p}', \vec{p}}] \, d\vec{p}'
\tag{10.8}
$$

10.1.2 *Quantum formulation*

We consider the case of electrons. We replace in (10.8) n by a distribution function $f(\vec{r}, \vec{k})$ which depends on the position $\vec{r}$ and on the wave vector $\vec{k} = \vec{p}/\hbar$. This function is the Fermi-Dirac distribution function f_0 when the system is at equilibrium (see Appendix A). We also replace

$$
\begin{aligned}
n(\vec{r}, \vec{p}')P_{\vec{p}, \vec{p}'} \quad &\text{by} \quad f(\vec{r}, \vec{k}')[1 - f(\vec{r}, \vec{k})]P_{\vec{k}, \vec{k}'} \\
n(\vec{r}, \vec{p})P_{\vec{p}', \vec{p}} \quad &\text{by} \quad f(\vec{r}, \vec{k})[1 - f(\vec{r}, \vec{k}')]P_{\vec{k}', \vec{k}}
\end{aligned}
$$

The factor $[1 - f(\vec{r}, \vec{k})]$ is introduced to make sure that the state $f(\vec{r}, \vec{k})$ is empty, namely $f(\vec{r}, \vec{k}) = 0$, so that the transition $\vec{k}' \to \vec{k}$ is possible. The same thing is for the inverse transition. The Boltzmann's equation in the quantum case is thus written as

$$\frac{\hbar \vec{k}}{m^*} \cdot \vec{\nabla}_{\vec{r}} f(\vec{k}) + \frac{\vec{F}}{\hbar} \cdot \vec{\nabla}_{\vec{k}} f(\vec{k}) = \frac{\Omega}{(2\pi)^3} \int [f(\vec{k}')(1 - f(\vec{k})) P_{\vec{k},\vec{k}'}$$
$$-f(\vec{k})(1 - f(\vec{k}')) P_{\vec{k}',\vec{k}}] d\vec{k}' \qquad (10.9)$$

where m^* is the electron effective mass and Ω the system volume. For simplicity, $f(\vec{r}, \vec{k})$ was noted by $f(\vec{k})$ and for a large system the sum over all $\vec{k}'$ was replaced by an integral.

At equilibrium, the right-hand side of (10.9), $(\frac{\partial n}{\partial t})_C$, is zero, we have $f(\vec{k}) = f_0(\vec{k})$:

$$f_0(\vec{k}) = \frac{1}{e^{\beta(E_{\vec{k}} - \mu)} + 1}$$

The fluxes of in-going and out-going particles are equal at equilibrium, we have

$$f_0(\vec{k}')(1 - f_0(\vec{k})) P_{\vec{k},\vec{k}'} = f_0(\vec{k})(1 - f_0(\vec{k}')) P_{\vec{k}',\vec{k}}$$
$$P_{\vec{k},\vec{k}'} \left(\frac{1}{e^{\beta(E_{\vec{k}'} - \mu)} + 1} \right) \left(1 - \frac{1}{e^{\beta(E_{\vec{k}} - \mu)} + 1} \right) = P_{\vec{k}',\vec{k}} \left(\frac{1}{e^{\beta(E_{\vec{k}} - \mu)} + 1} \right)$$
$$\times \left(1 - \frac{1}{e^{\beta(E_{\vec{k}'} - \mu)} + 1} \right)$$
$$P_{\vec{k},\vec{k}'} e^{-\beta E_{\vec{k}'}} = P_{\vec{k}',\vec{k}} e^{-\beta E_{\vec{k}}} \qquad (10.10)$$

For an elastic collision, namely $E_{\vec{k}'} = E_{\vec{k}}$, we have $P_{\vec{k},\vec{k}'} = P_{\vec{k}',\vec{k}}$. The right-hand side of (10.9) is reduced to

$$\left(\frac{\partial f}{\partial t}\right)_C = \frac{\Omega}{(2\pi)^3} \int P_{\vec{k}',\vec{k}} \left[f(\vec{k}')(1 - f(\vec{k})) - f(\vec{k})(1 - f(\vec{k}')) \right] d\vec{k}'$$
$$= \frac{\Omega}{(2\pi)^3} \int P_{\vec{k}',\vec{k}} \left[f(\vec{k}') - f(\vec{k}) \right] d\vec{k}'$$
$$= \frac{\Omega}{(2\pi)^3} \int P_{\vec{k}',\vec{k}} \left[\varphi(\vec{k}') - \varphi(\vec{k}) \right] d\vec{k}' \qquad (10.11)$$

where

$$f(\vec{k}) \equiv f_0(\vec{k}) + \varphi(\vec{k}) \qquad (10.12)$$

$\varphi(\vec{k})$ represents the out-of-equilibrium contribution.

The Boltzmann's equation in its general form (10.9) is not possible to solve. When the system is not far from equilibrium, we can linearize and solve it without difficulty as seen below.

10.2 Linearized Boltzmann's equation

10.2.1 *Explicit linearized terms*

We linearize (10.9) term by term as follows.

1. If the external force $\vec{F}$ is weak, f is not far from f_0, then

$$\vec{\nabla}_{\vec{k}} f \simeq \vec{\nabla}_{\vec{k}} f_0 = \frac{\partial f_0}{\partial E} \vec{\nabla}_{\vec{k}} E_{\vec{k}}$$

$$= \frac{\partial f_0}{\partial E} \frac{\hbar^2 \vec{k}}{m^*} \tag{10.13}$$

2. The term $\vec{\nabla}_{\vec{r}} f$ is linearized as

$$\vec{\nabla}_{\vec{r}} f \simeq \vec{\nabla}_{\vec{r}} f_0 = \frac{\partial f_0}{\partial T} \vec{\nabla}_{\vec{r}} T \tag{10.14}$$

We have

$$\frac{\partial f_0}{\partial T} = \frac{\partial f_0}{\partial \beta} \frac{\partial \beta}{\partial T} + \frac{\partial f_0}{\partial \mu} \frac{\partial \mu}{\partial T}$$

$$= \frac{1}{k_B T^2} (E_{\vec{k}} - \mu) \frac{e^{\beta(E_{\vec{k}} - \mu)}}{(e^{\beta(E_{\vec{k}} - \mu)} + 1)^2} - \frac{\partial f_0}{\partial E_{\vec{k}}} \frac{\partial \mu}{\partial T}$$

$$= -\frac{(E_{\vec{k}} - \mu)}{T} \left[-\beta f_0 (1 - f_0) \right] - \frac{\partial f_0}{\partial E_{\vec{k}}} \frac{\partial \mu}{\partial T}$$

$$= -\left[\frac{E_{\vec{k}} - \mu}{T} + \frac{\partial \mu}{\partial T} \right] \frac{\partial f_0}{\partial E_{\vec{k}}} \tag{10.15}$$

where we have used the identity $\frac{\partial f_0}{\partial \mu} = -\frac{\partial f_0}{\partial E_{\vec{k}}}$. We finally have

$$\vec{\nabla}_{\vec{r}} f \simeq -\frac{\partial f_0}{\partial E_{\vec{k}}} \left[\frac{E_{\vec{k}} - \mu}{T} + \frac{\partial \mu}{\partial T} \right] \vec{\nabla}_{\vec{r}} T \tag{10.16}$$

Equation (10.9) becomes

$$\left(\frac{\partial f}{\partial t} \right)_C = -\left(\frac{\partial f}{\partial t} \right)_F \simeq \frac{\hbar \vec{k}}{m^*} \cdot \frac{\partial f_0}{\partial E_{\vec{k}}} \vec{A} \tag{10.17}$$

where

$$\vec{A} = -e\vec{\varepsilon} - \left[\frac{E_{\vec{k}} - \mu}{T} + \frac{\partial \mu}{\partial T} \right] \vec{\nabla}_{\vec{r}} T \tag{10.18}$$

Note that we have replaced $\vec{F} = -e\vec{\varepsilon}$, $\vec{\varepsilon}$ being the applied electric field. We can also write $\vec{A}$ under the form:

$$\vec{A} = -e\vec{\varepsilon} - (\frac{E_{\vec{k}} - \mu}{T})\vec{\nabla}_{\vec{r}}T - \vec{\nabla}_{\vec{r}}\mu \tag{10.19}$$

Equation (10.17) is the linearized Bolzmann's equation. For $(\frac{\partial f}{\partial t})_C$, we introduce the relaxation-time approximation shown below.

10.2.2 *Relaxation-time approximation*

If the external perturbation is weak, we can replace $(\frac{\partial f}{\partial t})_C$ by

$$(\frac{\partial f}{\partial t})_C \simeq -\frac{f - f_0}{\tau} = -\frac{\varphi}{\tau} \tag{10.20}$$

where τ is the relaxation time between two collisions and (10.12) has been used. To be valid, τ should be independent of the external cause. We have to verify this condition when using this approximation. Writing

$$(\frac{\partial f}{\partial t})_C = -\frac{\varphi}{\tau} \equiv +\frac{\phi}{\tau}\frac{\partial f_0}{\partial E} \tag{10.21}$$

we obtain the linearized Boltzmann's equation in the relaxation-time approximation:

$$\frac{\hbar\vec{k}}{m^*} \cdot \frac{\partial f_0}{\partial E}\vec{A} = \frac{\phi}{\tau}\frac{\partial f_0}{\partial E} \tag{10.22}$$

where $\vec{A}$ is given by (10.18). We have

$$\phi = \tau\vec{A} \cdot \frac{\hbar\vec{k}}{m^*} = \tau\vec{A} \cdot \vec{v} \tag{10.23}$$

Note that when we apply a magnetic field $\vec{B}$, the force becomes $\vec{F} = -e[\vec{\varepsilon} + \vec{v} \wedge \vec{B}]$. We cannot therefore replace f by f_0 in $\frac{\vec{F}}{\hbar} \cdot \vec{\nabla}_{\vec{k}}f$ because the field $\vec{B}$ will not appear in the final equation. To see the effect of $\vec{B}$, we have to go to the second order: we have to replace f by $f_0 + \varphi$, not by f_0. The solution (10.23) becomes (see demonstration in Problem 1 in section 10.6)

$$\phi = \tau\vec{A} \cdot \vec{v} + \tau\frac{e}{\hbar^2}\vec{B} \cdot [\vec{\nabla}_{\vec{k}}\phi \wedge \vec{\nabla}_{\vec{k}}E] \tag{10.24}$$

This equation is a differential equation of ϕ because of $\vec{B}$.

10.3 Spin-independent transport — Ohm's law

We consider the effect of an electric field: we assume in (10.19) that $\vec{B} = 0$, $\vec{\nabla}_{\vec{r}} T = 0$ (no temperature gradient) and $\vec{\nabla}_{\vec{r}} \mu = 0$ (no gradient of chemical potential). From (10.24), we have

$$\phi = -\tau e \vec{\varepsilon} \cdot \vec{v} \tag{10.25}$$

We obtain

$$f = f_0 + \varphi = f_0 - \phi \frac{\partial f_0}{\partial E}$$

$$= f_0 - \tau e \vec{\varepsilon} \cdot \vec{v} \frac{\partial f_0}{\partial E}$$

$$= f_0 - \tau e \varepsilon_z v_z \frac{\partial f_0}{\partial E} \tag{10.26}$$

where we supposed $\vec{\varepsilon} \parallel \vec{O}z$. The current density ($\Omega = 1$, volume unit) in the z direction is

$$j_z = -e \int \varphi v_z d^3 v = e \int \phi \frac{\partial f_0}{\partial E} v_z d^3 v$$

$$= -e^2 \varepsilon_z \int d^3 v v_z \tau \frac{\partial f_0}{\partial E}$$

or, in quantum version, $\int d^3 v ... = \frac{2\Omega}{(2\pi)^3} \int d\vec{k} ...$ (factor 2 is spin degeneracy),

$$j_z = -e^2 \varepsilon_z \frac{2\Omega}{(2\pi)^3} \int \tau \frac{\hbar}{m^*} d\vec{k} \hbar k_z m^* \frac{\partial f_0}{\partial E}$$

$$= -\frac{e^2 \varepsilon_z \hbar^2}{(m^*)^2} \frac{\Omega}{4\pi^3} \int \tau 2\pi k^2 dk \frac{\partial f_0}{\partial E} \int_0^\pi k^2 \cos^2 \theta \sin \theta d\theta$$

$$= -\frac{e^2 \varepsilon_z \hbar^2}{(m^*)^2} \frac{\Omega}{4\pi^3} \int \tau 2\pi k^4 dk \frac{\partial f_0}{\partial E} [-\frac{\cos^3 \theta}{3}]_0^\pi$$

$$= -\frac{e^2 \varepsilon_z \hbar^2}{(m^*)^2} \frac{\Omega}{4\pi^3} \frac{2}{3} \int \tau 2\pi k^4 dk \frac{\partial f_0}{\partial E} \tag{10.27}$$

Transforming this integral into an integral on E, $\frac{\Omega}{4\pi^3} \int 4\pi k^2 dk ... = \int dE \rho(E) ...$, we get

$$j_z = -\frac{e^2 \varepsilon_z \hbar^2}{(m^*)^2} \frac{1}{3} \int \tau \rho(E) (\frac{2m^* E}{\hbar^2}) dE \frac{\partial f_0}{\partial E}$$

$$= -\frac{2e^2 \varepsilon_z}{3m^*} \int \tau \rho(E) E \frac{\partial f_0}{\partial E} dE \tag{10.28}$$

where $\rho(E)$ is the density of states given by (A.41) with $\Omega = 1$.

We suppose now that the electric field is weak, τ does not depend on the energy E of the electron. We can replace τ by an average value $< \tau >$. Using the expression (A.41) for $\rho(E)$, and integrating by parts, we obtain (see demonstration in Problem 2 in section 10.6)

$$j_z = \frac{e^2 \varepsilon_z < \tau >}{m^*} n = \sigma \varepsilon_z \qquad (10.29)$$

where n is the number of electrons per volume unit [see (2.1)] and

$$\sigma \equiv \frac{e^2 < \tau >}{m^*} \qquad (10.30)$$

σ is called "electric conductivity". Equation (10.29) is known as the "Ohm's law" which can be found by a simpler calculation. However, the method using the solution of the Boltzmann's equation allows us to treat the case of strong fields or when τ depends on E. In these cases, the Ohm's law is no more valid.

In the general case when the electric field $\vec{\varepsilon}$ is applied in an arbitrary direction the conductivity is a tensor. We write the current density in the i direction as

$$\begin{aligned} j_i = -e \int \varphi v_i d^3v &= e \int \phi \frac{\partial f_0}{\partial E} v_i d^3v \\ &= -e^2 \int d^3v \tau v_i \sum_j v_j \varepsilon_j \frac{\partial f_0}{\partial E} \\ &\equiv \sum_{j=x,y,z} \sigma_{ij} \varepsilon_j \end{aligned} \qquad (10.31)$$

where

$$\sigma_{ij} = -\frac{e^2}{4\pi^3} \int \frac{\partial f_0}{\partial E} \tau v_i v_j d\vec{k} \qquad (10.32)$$

10.4 Spin resistivity

The study of the resistivity is one of the fundamental tasks in materials science. This is because the transport properties are used in electronic devices and applications. The resistivity has been studied since the discovery of the electron a century ago by the simple Drude theory using the classical free particle model with collisions due to atoms in the crystal. The following relation is established between the conductivity σ and the electronic parameters e (charge) and m (mass):

$$\sigma = \frac{ne^2\tau}{m} \tag{10.33}$$

where τ is the electron relaxation time, namely the average time between two successive collisions. In more sophisticated treatments of the resistivity where various interactions are taken into account, this relation is still valid provided two modifications (i) the electron mass is replaced by its effective mass which includes various effects due to interactions with its environment (ii) the relaxation time τ is not a constant but dependent on collision mechanisms. The first modification is very important, the electron can have a "heavy" or "light" effective mass which modifies its mobility in crystals. The second modification has a strong impact on the temperature dependence of the resistivity: τ depends on some power of the electron energy, this power depends on the diffusion mechanisms such as collisions with charged impurities, neutral impurities, magnetic impurities, phonons, magnons, etc. As a consequence, the relaxation time averaged over all energies, $<\tau>$, depends differently on T according to the nature of the collision source. The properties of the total resistivity stem thus from different kinds of diffusion processes. Each contribution has in general a different temperature dependence. Let us summarize the most important contributions to the total resistivity $\rho_t(T)$ at low temperatures in the following expression

$$\rho_t(T) = \rho_0 + A_1 T^2 + A_2 T^5 + A_3 \ln \frac{\mu}{T} \tag{10.34}$$

where A_1, A_2 and A_3 are constants. The first term is T-independent, the second term proportional to T^2 represents the scattering of itinerant spins at low T by lattice spin-waves. Note that the resistivity caused by a Fermi liquid is also proportional to T^2. The T^5 term corresponds to a low-T resistivity in metals. This is due to the scattering of itinerant electrons by phonons. Note that at high T, metals show a linear-T dependence. The logarithm term is the resistivity due to the quantum Kondo effect caused by a magnetic impurity at very low T.

We are interested here in the spin resistivity ρ of magnetic materials. This subject has been investigated intensively both experimentally and theoretically for more than five decades. The rapid development of the field is due mainly to many applications in particular in spintronics.

10.4.1   *Experiments*

Experiments have been performed to investigate many magnetic materials including metals, semiconductors and superconductors. One of the interesting aspects of magnetic materials is the effect on the resistivity behavior due to the magnetic phase transition from a magnetically ordered phase to the paramagnetic state. Very recent experiments such as those performed on the following compounds show different forms of anomaly of the magnetic resistivity at the magnetic phase transition temperature: ferromagnetic $SrRuO_3$ thin films [145], Ru-doped induced ferromagnetic $La_{0.4}Ca_{0.6}MnO_3$ [87], antiferromagnetic ϵ-$(Mn_{1-x}Fe_x)_{3.25}Ge$ [43], semiconducting $Pr_{0.7}Ca_{0.3}MnO_3$ thin films [149], superconducting $BaFe_2As_2$ single crystals [137], and $La_{1-x}Sr_xMnO_3$ [119]. Depending on the material, ρ can show a sharp peak at the magnetic transition temperature T_c [93] or just only a change of its slope, or an inflexion point. The latter case gives rise to a peak of the differential resistivity $d\rho/dT$ [110, 121].

10.4.2   *Theories*

As for theories, the T^2 magnetic contribution in Eq. (10.34) has been obtained from the magnon scattering by Kasuya [74]. The theory was made by taking into account the exchange interaction between s and d electrons for transition metals, or s and f electrons for rare-earth metals. Using a mean-field treatment, the following expression was obtained

$$\rho = (\frac{3\pi m^{*2}}{Ne^2\hbar^2})(S - \sigma)(S + \sigma + 1)\frac{J_{eff}^2}{\varepsilon_0} \text{ Ohm} \qquad (10.35)$$

where $\sigma = < S^z >$ and J_{eff} is the effective exchange interaction. This relation shows that the temperature dependence of ρ is due to σ for $T < T_c$. For $T > T_c$, ρ is a constant because $\sigma = 0$ (see mean-field theory in chapter 4). Obviously, due to the mean-field treatment, this result does not correctly describe the behavior of ρ in particular at and around T_c.

De Gennes and Friedel have related [31] the spin resistivity to the spin-spin correlation function. The magnetic resistivity should therefore behave more or less as the magnetic susceptibility: it should diverge at T_c. Let us consider a system composed of interacting spins on a lattice. The conduction spin $\vec{S}_p$ interacts with the lattice spins $\vec{S}_R$ via a contact interaction G

$$H_I = \sum_{R,p} G\delta(R - R_p)\vec{S}_R \cdot \vec{S}_p \qquad (10.36)$$

The result of the model in the Ornstein-Zernike approximation for the relaxation time with the sum in the correlation function taken over all R shows a cusp of the resistivity at T_c. Below T_c, by neglecting the correlation between spins (because the Fermi wave length is rather short), one has

$$\frac{\tau_0}{\tau} = 1 - \frac{<S^z>^2}{S(S+1)} \tag{10.37}$$

This result shows the importance of the long-range ordering through the quantity $<S^z>^2$. We show in Fig. 10.1 the function $\frac{\tau_0}{\tau}$, which is proportional to ρ, with $<S^z>$ calculated by using, for example, Eq. (6.41) of the Green's function method. We observe the T^2 behavior at low T due to the scattering by spin-waves. If the spin-spin correlation is to be taken

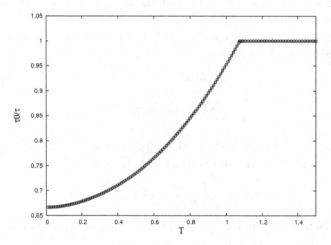

Fig. 10.1 The relaxation time $\frac{\tau_0}{\tau}$ versus T/T_c, with $S = 1/2$.

into account to any distance, then the following relaxation time for $T > T_c$ is obtained [31]

$$\frac{\tau_0}{\tau} = \frac{1}{4} \int_0^2 \frac{x^3 dx}{1 - \frac{T_c}{T} \frac{\sin(k_0 d\, x)}{(k_0 d\, x)}} \tag{10.38}$$

where $x^2 = 2(1 - \cos\theta)$, d is the lattice constant and k_0 the wave vector of the conduction electron. This function is shown in Fig. 10.2 for various values of the correlation range $k_0 d$. It indicates a decreasing $\frac{\tau_0}{\tau}$

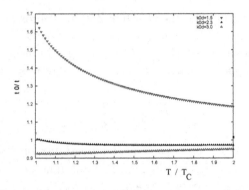

Fig. 10.2 $\dfrac{\tau_0}{\tau}$ versus T above T_c for different values of $k_0 d$.

with increasing T above T_c. Note that the maximum at T_c decreases with decreasing correlation length.

Fisher and Langer [51] have introduced a screening function in order to reduce the range of the correlation length as follows

$$p(R) = e^{-R/l} \tag{10.39a}$$

$$\frac{\tau_0}{\tau} = \sum_{s=0}^{\infty} \nu_s f(R_s) p(R_s) \Gamma(R_s) \tag{10.39b}$$

where l is the mean free path, ν_s the number of spins in the sth sphere of radius R_s, and $f(R_s)$ denotes the following decreasing function

$$f(R) = \frac{1}{4k_F^4 R} \frac{d^2}{dR^2} \left[\frac{\cos(2k_F R) - 1}{R} \right] \tag{10.40}$$

The last factor $\Gamma(R_s)$ in Eq. (10.39b) is the real-space spin-spin correlation function. The relaxation time is given by

$$\frac{\tau_0}{\tau} = \Gamma(0, T) + \frac{1}{8k_F^4} \int_0^{2k_F} \hat{\Gamma}(K, T) K^3 dK \tag{10.41}$$

$$\hat{\Gamma}(K, T) = \sum_{\vec{R}_i \neq 0} \Gamma(\vec{R}_i, T) p(\vec{R}_i) \exp(i\vec{K} \cdot \vec{R}_i) \tag{10.42}$$

The function $\hat{\Gamma}(K, T)$ is given by $\simeq 1/K^{2-\eta}$ (η being the usual critical exponent). Its singularity is overwhelmed by the factor K^3 in the above integral so that the divergence of $\frac{\tau_0}{\tau}$ is suppressed.

$\hat{\Gamma}(K, T)$ is shown in Fig. 10.3 for various values of Ka (K: wave vector, a: lattice constant). For small Ka (long-range correlation), the peak tends to infinity, while for large Ka (short range) the peak is reduced and rounded.

Note that Fisher and Langer have also suggested in their paper that above T_c, the short-range correlation dominates so that ρ behaves as the internal energy. Consequently, the resistivity derivative $d\rho/dT$ behaves as the heat capacity with a sharp peak at T_c. This picture seems to be consistent with what observed in some materials such as Ni [51].

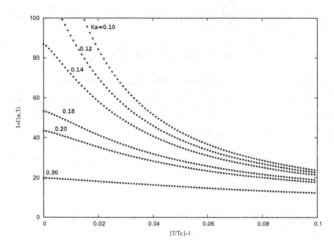

Fig. 10.3 Spin-spin correlation function $\hat{\Gamma}(K, T)$ versus T for different values of Ka.

We introduce now the model of Kataoka [75] by following his paper. The Hamiltonian is given by

$$H_e = \sum_{\vec{k},\sigma} \epsilon_{\vec{k},\sigma} a^+_{\vec{k},\sigma} a_{\vec{k},\sigma} - \frac{J_{sd}}{N} \sum_{\vec{k},\sigma} [\ \overline{S}_z(\vec{q})(a^+_{\vec{k}+\vec{q}\uparrow} a_{\vec{k}\uparrow}$$

$$-a^+_{\vec{k}+\vec{q}\downarrow} a_{\vec{k}\downarrow}) + S_-(\vec{q})a^+_{\vec{k}+\vec{q}\uparrow} a_{\vec{k}\downarrow} + S_+(\vec{q})a^+_{\vec{k}+\vec{q}\downarrow} a_{\vec{k}\uparrow}] \quad (10.43)$$

where J_{sd} represents the $s-d$ exchange interaction, $S(\vec{q})$ the Fourier transform of the lattice spin $\vec{S}_i$, $\sigma_z = \pm 1$ denotes the itinerant spin, and $\epsilon_{k,\sigma}$ is the mean-field electron energy given by

$$\epsilon_{\vec{k},\sigma} = \frac{\hbar^2 k^2}{2m^*_\sigma} - J_{sd} < S_z > \sigma_z - \frac{1}{2} g\mu_B H_{ext}\sigma_z. \quad (10.44)$$

and

$$\vec{S}(\vec{q}) = \sum_i \vec{S}_i \exp(-i\vec{q} \cdot \vec{R}_i) \quad (10.45)$$

$$\overline{S}_z(\vec{q}) = S_z(\vec{q}) - < S_z(0) > \delta_{\vec{q},0} \quad (10.46)$$

$$S_\pm = S_x(\vec{q}) \pm i S_y(\vec{q}) \quad (10.47)$$

The collision term of the Boltzmann's equation is given by

$$
(\frac{df_\sigma(\vec{k})}{dt})_c = (\frac{J_{sd}^2}{\hbar N^2}) \sum_{\vec{k}'} [f_\sigma(\vec{k}')(1 - f_\sigma(\vec{k}))W(\vec{k}'\sigma, \vec{k}\sigma)
$$

$$
-f_\sigma(\vec{k})(1 - f_\sigma(\vec{k}'))W(\vec{k}\sigma, \vec{k}'\sigma)]
$$

$$
+(\frac{J_{sd}^2}{\hbar N^2}) \sum_{\vec{k}'} [f_{\sigma^-}(\vec{k}')(1 - f_\sigma(\vec{k}))W(\vec{k}'\sigma^-, \vec{k}\sigma)
$$

$$
-f_\sigma(\vec{k})(1 - f_{\sigma^-}(\vec{k}'))W(\vec{k}\sigma, \vec{k}'\sigma^-)] \tag{10.48}
$$

where f is the electron Fermi distribution function and W the transition probability. The first sum corresponds to the spin-conserved collisions and the second one to the spin-flip collisions. The corresponding transition probabilities are expressed in terms of the spin-spin correlation as follows

$$
W(\vec{k}\sigma, \vec{k}'\sigma) = (\frac{J_{sd}^2}{\hbar N^2}) \int <\overline{S}_z(\vec{k} - \vec{k}', 0)\overline{S}_z(\vec{k}' - \vec{k}, t) > e^{iT_{\sigma\sigma}(\vec{k}, \vec{k}')t} dt
$$

$$
\tag{10.49a}
$$

$$
W(\vec{k}\sigma, \vec{k}'\sigma^-) = (\frac{J_{sd}^2}{\hbar N^2}) \int <S_{\sigma^-\sigma}(\vec{k} - \vec{k}', 0)S_{\sigma\sigma^-}(\vec{k}' - \vec{k}, t) > e^{iT_{\sigma\sigma^-}(\vec{k}, \vec{k}')t} dt
$$

$$
\tag{10.49b}
$$

where $\sigma^- = -\sigma$ and $T_{\sigma\sigma^-}(\vec{k}, \vec{k}') = \epsilon_{\vec{k}\sigma} - \epsilon_{\vec{k}'\sigma^-}$ is the transferred energy during the transition $|\vec{k}\sigma\rangle \to |\vec{k}'\sigma^-\rangle$ and $S_{\downarrow\uparrow}$ (resp. $S_{\uparrow\downarrow}$) represents S_- (resp. S_+).

Using the relaxation-time approximation, Kataoka obtained

$$
\rho = \frac{3}{2e^2} [\frac{<< \tau_\uparrow >>}{m_\uparrow^*} + \frac{<< \tau_\downarrow >>}{m_\downarrow^*}]^{-1} \tag{10.50}
$$

where the relaxation times have been averaged over the two bands separated by $2\Delta = J_{sd} < S_z > + g\mu_B H_{ext}/2$.

Now, in order to see the effect of the magnetic instability, an antiferromagnetic interaction J_2 between the next nearest neighbors is introduced:

$$
H_{spin} = -\frac{1}{N} \sum_{\vec{q}} [\overline{S}_z(\vec{q})\overline{S}_z(-\vec{q}) + \frac{1}{2}[S_+(\vec{q})S_-(-\vec{q}) + S_-(\vec{q})S_+(-\vec{q})]]
$$

$$
- \sum_i [2J^{eff}(0) < S_z > + H_{ext}]S_{iz}
$$

where

$$
J^{eff}(q) = J_1 \cos(\frac{aq}{2}) - J_2 \cos(aq) \tag{10.51}
$$

J_1 and J_2 are positive, a is the lattice constant, $\vec{q}$ the wave vector. For a simple cubic lattice, the ferromagnetic state is no more stable for $J_2/J_1 > x_c = 1/4$ (see section 10.4.7). For convenience, we define the following stability parameter

$$\xi \equiv \frac{1}{\sqrt{2}} \frac{\sqrt{J_1 - 4J_2}}{J_1 - J_2}$$

When ξ is small, the ferromagnetic state becomes unstable. The correlation can be calculated and the resistivity is finally obtained by solving the Boltzmann's equation for $T > T_c$:

$$\rho = \rho_0 S(S + 1)(Q/\epsilon_F)r(t)$$
$$r(t) = (t - t_c + t_s)[1 - t\ln(1 + t^{-1})]$$
$$t = t_s(\frac{T}{T_c} - 1) + t_c$$
$$t_s = (2ak_F\xi)^{-2}$$
$$t_c = (2k_F l_0)^{-2}$$

where Q is the conduction band width, $t_s(\xi)$ a function of the ferromagnetic stability and $t_c(l_0)$ expresses the effect of the mean free path l_0. The effects of the conduction electron density n [which is contained in k_F: $k_F \propto n^{1/3}$, see (2.6)], the ordering stability ξ and the mean free path l_0 are shown in Figs. 10.4, 10.5 and 10.6, respectively: the peak of the resistivity R (Kataoka's notation) decreases with increasing n, it also decreases with increasing ξ, but it increases with increasing l_0. Note that in Fig. 10.5 the peak moves to a higher temperature when the stability ξ increases: this is easily understood because the system needs a higher temperature to destroy the more stable ordering.

The resistivity due to magnetic impurities has been calculated by Zarand et al. [148] as a function of the Anderson's localization length. This parameter expresses in fact a kind of the correlation sphere induced around each impurity. Their result shows the existence of a resistivity peak which is a consequence caused by the lattice magnetic transition. The peak is gradually reduced and pushed towards higher temperatures with an increasing applied magnetic field, in agreement with experiments.

After examining different theoretical possibilities, we see that the shape of the resistivity at T_c depends on several parameters. Such a situation has also been experimentally observed in various materials.

Let us consider numerical investigations on the spin transport. There has been recently a number of Monte Carlo simulations in the literature

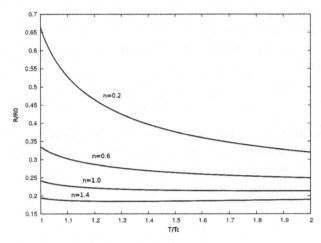

Fig. 10.4   Renormalized resistivity R/R_0 versus T ($> T_c$) for various conducting electron densities.

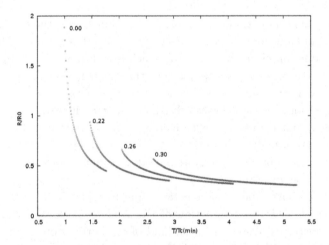

Fig. 10.5   Renormalized resistivity R/R_0 versus T for various values of ξ. See text for comments.

dealing with the spin resistivity. These works have studied the spin current in ferromagnetic [1–3] and antiferromagnetic [4, 88–90] materials. The behavior of ρ as a function of T has been shown to be in qualitative agreement with main experimental features and theoretical results mentioned above.

In the following, we give a review of these works, outline the most important aspects and results. We consider in some details the case of

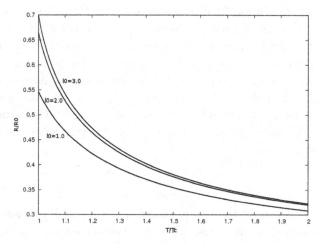

Fig. 10.6 Renormalized resistivity R/R_0 versus T for various l_0.

MnTe where simulation is in excellent agreement with experiments.

10.4.3 *Model for spin transport in Monte Carlo simulation*

The model used in the following Monte Carlo simulation is very general. The itinerant spins move in a crystal whose lattice sites are occupied by localized spins. The itinerant spins and the localized spins may be of Ising, XY or Heisenberg models. Their interaction is usually limited to nearest neighbors (NN) but this assumption is not necessary. It can be ferromagnetic or antiferromagnetic.

The objective is to study the effect of the magnetic transition on ρ. This transition occurs at a high temperature where it is known that the quantum nature of itinerant electron spins does not make significant additional effects with respect to the classical spin model. Therefore, to simplify the task, we consider here the classical spin model.

10.4.4 *Interactions*

We consider a crystal of a given lattice structure where each lattice site is occupied by a spin. The interaction between the lattice spins is given by the following Hamiltonian :

$$\mathcal{H}_l = -\sum_{(i,j)} J_{i,j}\mathbf{S}_i \cdot \mathbf{S}_j \tag{10.52}$$

where $\mathbf{S}_i$ is the spin localized at lattice site i of Ising, XY or Heisenberg model, $J_{i,j}$ the exchange integral between the spin pair $\vec{S}_i$ and $\vec{S}_j$ which is not limited to the interaction between nearest-neighbors (NN). Hereafter, except otherwise stated, we take $J_{i,j} = J$ for NN spin pairs, for simplicity. $J > 0$ (< 0) denotes ferromagnetic (antiferromagnetic) interaction. The system size is $L_x \times L_y \times L_z$ where $L_i(i = x, y, z)$ is the number of lattice cells in the i direction. Periodic boundary conditions (PBC) are used in all directions. We define the interaction between itinerant spins and localized lattice spins as follows

$$\mathcal{H}_r = -\sum_{i,j} I_{i,j} \sigma_i \cdot \mathbf{S}_j \tag{10.53}$$

where σ_i is the spin of the i-th itinerant electron and $I_{i,j}$ denotes the interaction which depends on the distance between electron i and spin $\mathbf{S}_j$ at lattice site j. For simplicity, we suppose the following interaction expression

$$I_{i,j} = I_0 e^{-\alpha r_{ij}} \tag{10.54}$$

where $r_{ij} = |\vec{r}_i - \vec{r}_j|$, I_0 and α are constants. We use a cut-off distance D_1 for the above interaction. In the same way, interaction between itinerant electrons is defined by

$$\mathcal{H}_m = -\sum_{i,j} K_{i,j} \sigma_i \cdot \sigma_j \tag{10.55}$$

$$K_{i,j} = K_0 e^{-\beta r_{ij}} \tag{10.56}$$

with $K_{i,j}$ being the interaction between electrons i and j, limited in a sphere of radius D_2. The choice of the constants K_0 and β will be discussed below. Note that the choice of an exponential law does not affect the general feature of our results presented here because the short cut-off distance used here limits the interaction to a small number of neighbors, typically to next nearest neighbors (NNN), so the choice of another law such as a power law, or even discrete interaction values, for such a small cut-off will not make a qualitative difference in the results. Itinerant electrons move under an electric field applied along the x axis. The PBC ensure that the electrons who leave the system at one end are to be reinserted at the other end. These boundary conditions are used in order to conserve the average density of itinerant electrons. One has

$$\mathcal{H}_E = -e\vec{\epsilon}.\vec{\ell} \tag{10.57}$$

where e is the electronic charge, $\vec{\epsilon}$ an applied electric field and $\vec{\ell}$ a displacement vector of an electron. Since the interaction between itinerant electron spins is attractive, we need to add a kind of "chemical potential" in order to avoid a possible collapse of electrons into some points in the crystal and to ensure a homogeneous spatial distribution of electrons during the simulation. The chemical potential term is given by

$$\mathcal{H}_c = D[n(\vec{r}) - n_0] \qquad (10.58)$$

where $n(\vec{r})$ is the concentration of itinerant spins in the sphere of D_2 radius, centered at $\vec{r}$, n_0 the average concentration, and D a constant parameter.

10.4.4.1 *Choice of parameters and units*

The above model is very general. Several kinds of materials such as metals, semiconductors, insulating magnetic materials etc. can be studied with this model, provided an appropriate choice of the parameters. For example, non magnetic metals correspond to $I_{i,j} = K_{i,j} = 0$ (free conduction electrons). Magnetic semiconductors correspond to the choice of parameters K_0 and I_0 so as the energy of an itinerant electron due to the interaction $\mathcal{H}_r$ should be much lower than that due to $\mathcal{H}_m$, namely itinerant electrons are more or less bound to localized atoms. Note that $\mathcal{H}_m$ depends on the concentration of itinerant spins: for example the dilute case yields a small $\mathcal{H}_m$. We make simulations for typical values of parameters which correspond to semiconductors. The choice of the parameters has been made after numerous test runs. We describe the principal requirements which guide the choice: (i) We choose the interaction between lattice spins as unity, i. e. $|J| = 1$, (ii) We choose interaction between an itinerant and its surrounding lattice spins so as its energy E_i in the low T region is the same order of magnitude with that between lattice spins. To simplify, we take $\alpha = 1$. This case corresponds to a semiconductor, (iii) Interaction between itinerant spins is chosen so that this contribution to the itinerant spin energy is smaller than E_i in order to highlight the effect of the lattice ordering on the spin current. To simplify, we take $\beta = 1$, (iv) The choice of D is made in such a way to avoid the formation of clusters of itinerant spins (agglomeration) due to their attractive interaction [Eq. (10.56)], (v) The electric field is chosen not so strong in order to avoid its dominant effect that would mask the effects of thermal fluctuations and of the magnetic ordering, (vi) The density of the itinerant spins is chosen in a way that the contribution of interactions between themselves is much weaker than E_i, as said above in the case of semiconductors. Within the above requirements, a variation of each parameter does not change qualitatively the results shown below. Only the variation of D_1 in some antiferromagnets does change the results (see Ref. [89]). The energy is measured in the unit of $|J|$. The temperature is expressed in the unit of $|J|/k_B$. The distance (D_1 and D_2) is in the unit of the lattice constant a. Real units will be used in subsect. 10.4.9 for comparison with experiments.

10.4.5 *Simulation method*

Using the Metropolis algorithm, we first equilibrate the lattice at a given temperature T without itinerant electrons. When equilibrium is reached, we randomly add N_0 polarized itinerant spins into the lattice. Each itinerant electron interacts with lattice spins in a sphere of radius D_1 centered at its position, and with other itinerant electrons in a sphere of radius D_2. We next equilibrate the itinerant spins using the following updating. We calculate the energy E_{old} of an itinerant electron taking into account all interactions described above. Then we perform a trial move of length ℓ taken in an arbitrary direction with random modulus in the interval $[R_1, R_2]$ where $R_1 = 0$ and $R_2 = a$ (NN distance), a being the lattice constant. Note that the move is rejected if the electron falls in a sphere of radius r_0 centered at a lattice spin or at another itinerant electron. This excluded space emulates the Pauli exclusion. We calculate the new energy E_{new} and use the Metropolis algorithm to accept or reject the electron displacement. We choose another itinerant electron and begin again this procedure. When all itinerant electrons are considered, we say that we have made a Monte Carlo sweeping, or one Monte Carlo step/spin. We have to repeat a large number of Monte Carlo steps/spin to reach a stationary transport regime. We then perform the averaging to determine physical properties such as magnetic resistivity, electron velocity, energy etc. as functions of temperature. We define the dimensionless spin resistivity ρ as:

$$\rho = \frac{1}{n_e} \tag{10.59}$$

where n_e is the number of itinerant electron spins crossing a unit slice perpendicular to the x direction per unit of time. An example with real units is shown in subsection 10.4.9. In order to have sufficient statistical averages on microscopic states of both the lattice spins and the itinerant spins, we use what we call "multi-step averaging procedure": after averaging the resistivity over N_1 steps for "each" lattice spin configuration, we thermalize again the lattice with N_2 steps in order to take another disconnected lattice configuration. Then we take back the averaging of the resistivity for N_1 steps for the new lattice configuration. We repeat this cycle for N_3 times, usually several hundreds of thousands times. The total Monte Carlo steps for averaging is about 4×10^5 steps per spin in our simulations. This procedure reduces strongly thermal fluctuations observed in our previous work [2]. Of course, the larger N_2 and N_3 are, the better the statistics become. The question is what is the correct value of N_1 for averaging with one lattice spin configuration at a given T? This question is important because this is related to the relaxation time τ_L of the lattice spins compared to that of the itinerant spins, τ_I. The two extreme cases are (i) $\tau_L \simeq \tau_I$, one should take $N_1 = 1$, namely the lattice spin configuration should change with each move of itinerant spins (ii) $\tau_L \gg \tau_I$, in this case, itinerant spins can travel in the same lattice configuration for many times during the averaging. In order to choose a right value of N_1, we consider the following temperature dependence of τ_L in non frustrated spin systems. The relaxation time is expressed in this case as [65, 108]

$$\tau_L = \frac{A}{|1 - T/T_c|^{z\nu}} \tag{10.60}$$

where A is a constant, ν the correlation critical exponent, and z the dynamic exponent which depend on the spin model and space dimension. For 3D Ising model, $\nu = 0.638$ and $z = 2.02$. From this expression, we see that as T tends to T_c, τ_L diverges. In the critical region around T_c the system encounters thus the so-called "critical slowing down": the spin relaxation is extremely long due to the divergence of the spin-spin correlation. When we take into account the temperature dependence of τ_L, the shape of the resistivity is modified strongly at T_c where τ_L is very long, and in the paramagnetic phase where the relaxation time is very short due to rapid thermal fluctuations. On the other hand, at low T, τ_L does not modify ρ because in the ordered phase the spin landscape from one microscopic state to another does not change significantly to affect the motion of the itinerant spin (see discussion in Ref. [90]).

10.4.6   *Ferromagnets and antiferromagnets*

In ferromagnets, experimental data mentioned above show a peak at T_c. The peak is related to the critical slowing-down where the relaxation time diverges. Direct Monte Carlo simulations in the case of Ising spin give a pronounced peak at T_c as shown in Fig. 10.7 in agreement with experiments. Note that ρ increases at low T. The reason for this is multiple: it can stem from the freezing or crystallization of itinerant spins at low T or just from the smallness of the number of conduction electrons in such a low-T region. The shape of ρ depends on many factors: lattice structure, various interactions encountered by itinerant spins, electron concentration, relaxation time, spin model, magnetic-field amplitude etc. For example, a decrease in the interaction between itinerant spins K_0 will reduce the increase of ρ as $T \to 0$, an applied magnetic field will decrease the peak height, the larger carrier concentration will reduce ρ in particular at T_c. All of these have been discussed in Ref. [88]. We note a strong effect of the temperature dependence of τ_L on ρ for $T \geq T_c$. This is very important because τ_L depends intrinsically on the material via ν and z. For a quantitative comparison with experiments for a given material, it is necessary to take into account the specific parameters of that material. This is what we do in subsection 10.4.9.

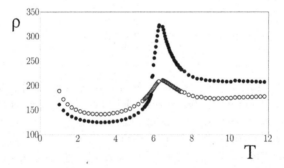

Fig. 10.7   BCC ferromagnetic and antiferromagnetic films: Resistivity ρ with temperature-dependent relaxation for ferro- (black circles) and antiferromagnet (white circles) in arbitrary unit versus temperature T, in zero magnetic field, with electric field $\epsilon = 1$, $I_0 = 2$, $K_0 = 0.5$, $A = 1$.

In antiferromagnets much less is known because there have been very few theoretical investigations which have been carried out. Haas [57] has shown that while in ferromagnets the resistivity ρ shows a sharp peak at the magnetic transition of the lattice spins, in antiferromagnets there is no such a peak. We found that the peak exists in antiferromagnets but it is less pronounced as seen in Fig. 10.7. The alternate change of sign of the spin-spin correlation with distance may have something to do with the absence of a sharp peak. We have tested for example the effect of the cut-off distance D_1 [89]: when D_1 increases, it will include successively up-spin shells and down-spin shells in the sphere of radius D_1. As a consequence, the difference between the numbers of up and down spins in the sphere oscillates with varying D_1, making an oscillatory behavior of ρ at small D_1, unlike in ferromagnets. It is interesting to note that in the presence of an itinerant spin, the ferromagnet and its antiferromagnet counterpart are no more invariant by the local Mattis transformation ($J_{ij} \rightarrow -J_{ij}, \mathbf{S}_j \rightarrow -\mathbf{S}_j$).

10.4.7 *Frustrated systems*

We consider the simple cubic lattice shown in Fig. 10.8. The Hamiltonian is given by

$$\mathcal{H} = -J_1 \sum_{(i,j)} \mathbf{S}_i \cdot \mathbf{S}_j - J_2 \sum_{(i,m)} \mathbf{S}_i \cdot \mathbf{S}_m \qquad (10.61)$$

where $\mathbf{S}_i$ is the Ising spin at the lattice site i, $\sum_{(i,j)}$ is made over the NN spin pairs with interaction J_1, while $\sum_{(i,m)}$ is performed over the NNN pairs with interaction J_2. We are interested in the frustrated regime. Therefore, hereafter we suppose that $J_1 = J$ ($J > 0$, antiferromagnetic interaction), and $J_2 = -\eta J$ where η is a positive parameter. The ground state (GS) of this system is easy to obtain either by minimizing the energy, or by comparing the energies of different spin configurations, or just by a numerical minimizing with a steepest descent method shown in Problem 4 of chapter 8 [103]. We obtain the antiferromagnetic configuration shown by the upper figure of Fig. 10.9 for $|J_2| < 0.25|J_1|$, or the configuration shown in the lower figure for $|J_2| > 0.25|J_1|$. Note that this latter configuration is 3-fold degenerate by choosing the parallel NN spins on x, y or z axis. With the permutation of black and white spins, the total degeneracy is thus 6. The phase transition in the case of the Heisenberg model in the frustrated region ($|J_2| > 0.25|J_1|$) has been found to be of first order [111]. The system is very unstable due to its large degeneracy. We find that the case of the Ising spin shows an even stronger first-order transition [63]. It is interesting to note that the resistivity of itinerant spins in systems with a first-order transition undergoes a discontinuity at T_c just as the system energy and the order parameter. We show ρ in Fig. 10.10 for several cut-off distance D_1. One observes here that ρ can jump or fall at the transition depending on the interaction range D_1. The resistivity discontinuity has been confirmed in another system with first-order transition, the frustrated FCC antiferromagnet [89]. This seems to be a general rule.

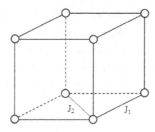

Fig. 10.8 Simple cubic lattice with nearest and next-nearest neighbor interactions, J_1 and J_2, indicated.

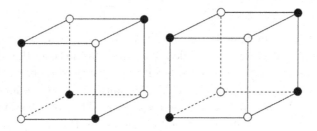

Fig. 10.9 Simple cubic lattice. Up-spins: white circles, down-spins: black circles. Left: Ground state when $|J_2| < 0.25|J_1|$. Right: Ground state when $|J_2| > 0.25|J_1|$.

10.4.8 *Thin films — Surface effects*

We see so far that when there is a magnetic phase transition, the spin resistivity undergoes an anomaly. In magnetic thin films, when there is a surface phase transition at a temperature T_S different from that of the bulk one (T_c), we expect two peaks of ρ one at T_S and the other at T_c. We show here an example of a thin film composed of three sub-films: the middle film of 4 atomic layers between two surface films of 5 layers. The lattice sites are occupied by Ising spins interacting with each other via nearest-neighbor

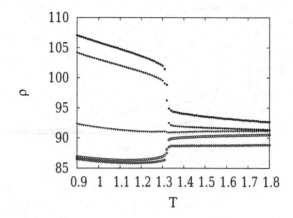

Fig. 10.10   Spin resistivity versus T for $|J_2| = 0.26|J_1|$ for several values of D_1: from up to down $D_1 = 0.7a$, $0.8a$, $0.94a$, a, $1.2a$. Other parameters are $L_x = L_y = 20, L_z = 6$, $I_0 = K_0 = 0.5$, $D_2 = a$, $D = 1$, $\epsilon = 1$.

ferromagnetic interaction. Let us suppose the interaction between spins in the outside films be J_S and that in the middle film be J. The inter-film interaction is J. The lattice structure is face-centered cubic. In order two enhance surface effects we suppose in addition $J_S \ll J$. We show in Fig. 10.11 the layer magnetization as well as the layer susceptibility. We observe that the outside films undergo a phase transition at a temperature far below the transition temperature of the middle film. As stated above, a phase transition induces an anomaly in the spin resistivity: the two phase transitions observed in Fig. 10.11 give rise to two peaks of ρ shown in Fig. 10.12. The surface peak of ρ has been also seen in a frustrated film [89].

10.4.9   *The case of MnTe*

The pure MnTe has either the zinc-blend structure [61] or the hexagonal NiAs one shown in Fig. 10.13. We confine ourselves in the latter case where the Néel temperature is $T_N = 310$ K [127]. Hexagonal MnTe is a crossroad semiconductor with a big gap (1.27 eV) and a room-temperature carrier concentration of $n = 4.3 \times 10^{17} \mathrm{cm}^{-3}$ [99, 5]. Without doping, MnTe is non degenerate. The behavior of ρ in MnTe as a function of T has been experimentally shown [27, 84, 9, 44, 59]. The system is composed of ferromagnetic xy hexagonal planes antiferromagnetically stacked in the c direction. The NN distance in the c direction is $c/2 \simeq 3.36$ Å shorter

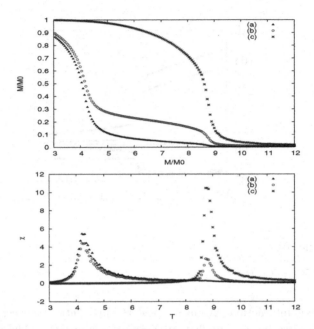

Fig. 10.11 Top: Magnetization versus T in the case where the system is made of three films: the first and the third have 5 layers with a weaker interaction J_s, while the middle has 4 layers with interaction $J = 1$. We take $J_s = 0.2J$. Black triangles: magnetization of the surface films, stars: magnetization of the middle film, void circles: total magnetization. Bottom: Susceptibility versus T. Black triangles: susceptibility of the surface films, stars: susceptibility of the middle films, void circles: total susceptibility. See text for comments.

than the in-plane NN distance which is $a = 4.158\text{Å}$. Neutron scattering experiments show that the main exchange interactions between Mn spins in MnTe are i) interaction between NN along the c axis with the value $J_1/k_B = -21.5 \pm 0.3$ K, (ii) ferromagnetic exchange $J_2/k_B \approx 0.67 \pm 0.05$ K between in-plane neighboring Mn (they are next NN by distance), (iii) third NN antiferromagnetic interaction $J_3/k_B \simeq -2.87 \pm 0.04$ K. The spins are lying in the xy planes perpendicular to the c direction with a small in-plane easy-axis anisotropy D_a [127]. We note that the values of the exchange integrals given above have been deduced from experimental data by fitting with a formula obtained from a free spin-wave theory [127]. Other fittings with mean-field theories give slightly different values: $J_1/k_B = -16.7$ K, $J_2/k_B = 2.55$ K and $J_3/k_B = -0.28$ K [99]. The lattice Hamiltonian is given by

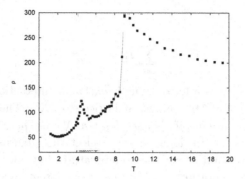

Fig. 10.12 Resistivity ρ in arbitrary unit versus T of the system described in the previous figure's caption.

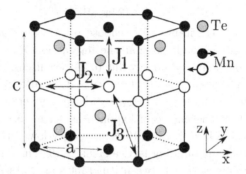

Fig. 10.13 Structure of MnTe of NiAs type is shown. Antiparallel spins are shown by black and white circles. NN interaction is marked by J_1, next NN interaction by J_2, and third NN one by J_3.

$$\mathcal{H} = -J_1 \sum_{(i,j)} \mathbf{S}_i \cdot \mathbf{S}_j - J_2 \sum_{(i,m)} \mathbf{S}_i \cdot \mathbf{S}_m - J_3 \sum_{(i,k)} \mathbf{S}_i \cdot \mathbf{S}_k$$
$$-D_a \sum_i (S_i^x)^2 \tag{10.62}$$

where $\mathbf{S}_i$ is the Heisenberg spin at the lattice site i, $\sum_{(i,j)}$ is made over the NN spin pairs $\mathbf{S}_i$ and $\mathbf{S}_j$ with interaction J_1, while $\sum_{(i,m)}$ and $\sum_{(i,k)}$ are made over the NNN and third NN neighbor pairs with interactions J_2 and J_3, respectively. $D_a > 0$ is an anisotropy constant which favors the in-plane x easy-axis spin configuration. The Mn spin is experimentally known to be of the Heisenberg model with magnitude $S = 5/2$ [127]. The interaction between an itinerant spin and surrounding Mn spins in semiconducting

MnTe is written as

$$\mathcal{H}_i = -\sum_n I(\vec{r} - \vec{R}_n)\sigma \cdot \mathbf{S}_n \qquad (10.63)$$

where $I(\vec{r} - \vec{R}_n) > 0$ is a ferromagnetic exchange interaction between itinerant spin σ at $\vec{r}$ and Mn spin $\mathbf{S}_n$ at lattice site $\vec{R}_n$. The sum on lattice spins $\mathbf{S}_n$ is limited at cut-off distance $D_1 = a$. We use here the Ising model for the electron spin. In doing so, we neglect the quantum effects which are of course important at very low temperatures but not in the transition region at room temperature where we focus our attention. We suppose the following distance dependence of $I(\vec{r} - \vec{R}_n)$:

$$I(\vec{r} - \vec{R}_n) = I_0 \exp[-\alpha(\vec{r} - \vec{R}_n)] \qquad (10.64)$$

where I_0 and α are constants. We choose $\alpha = 1$ for convenience. The choice of I_0 should be made so that the interaction $\mathcal{H}_i$ yields an energy much smaller than the lattice energy due to $\mathcal{H}$ (see discussion on the choice of variables given above). Note that the cut-off distance is rather short so that the obtained results shown below still keep a general character which does not depend on the choice of exponential form. Since in MnTe the carrier concentration is $n = 4.3 \times 10^{17} \text{cm}^{-3}$, very low with respect to the concentration of its surrounding lattice spins $\simeq 10^{22} \text{cm}^{-3}$, we do not take into account the interaction between itinerant spins. As said before, the values of the exchange interactions deduced from experimental data depend on the model Hamiltonian, in particular the spin model, as well as the approximations. Furthermore, in semiconductors, the carrier concentration is a function of T. In our model, there is however no interaction between itinerant spins. Therefore, the number of itinerant spins used in the simulation is important only for statistical average: the larger the number of itinerant spins the better the statistical average. The current obtained is proportional to the number of itinerant spins but there are no extra physical effects. Using the exchange integrals slightly modified with respect to the ones given above, we have calculated ρ of the hexagonal MnTe. The result of ρ is shown in Fig. 10.14. Note that with J_3 slightly larger in magnitude than the value deduced from experiments, we find $T_N = 310$ K in excellent agreement with experiments. Furthermore, we observe that ρ shows a pronounced peak and coincides with the experimental data. The values we used to obtain that agreement are $A = 1$ and Heisenberg critical exponents $\nu = 0.707$, $z = 1.97$ [108]. In the temperature regions below $T < 140$ K and above T_N the Monte Carlo result is also in excellent agreement with experiment, unlike in our previous work [4] using the Boltzmann's equation. Using the value of ρ, we obtain the relaxation time of itinerant spin equal to $\tau_I \simeq 0.1$ ps, and the mean free path equal to $\bar{l} \simeq 20$ Å, at the critical temperature.

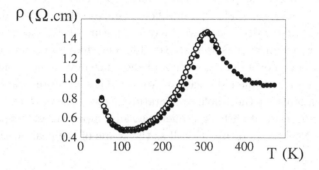

Fig. 10.14 Spin resistivity ρ versus temperature T. Black circles are from Monte Carlo simulation, white circles are experimental data taken from He *et al.*. The parameters used in the simulation are $J_1 = -21.5$K, $J_2 = 2.55$ K, $J_3 = -9$ K, $I_0 = 2$ K, $D_a = 0.12$ K, $D_1 = a = 4.148$ Å, $\epsilon = 2 \times 10^5$ V/m, $L = 30a$ (lattice size L^3).

10.4.10 *Calculation with the Boltzmann's equation*

We can combine the Boltzmann's equation with the Monte Carlo simulation to calculate the spin resistivity. The principle consists of two steps:

i) to write the Boltzmann's equation in terms of the cluster sizes and cluster numbers at a given T,

ii) to calculate the cluster numbers and the cluster sizes by Monte Carlo simulations using the Hoshen-Kopelmann's algorithm [66]. Inserting the results into the Boltzmann's equation we obtain the relaxation time which allows us to calculate the resistivity. This has been done in Refs. [3, 4].

10.5 Conclusion

We have shown above how Monte Carlo simulations can be used to produce properties of spin transport in magnetic materials. The method is very general, it can be easily applied to a wide range of materials from ferromagnets to antiferromagnets of different lattices and spin models. The results of the spin resistivity ρ as a function of temperature under different situations can be obtained and compared to experiments. We were concentrated in the magnetic phase transition region where theories failed to predict correct behaviors of ρ. This is due to the fact that the magnetic resistivity is intimately related to the spin-spin correlation which is very different from one material to another. This correlation, as we know in the domain of phase transitions and critical phenomena, governs the nature of the transition: phase transitions of second order of different universality classes and phase transitions of first-order. Needless to say, the nature of the phase transition affects the behavior of ρ as seen above: different shapes of ρ and discontinuity at T_c, etc. For a demonstration of the efficiency of the Monte Carlo method, we have studied the case of MnTe where experimental data are recently available for the whole temperature range. The obtained result is in excellent agreement with experiments at all temperatures: it reproduces the correct Néel temperature as well as the shape of the peak at the phase transition.

10.6   Problems

Problem 1. Effect of magnetic field: demonstrate Eq. (10.24).

Problem 2. Ohm's law: demonstrate Eq. (10.29).

Problem 3. Hall effect - Magneto-resistance:

The general expression of the current density in a system under an applied electric field $\vec{\epsilon}$ and an external magnetic field $\vec{B}$ can be written as a series of $\vec{\epsilon}$ and $\vec{B}$:

$$j_i = \sum_j \sigma_{ij}\epsilon_j + \sum_{j,l} \sigma_{ijl}\epsilon_j B_l + \sum_{j,l,m} \sigma_{ijlm}\epsilon_j B_l B_m$$

where σ_{ij} is the "normal" or "ordinary" electric conductivity tensor, and σ_{ijl} denotes the conductivity tensor due to the interaction between $\vec{\epsilon}$ and $\vec{B}$. When $\vec{\epsilon} \cdot \vec{B} = 0$, we have the geometry of the Hall effect. σ_{ijlm} is the conductivity tensor due to the interaction between $\vec{\epsilon}$ and $\vec{B}$ at the second order. This is at the origin of the magneto-resistance.

a) Weak fields: linear approximation:

This approximation is used when $\omega_c \tau \ll 1$ where ω_c is the cyclotron frequency and τ the relaxation temps. One supposes that $\vec{B}$ is parallel to $\vec{Oz}$ and $\vec{\epsilon}$ parallel to $\vec{Ox}$. One supposes in addition that τ is independent of the electron energy and that the electron effective mass is isotropic.

i) Calculate the Hall coefficient R_e defined as the ratio ϵ_y to $j_x B$ where ϵ_y is the y component of the electric field induced by the magnetic field and j_x the current density due to the conducting electrons.

ii) What is the coefficient R_h for the hole current?

iii) Give the expression of the total coefficient R when both electrons and holes participate in the conduction. We denote $R = R_0$. We express the results as functions of n, p, σ, μ_e and μ_h [respectively, electron and hole densities, electron and hole conductivity $\sigma_e = \sigma_h = \sigma$, electron and hole mobilities $\mu_e = \sigma_e/(ne)$, $\mu_h = \sigma_h/(pe)$].

b) Moderate fields:

i) Write down the equations of motion of an electron in the x and y directions. One introduces the following complex quantity $Z = v_x + iv_y$ where v_x and v_y are the components of the electron velocity. Show that the solution of the equations of motion is of the form

$$Z(t) = Z_0 + \frac{e}{i\omega_c m_e}(\epsilon_x + i\epsilon_y)(1 - e^{i\omega_c t})$$

where $Z_0 = Z(t = 0)$.

ii) Show that the average value of Z taken over all collisions is given by

$$\overline{Z} = \frac{e}{m_e}\frac{\tau}{i\omega_c \tau - 1}.$$

Deduce that

$$j_x = \frac{ne^2}{m_e}\left[\frac{\tau}{1 + \omega_c^2\tau^2}\epsilon_x - \frac{\omega_c\tau^2}{1 + \omega_c^2\tau^2}\epsilon_y\right]$$

$$j_y = \frac{ne^2}{m_e}\left[\frac{\tau}{1 + \omega_c^2\tau^2}\epsilon_y + \frac{\omega_c\tau^2}{1 + \omega_c^2\tau^2}\epsilon_x\right]$$

iii) Show that when both electrons and holes participate in the conduction, the Hall coefficient is given by

$$R = \frac{\sigma_e^2 R_e + \sigma_h^2 R_h + \sigma_e^2\sigma_h^2 R_e R_h (R_e + R_h)B^2}{(\sigma_e + \sigma_h)^2 + \sigma_e^2\sigma_h^2(R_e + R_h)^2 B^2}.$$

Give comments on the case where the conduction is due to only one type of impurity (p or n) and on the case of an intrinsic conductor ($p = n$).

iv) The above results show that the presence of a magnetic field has no effect on the resistance when the conduction is due to only one type of carriers (n or p) and when the relaxation time is constant and when the effective mass is isotropic (spherical iso-energy surfaces). If one of these three conditions is not fulfilled, there is a correction to the initial resistivity ρ_0. Show that when both electrons and holes participate to the conduction, keeping isotropic effective mass and constant τ, the correction is given by

$$\frac{\Delta\rho}{\rho_0} = \frac{np\mu_e\mu_h(\mu_e + \mu_h)^2 B^2}{(n\mu_e + p\mu_h)^2} \equiv \xi R_0^2 \sigma_0^2 B^2$$

where the coefficient of the transverse magneto-resistance ξ is

$$\xi = \frac{np\mu_e\mu_h(\mu_e + \mu_h)^2}{(p\mu_h^2 - n\mu_e^2)^2}$$

and $\sigma_0 \equiv \sigma_e + \sigma_h$ when $B = 0$, R_0 being given by the above weak-field approximation. Numerical application: in order to have $\frac{\Delta\rho}{\rho_0} \simeq 0.1$ in an applied field of 10^3 Gauss, what is the total mobility $\mu_e + \mu_h$?

c) Effects of collisions:

One supposes again that the effective mass m is isotropic, but the relaxation time τ depends on the electron energy under the form $\tau = aE^{-s}$. This form represents several types of collision. The results for j_x and j_x in the weak-field approximation shown above are still valid provided that the quantities $\frac{\tau}{1+\omega_c^2\tau^2}$ and $\frac{\omega_c\tau^2}{1+\omega_c^2\tau^2}$ are replaced by their values averaged over all energies.

i) Show that in the present case the Hall coefficient is written as $R = -\frac{K}{ne}$ where K is given by

$$K = \frac{< \frac{\tau^2}{1+\omega_c^2\tau^2} >}{< \frac{\tau}{1+\omega_c^2\tau^2} >^2 + \omega_c^2 < \frac{\tau^2}{1+\omega_c^2\tau^2} >^2} \qquad (10.65)$$

Show that $K = \frac{<\tau^2>}{<\tau>^2}$ for weak fields and strong enough collisions.

ii) Calculate $< \tau^2 >$ and $< \tau >$. Show that

$$< \tau > = a(k_B T)^{-s} \frac{\Gamma(5/2 - s)}{\Gamma(5/2)}$$

and

$$< \tau^2 > = a^2 (k_B T)^{-2s} \frac{\Gamma(5/2 - 2s)}{\Gamma(5/2)^2}.$$

Estimate K for $s = 1/2$ (collisions with phonons) and $s = -3/2$ (collisions with impurities).

iii) One supposes there is only one kind of carriers, calculate $\frac{\Delta \rho}{\rho_0}$.

Problem 4. Boltzmann's equation in the case of a strong field:

One applies an electric field $\vec{\varepsilon}$ parallel to the $\vec{Oz}$ axis. The distribution function f depends on the electron energy E and the angle θ between the wave vector $\vec{k}$ and $\vec{\varepsilon}$. One expresses f as a series of the Legendre polynomials P_n as follows

$$f(E, \cos \theta) = \sum_{n=0}^{\infty} g_n(E) P_n(\cos \theta)$$

If one retains only the first two terms,

a) Show that the field term in the Boltzmann's equation is given by

$$-\left(\frac{\partial f}{\partial t}\right)_F = \frac{q\varepsilon}{\hbar}\left(g + \frac{2}{3}Eg' + \frac{\hbar^2 k_z}{m^*}g_0'\right)$$

where $g = g_1/k$ ($k_z = k\cos\theta$), g' and g_0' are the first derivatives with respect to E

b) Show that the collision term is written as

$$\left(\frac{\partial f}{\partial t}\right)_C = \phi_0 + \phi_1$$

where

$$\phi_0 = \frac{\Omega}{(2\pi)^3} \int \{w(\vec{k}, \vec{k}')g_0(E') - w(\vec{k}', \vec{k})g_0(E)\} d^3 k'$$

$$\phi_1 = \frac{\Omega}{(2\pi)^3} \int \{w(\vec{k}, \vec{k}')k_z'g(E') - w(\vec{k}', \vec{k})k_z g(E)\} d^3 k'$$

$w(\vec{k}, \vec{k}')$ is the probability per time unit of the transition $\vec{k} \to \vec{k}'$.

c) Taking into account the probability of an electron in the initial state $\vec{k}$ and that of the final state $\vec{k}'$, the transition probability is written as

$$\bar{w} = w(\vec{k}, \vec{k}')f(E_k)[1 - f(E_{k'})]$$

with

$$w(\vec{k}, \vec{k}') = \frac{\pi}{NM}\frac{|I|^2}{w_{\vec{q}_j}}\left[s(w_{\vec{q}_j}) + \frac{1}{2} \pm \frac{1}{2}\right]\delta(E_{k'} - E_k \pm \hbar w_{\vec{q}_j})$$

where I is the electron-phonon coupling, N the total number of electrons and M the mass of ions. The signs $\pm$ correspond respectively to emission and absorption of a phonon of energy $\hbar\omega_{\vec{q}_j}$ of the phonon branch j, of wave vector $\vec{q}$. $s(\omega_{\vec{q}_j})$ is the occupation number of mode $\omega_{\vec{q}_j}$. One supposes that $f[1 - f] \simeq f$ (high temperatures). For acoustic phonons, one shall use $\omega_{(\vec{q})} = v_s q$ (v_s: sound velocity) and $I = \pm iCq$ (C: deformation potential). Without demonstration (see Ref. [58] for details), we admit the following result

$$\phi_0 = \frac{2m^* v_s^2}{\tau}\left[Eg_0'' + \left(\frac{E}{k_B T} + 2\right)g_0' + \frac{2g_0}{k_B T}\right]$$

$$\phi_1 = -\frac{k_z g}{\tau}$$

where τ is the relaxation time calculated for weak fields:

$$\tau = \frac{2\pi M}{\Omega}\frac{\hbar v_s^2}{C^2}\left(\frac{\hbar^2}{2m^*}\right)^{3/2}\frac{1}{\sqrt{E}}\frac{1}{k_B T} \tag{10.66}$$

Write the Boltzmann's equation and show that

$$g = -\frac{q\varepsilon\hbar\ell}{\sqrt{2m^*}}\frac{g_0'}{\sqrt{E}}$$

where ℓ is the mean free path. Deduce the following differential equation of g_0:

$$(E + \lambda k_B T)g_0'' + +\left(\frac{E}{k_B T} + 2 + \frac{\lambda k_B T}{E}\right)g_0' + \frac{2}{k_B T}g_0 = 0$$

where

$$\lambda = \frac{q^2\ell^2}{6m^* v_s^2 k_B T}\varepsilon^2 \equiv \frac{\varepsilon^2}{\varepsilon_0^2}$$

d) In the strong-field limit $\lambda = \varepsilon^2/\varepsilon_0^2 \gg E/k_B T$, show that the solution of the above equation is of the form

$$g_0 = Ae^{-E^2/2\lambda k_B^2 T^2}$$

where A is a constant.

e) Show that A is given by

$$A = \frac{4\pi^2}{\Gamma(3/4)}\frac{n}{(2\lambda)^{3/4}}\left(\frac{\hbar^2}{2m^* k_B T}\right)^{3/2}$$

with $n = \frac{N}{\Omega}$ (electron density).

f) Calculate the current j and show that $j = \gamma\sqrt{\varepsilon}$ where

$$\gamma = en\frac{\sqrt{e\ell v_s}}{(m^* k_B T)^{1/4}}\frac{\sqrt{2\pi}}{3^{3/4}\Gamma(3/4)}$$

PART 3

Solutions of Problems

Chapter 11

Solutions of Problems

11.1 Solutions of problems of chapter 1

Problem 1. Orbital and spin moments of an electron:

Guide: The state of an electron in an atomic orbital is defined by four quantum numbers n, l, m_l, m_s where $n = 1, 2, 3, ...$; $l = 0, 1, 2, ..., n-1$; $m_l = -l, -l+1, ..., l-1, l$; $m_s = -s, s$ with $s = 1/2$. The orbital angular momentum is $\vec{L}$ with eigenvalues $\hbar m_l$, the spin angular momentum is $\vec{S}$ with eigenvalue $\hbar m_s$. The orbital magnetic moment is $\vec{M}_l = -\mu_B \vec{L}$, and the spin magnetic moment is $\vec{M}_s = -g\mu_B \vec{S}$ where $g = 2.0023 \simeq 2$ (Landé factor or electron spectroscopic factor). The total magnetic moment of an electron is thus $\vec{M}_t = -\mu_B(\vec{L} + g\vec{S})$.

Problem 2. Zeeman effect:

Guide:

a) The number of Fe atoms in 1 m^3: $N = \frac{7970 \times 6.025 \times 10^{26}}{56} = 8.58 \times 10^{28}$ (the Avogadro number per kilogram is 6.025×10^{26})

The magnetic moment of a Fe atom is thus

$M = \frac{1.7 \times 10^6}{8.58 \times 10^{28}}$ Am^2=$1.98 \times 10^{-23} = 2.49 \times 10^{-29}$ JA/m=$2.14\mu_B$ (Bohr magneton $\mu_B = 9.27 \times 10^{-24}$ Joule/Tesla $= 1.16 \times 10^{-29}$ JA/m).

b) The energy due to the Zeeman effect is $\Delta E = \mu_B B = \mu_B \mu_0 H$ ($B = \mu_0 H$: magnetic field). We have $\Delta E = 0.9273 \times 10^{-23}\mu_0 H$ Joules (μ_0, vacuum permeability, $= 1.257 \times 10^{-6}$ H/m).

For $\mu_0 H = 0.5$ Tesla, $\Delta E = 0.464 \times 10^{-23}$ J

For $\mu_0 H = 1$ Tesla, $\Delta E = 0.927 \times 10^{-23}$ J

For $\mu_0 H = 2$ Tesla, $\Delta E = 1.85 \times 10^{-23}$ J.

The variation of the frequency: $\Delta \nu = \nu - \nu_0 = \Delta E/h$.

Problem 3. Density of states:

Solution: For a free electron in a box of linear dimension L, the Schrödinger equation with the periodic boundary conditions gives the following solution for the energy

$$E(\vec{k}) = \frac{\hbar^2 k^2}{2m} \qquad (11.1)$$

with

$$k_i = \frac{2\pi n_i}{L} \quad (i = x, y, z) \qquad (11.2)$$

where $n_i = 0, \pm 1, \pm 2, ...$ Let $\mathcal{N}(E)$ be the number of microscopic states of energy $\leq E$. The number of states between E and $E + \Delta E$ is thus $\mathcal{N}(E + \Delta E) - \mathcal{N}(E)$. The density of states $\rho(E)$ is defined by

$$\rho(E) = lim_{\Delta E \to 0} \frac{\mathcal{N}(E + \Delta E) - \mathcal{N}(E)}{\Delta E} = \frac{d\mathcal{N}}{dE} \qquad (11.3)$$

This is the number of states of energy E, namely the degeneracy in the continuous energy case. We calculate $\mathcal{N}(E)$: in the phase space defined by (k_x, k_y, k_z) where each state is defined by a point (k_x, k_y, k_z), the volume of each state is $(2\pi/L)^3$ [see Eq. (11.2)]. The number of microscopic states of energy $\leq E$ is the volume of the sphere of radius $k = \sqrt{\frac{2mE}{\hbar^2}}$ divided by the volume of a state:

$$\mathcal{N}(E) = \frac{4\pi}{3} k^3 \frac{L^3}{8\pi^3} = \frac{L^3}{6\pi^2} \left(\frac{2m}{\hbar^2}\right)^{3/2} E^{3/2} \qquad (11.4)$$

We deduce

$$\rho(E) = \frac{L^3}{4\pi^2} \left(\frac{2m}{\hbar^2}\right)^{3/2} E^{1/2} \qquad (11.5)$$

If we take into account the spin degeneracy $(2s + 1)$, we have

$$\rho(E) = (2s + 1) \frac{L^3}{4\pi^2} \left(\frac{2m}{\hbar^2}\right)^{3/2} E^{1/2} \qquad (11.6)$$

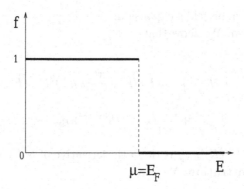

Fig. 11.1   Fermi-Dirac distribution function at $T = 0$ versus energy E. The scale of E is arbitrary, μ is taken equal to 1.

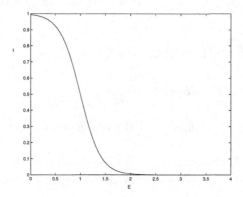

Fig. 11.2   Fermi-Dirac distribution function at $T \neq 0$ versus energy E. The scale of E is arbitrary, μ is taken equal to 1.

Problem 4. Fermi-Dirac distribution for free-electron gas:

   Solution: At $T = 0$, namely $\beta = \infty$, we see that $f = 1$ for $E \leq \mu$, and $f = 0$ for $E > \mu$: the electrons occupy all energy levels up to μ. Each energy level has two electrons, one with up spin and the other with down spin. One defines the Fermi level E_F by $E_F = \mu$, namely the highest energy level which is occupied at $T = 0$. The function f at $T = 0$ is shown in Fig. 11.1.

   At $T \neq 0$, the Fermi-Dirac distribution function is shown in Fig. 11.2. Electrons occupy all levels with decreasing f for increasing E.

Problem 5. Sommerfeld's expansion:
Solution: We show that

$$I = \int_0^\mu h(E)dE + \frac{\pi^2}{6}(k_BT)^2 h^{(1)}(E)|_{E=\mu}$$

$$+ \frac{7\pi^4}{360}(k_BT)^4 h^{(3)}(E)|_{E=\mu} + ... \tag{11.7}$$

where $h^{(n)}(E)|_{E=\mu}$ is the n-th derivative of $h(E)$ at $E = \mu$.
Demonstration: We have

$$I = \int_{-\infty}^{\infty} h(E)f(E)dE \tag{11.8}$$

We define $g(E) = \int_{-\infty}^{E} dE h(E)f(E)$. Integration of I by parts gives

$$I = \int_{-\infty}^{\infty} h(E)f(E)dE = [g(E)f(E)]_{-\infty}^{\infty} - \int_{-\infty}^{\infty} dE g(E)\frac{\partial f(E)}{\partial E}$$

$$= 0 + \int_{-\infty}^{\infty} dE g(E)(-\frac{\partial f(E)}{\partial E}) \tag{11.9}$$

where we have used $f \to 0$ when $E \to \infty$, and $g(E) \to 0$ when $E \to -\infty$.
At low T, the function $-\frac{\partial f(E)}{\partial E}$ is significant only near μ (slope of $f(E)$). This justifies an expansion of $g(E)$ around μ:

$$g(E) = g(\mu) + \sum_{n=1}^{\infty} \left[\frac{(E-\mu)^n}{n!}\right]\left[\frac{d^n g(E)}{dE^n}\right]_{E=\mu} \tag{11.10}$$

Replacing this series in (11.9), we get

$$I = g(\mu) + \sum_{n=1}^{\infty} \int_{-\infty}^{\infty} dE \frac{(E-\mu)^n}{n!}\left[\frac{d^n g(E)}{dE^n}\right]_{E=\mu}(-\frac{\partial f(E)}{\partial E})$$

$$= g(\mu) + \sum_{n=1}^{\infty} \int_{-\infty}^{\infty} dE \frac{(E-\mu)^{2n}}{(2n)!}\left[\frac{d^{2n-1} h(E)}{dE^{2n-1}}\right]_{E=\mu}(-\frac{\partial f(E)}{\partial E})$$

where only terms of even power are non zero, odd terms being zero because the integrands are odd functions with symmetric limits [note that $\frac{\partial f(E)}{\partial E}$ is an even function with respect to $(E - \mu)$]. Putting $x = \beta(E - \mu)$, we obtain

$$I = g(\mu) + \sum_{n=1}^{\infty} c_{2n}(k_B T)^{2n} \left[\frac{d^{2n-1}h(E)}{dE^{2n-1}} \right]_{E=\mu} \tag{11.11}$$

where

$$c_{2n} = \frac{1}{(2n)!} \int_{-\infty}^{\infty} x^{2n} [-\frac{d}{dx} \frac{1}{e^x + 1}] dx \tag{11.12}$$

Integration by parts gives

$$c_{2n} = \frac{1}{(2n)!} \left[-[x^{2n} \frac{1}{e^x + 1}]_{-\infty}^{\infty} + 2n \int_{-\infty}^{\infty} x^{2n-1} \frac{1}{e^x + 1} dx \right]$$

$$= \frac{1}{(2n)!} \left[0 + 4n(1 - 2^{1-2n})\Gamma(2n)\zeta(2n) \right] \tag{11.13}$$

where we have used the formula

$$\int_0^{\infty} x^{n-1} \frac{1}{e^x + 1} dx = (1 - 2^{1-n})\Gamma(n)\zeta(n) \tag{11.14}$$

We have

$$\Gamma(n+1) = \int_0^{\infty} t^n \exp(-t) dt = n!$$

$$\Gamma(n+1) = n\Gamma(n)$$

$$\zeta(x) = \sum_{n=1}^{\infty} \frac{1}{n^x}$$

with $\zeta(2) = \frac{\pi^2}{6}$ and $\zeta(4) = \frac{\pi^4}{90}$. Rarely, we need to go beyond $2n = 4$ in the Sommerfeld's expansion.

Problem 6. Pauli paramagnetism:

Solution: Energies of spin in a magnetic field $\vec{B}$ applied in the z direction: $E_\uparrow = E - \mu_B B$ and $E_\downarrow = E + \mu_B B$. The resulting magnetic moment is $M = \mu_B(N_\uparrow - N_\downarrow)$ which is written as

$$M = \mu_B \left[\int_{\mu_B B}^{\infty} dE \frac{\rho(E)}{e^{\beta(E-\mu_B B-\mu)} + 1} - \int_{-\mu_B B}^{\infty} dE \frac{\rho(E)}{e^{\beta(E+\mu_B B-\mu)} + 1} \right]$$

$$= \mu_B \int_0^{\infty} dE \left[\rho(E + \mu_B B) - \rho(E - \mu_B B) \right] f(E, T, \mu) \tag{11.15}$$

where we changed $E \to E \pm \mu_B B$. For weak fields, we can expand the density of states around E (energy in zero field): $\rho(E \pm \mu_B B) \simeq$

$\rho(E) \pm \mu_B B \left[\rho'(E)\right]_E$. We get $M \simeq 2\mu_B^2 B \int_0^\infty dE\rho'(E)f(E,T,\mu)$. We deduce

$$\chi = \frac{dM}{dB} \simeq 2\mu_B^2 \int_0^\infty dE\rho'(E)f(E,T,\mu) \qquad (11.16)$$

At low T, we can make a Sommerfeld's expansion [see (1.84)] for this integral

$$\chi \simeq 2\mu_B^2 \left[\int_0^\mu dE\rho'(E) + \frac{\pi^2}{6}\rho''(\mu)(k_BT)^2 \right]$$

$$= 2\mu_B^2 \left[\rho(\mu) - A\frac{\pi^2}{48\mu^{3/2}}(k_BT)^2 \right]$$

where we have used (A.40): $\rho(E) = AE^{1/2}$. Using (2.9), we obtain

$$\chi = 2\mu_B^2\rho(E_F)\left[1 - \frac{\pi^2}{12}(\frac{k_BT}{E_F})^2 \right] \qquad (11.17)$$

The first term is independent of T as we have found in (1.18). The second term depends on T^2.

At high T, $f \simeq e^{-\beta(E-\mu)}$, (11.16) becomes

$$\chi \simeq 2\mu_B^2 \int_0^\infty dE\rho'(E)e^{-\beta(E-\mu)}$$

$$= 2\mu_B^2 \left[\rho'(E)e^{-\beta(E-\mu)}|_0^\infty + \beta \int_0^\infty dE\rho(E)e^{-\beta(E-\mu)} \right]$$

$$= 2\mu_B^2\beta\left[0 + N/2\right] = \frac{N\mu_B^2}{k_BT} \qquad (11.18)$$

where N is the total number of electrons of spins $\uparrow$ and $\downarrow$. This results is called "Curie's law".

Remark: we have used (A.40) without factor 2 of the spin degeneracy because we distinguish in the calculation each kind of spin.

Problem 7. Paramagnetism of free atoms for arbitrary $\vec{J}$:

   Solution: The average total magnetic moment in the magnetic field $\vec{B}$ applied in the z direction is

$$M^z = g\mu_B \sum_{i=1}^N J_i^z \qquad (11.19)$$

where N is the total number of atoms. J_i^z is the z component of the moment $\vec{J}_i$ of the i-th atom. The Zeeman energy of the magnetic moment of the i-th atom in the field is $H_i = -\vec{M}_i \cdot \vec{B} = -M_i^z B$. The average value $< J_i^z >$ is calculated by the canonical description (see Appendix A) as follows

$$< J_i^z > = \frac{\sum_{J_i^z=-J}^{J} J_i^z e^{-\beta H_i}}{Z_i} \tag{11.20}$$

where $\beta = \frac{1}{k_B T}$ and Z_i the partition function defined by

$$Z_i = \sum_{J_i^z=-J}^{J} \exp(\beta B M_i^z) = \sum_{J_i^z=-J}^{J} \exp(\beta g \mu_B B J_i^z)$$

$$= \frac{\sinh[\beta g \mu_B B J(J+\frac{1}{2})]}{\sinh[\frac{1}{2}\beta g \mu_B B J]} \tag{11.21}$$

where we have used the formula of geometric series. We obtain

$$\sum_{J_i^z=-J}^{J} J_i^z e^{-\beta H_i} = \frac{\partial}{\partial \alpha} \sum_{J_i^z=-J}^{J} e^{\beta g \mu_B J_i^z} \quad (\alpha \equiv \beta g \mu_B)$$

$$= \frac{\partial}{\partial \alpha} Z_i$$

$$= \frac{(J+\frac{1}{2}) \cosh(J+\frac{1}{2})\alpha \sinh\frac{\alpha}{2} - \frac{1}{2} \sinh(J+\frac{1}{2})\alpha \cosh\frac{\alpha}{2}}{\sinh^2 \frac{\alpha}{2}}$$

Thus

$$< J^z > = J B_J(x) \tag{11.22}$$

$B_J(\cdots)$ is the Brillouin function defined by

$$B_J(x) = \frac{2J+1}{2J} \coth \frac{(2J+1)x}{2J} - \frac{1}{2J} \coth \frac{x}{2J} \tag{11.23}$$

where

$$x = \beta g \mu_B J B \tag{11.24}$$

We get

$$\overline{m} = \frac{< M^z >}{V} = g \mu_B \sum_{i=1}^{N} < J_i^z > = \frac{N g \mu_B J}{V} B_J\left(\frac{g \mu_B J B}{k_B T}\right) \tag{11.25}$$

At high temperatures, one has $B_J(x) \simeq \frac{J+1}{3J}x - \dots$. Thus,

$$\chi = \frac{m}{B} \simeq \frac{N}{V}(g\mu_B)^2 \frac{J(J+1)}{3} \frac{1}{k_BT} \qquad (11.26)$$

This is the Curie's $1/T$ law. At low T, one has $B_J(x) \simeq 1 - \frac{1}{J}\exp(-x/J)$. One gets

$$\overline{m} \simeq \frac{N}{V}g\mu_B J[1 - \frac{1}{J}\exp(-x/J)] \to_{T\to 0} \frac{N}{V}g\mu_B J \qquad (11.27)$$

The value at $T = 0$ corresponds to the saturated value of $\overline{m}$ ($= \frac{N}{V}g\mu_B J$).

Problem 8. Langevin's theory of diamagnetism:

Solution: The first explanation of the diamagnetism has been given by the theory of Langevin using the classical mechanics:

a) We have the relation $m = iA$

b) We have for an electron, $m = eA/\tau$ where τ is the period (time necessary to make a full circular motion), e electron charge. With $\tau = 2\pi r/v$ (r: radius, v: velocity) and $A = \pi r^2$, we have the orbital magnetic moment written as $m = evr/2$.

c) The variation of the magnetic flux ϕ induced by $\vec{B}$ gives rise to an electric field

$$E = -\frac{1}{L}\frac{d\phi}{dt} = -\frac{1}{L}\frac{d(AB)}{dt} = -\frac{A}{L}\frac{dB}{dt} \quad (L = 2\pi r).$$

Acceleration:

$$a = \frac{dv}{dt} = \frac{eE}{m_e} = -\frac{er}{2m_e}\frac{dB}{dt} = -\frac{\mu_0 er}{2m_e}\frac{dH}{dt}.$$

Integrating this relation, we have

$$\int_{v_1}^{v_2} dv = v_2 - v_1 = -\frac{\mu_0 er}{2m_e}H.$$

The variation of the magnetic moment of the electron is thus

$$\Delta m = er(v_2 - v_1)/2 = -\frac{\mu_0 e^2 r^2 H}{4m_e}.$$

The negative sign indicates the diamagnetic character.

d) We project the orbit of radius r on a plane perpendicular to the field: the radius of the projected orbit $R = r\cos\theta$. We replace, in the above result of Δm, r by $r\cos\theta$. The average on all directions is obtained by integrating on θ:

$$
\begin{aligned}
\Delta m &= -\frac{\mu_0 e^2 H}{4m_e} \int r^2 \cos^2\theta \sin\theta d\theta \\
&= -\frac{\mu_0 e^2 r^2 H}{4m_e} [-\frac{\cos^3\theta}{3}]_0^\pi \\
&= -\frac{\mu_0 e^2 r^2 H}{6m_e}
\end{aligned} \tag{11.28}
$$

e) If there are Z electrons in an atom:

$$
\Delta \mathcal{M} = -N \frac{\mu_0 Z e^2 r^2 H}{4m_e}
$$

where N is the number of atoms in a volume unit: $N = N_A \rho / M$ (N_A: Avogadro number). The resulting susceptibility is

$$
\chi = \Delta \mathcal{M} / H = -N \frac{\mu_0 Z e^2 r^2 H}{4m_e} < 0.
$$

Note: See section 1.4 for a quantum treatment.

Problem 9. Langevin's theory of paramagnetism:
   Solution: The case of discrete spins of magnitude $1/2$ has been studied in section 1.2. Here we study the case of continuous spins (Langevin's theory).
   Langevin's theory: The Maxwell-Boltzmann's probability for a state of energy $E = -\vec{m} \cdot \vec{B}$ (Zeeman energy) is

$$
p(E) = C \exp(-\beta E) = \exp(\beta \vec{m} \cdot \vec{B})
$$

where C is the normalization constant.
   In an isotropic material, magnetic moments $\vec{m}$ are distributed in random directions. The number of moments in an elementary volume is $dn = C 2\pi \sin\theta d\theta \exp(\beta m B \cos\theta)$. The total number of moments in a volume unit is

$$
N = C 2\pi \int_0^\pi \sin\theta d\theta \exp(\beta m B \cos\theta),
$$

thus

$$C = N/2\pi \int_0^\pi \sin\theta d\theta \exp(\beta m B \cos\theta).$$

The component along the z axis of the total resulting magnetic moment, namely magnetization, is

$$M = \int_0^\pi m \cos\theta dn$$

$$= \frac{Nm \int_0^\pi \cos\theta \sin\theta d\theta \exp(\beta m B \cos\theta)}{\int_0^\pi \sin\theta d\theta \exp(\beta m B \cos\theta)}$$

$$= Nm\mathcal{L}(\frac{mB}{k_B T})$$

$$= Nm\mathcal{L}(\frac{\mu_0 m H}{k_B T}) \tag{11.29}$$

where we used $x = \cos\theta$ $(dx = -\sin\theta d\theta)$ for integration, and $\mathcal{L}(y) \equiv \coth(y) - \frac{1}{y}$. μ_0, vacuum permeability, is equal to $1,257 \times 10^{-6}$ H/m. For weak fields, an expansion of the Langevin function $\mathcal{L}(y)$ gives $M = N\mu_0 m^2 H/(3k_B T)$, leading to the Curie's law $\chi = M/H = N\mu_0 m^2/(3k_B T) > 0$ (paramagnetism).

Problem 10. Variation of the band gap:

   Solution: In a strong applied magnetic field $\vec{B}$, the Landau's levels are given by Eq. (1.51): $E_n = (n + 1/2)\hbar\omega_c + \hbar^2 k_z^2/2m^*$ where $\omega_c = eB/m^*$. This is for electrons in the conduction band (CB) and valence band (VB) with $m^* = m_c^*$ (CB) and $m^* = m_v^*$ (VB). The zero-field CB and VB become multiple bands (Landau's levels) and the initial band gap becomes wider under the effect of the applied field. Its variation is given by the distance between the first Landau's levels of CB and VB, namely $\Delta E_g = \hbar eB/2m_c^* + \hbar eB/2|m_v^*|$.

Problem 11. Paramagnetic resonance:

   Solution: As the orbital angular momentum $\vec{L}$, the spin angular momentum $\vec{S}$ gives rise to a magnetic momentum $\vec{M}$

$$\vec{M} = h\gamma\vec{S} = -g\mu_B\vec{S}$$

with the electron magnetogyric ratio γ, the Bohr magneton $\mu_B = e\hbar/2m_e$ and the Landé factor $g \simeq 2$. In an applied magnetic field $\vec{H}_0$ along the z direction, the energy of the electron splits into two levels according to its direction with respect to the field: $E = -\vec{M} \cdot \vec{H}_0 = \pm\frac{1}{2}g\mu_B H_0$ (S=1/2). If we send a photon of energy $h\nu$ which matches the energetic difference ΔE between the two states, there is an absorption: $\Delta E = h\nu = \hbar\omega = g\mu_B H_0$. This phenomenon is called "electron paramagnetic resonance" (EPR) which is used to determine properties of molecules and atoms.

a) Equation of motion: the motion of a magnetic moment $\vec{M}$ in a magnetic field $\vec{H}_0$ is given by the following equation of motion

$$\frac{d\vec{M}}{dt} = \gamma\vec{H}_0 \wedge \vec{M} \qquad (11.30)$$

At equilibrium ($\vec{M}$ lies along $\vec{H}_0$), $d\vec{M}/dt = 0$ so a static magnetic field $\vec{H}_0$ applied along z cannot detect the magnetization vector $\vec{M}$. When $\vec{M}$ makes an angle θ with $\vec{H}_0$, the above equation gives rise to

$$\frac{dM_x}{dt} = -\gamma H_0 M_y$$
$$\frac{dM_y}{dt} = \gamma H_0 M_x$$
$$\frac{dM_z}{dt} = 0 \qquad (11.31)$$

The solution is M_z =constant, $M_x = A\cos\omega_0 t$ and $M_y = A\sin\omega_0 t$ where $\omega_0 = \gamma H_0$ (see Fig. 11.3). Therefore, $\vec{M}$ precesses around z with the Larmor frequency ω_0.

b) To detect an EPR signal, besides $\vec{H}_0$ an additional oscillating micro-wave field is necessary: $\vec{H}_1(t) = H_1[\cos(\omega_1 t)\vec{i} + \sin(\omega_1 t)\vec{j}]$ is applied in the xy plane. The resulting field is $\vec{H} = \vec{H}_1 + H_0\vec{k}$ $[(\vec{i},\vec{j},\vec{k})$: unit vectors on x, y and z axes].

Let us use the system of rotating coordinates $\zeta' = (x',y',z')$ with $z' = z$: the $x'y'$ plane rotates about z with frequency ω_1 (see Fig. 11.4). The rotating applied field $\vec{H}_1$ rotates with the $x'y'$ plane: it is thus "constant" in the system of rotating coordinates (see Fig. 11.5).

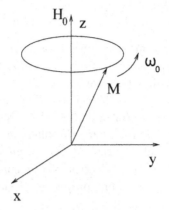

Fig. 11.3 A constant magnetic field $\vec{H}$ applied along the z direction: the magnetic moment $\vec{M}$ precesses around z with the Larmor frequency ω_0.

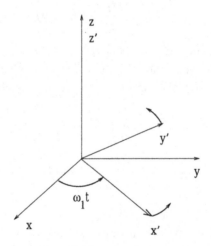

Fig. 11.4 System of rotating coordinates (x', y', z') with $z' = z$ rotates around the z axis with the applied field frequency ω_1.

In the laboratory coordinates $\zeta = (xyz)$, the equation of motion is written as

$$[\frac{d\vec{M}}{dt}]_\zeta = \gamma \vec{H} \wedge \vec{M} \tag{11.32}$$

The relation between quantities measured in the systems of fixed and rotating coordinates is

$$[d\vec{M}]_\zeta = [d\vec{M}]_{\zeta'} + \text{motion of } \zeta' \text{ with respect to } \zeta$$
$$= [d\vec{M}]_{\zeta'} + [\omega_1 dt]\vec{k}' \wedge \vec{M} \tag{11.33}$$

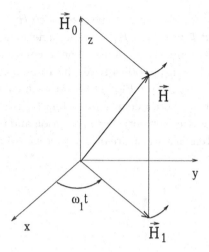

Fig. 11.5 A constant magnetic field $\vec{H}_0$ applied along the z direction and a rotating field $\vec{H}_1$ of frequency ω_1 applied in the xy plane. The resulting field $\vec{H} = \vec{H}_0 + \vec{H}_1$ rotates around the z axis with the frequency ω_1.

where $\vec{k}'$ is the unit vector on the z' axis. The equation of motion is thus

$$[\frac{d\vec{M}}{dt}]_\varsigma = [\frac{d\vec{M}}{dt}]_{\varsigma'} + \omega_1 \vec{k}' \wedge \vec{M} \tag{11.34}$$

The last term in the above equation is the "drift" term due to the rotation of $x'y'$, at the frequency ω_1, with respect to xy coordinates. Replacing (11.32) in the above equation, we have

$$[\frac{d\vec{M}}{dt}]_{\varsigma'} = [\gamma\vec{H} - \omega_1\vec{k}'] \wedge \vec{M} = \gamma\vec{H}_E \wedge \vec{M} \tag{11.35}$$

where $\vec{H}_E$ is the effective magnetic field in the rotating frame which is given by

$$\vec{H}_E = H_1\vec{i}' + (H_0 - \omega_1/\gamma)\vec{k}' \tag{11.36}$$

where $\vec{i}'$ is the unit vector on the x' axis. We have used $\vec{H} = H_1\cos(\omega_1 t)\vec{i} + H_1\sin(\omega_1 t)\vec{j} + H_0\vec{k}$ with $\vec{i}$ and $\vec{j}$ replaced by $\vec{i} = \cos(\omega_1 t)\vec{i}' - \sin(\omega_1 t)\vec{j}'$ and $\vec{j} = \sin(\omega_1 t)\vec{i}' + \cos(\omega_1 t)\vec{j}'$.

The whole system is shown in Fig. 11.6: $\vec{H}_E$ rotates around the z axis with a frequency ω_1 and the magnetic moment rotates around the effective field $\vec{H}_E$ with the frequency $\Omega = \gamma H_0 - \omega_1 = \omega_0 - \omega_1$. The resonance occurs when ω_1 is chosen so that the deviation of $\vec{M}$ corresponds to the annihilation of the Larmor frequency by ω_1, namely $\Omega = 0$. We have at the resonance $\vec{H}_E = \vec{H}_1$ perpendicular to $\vec{H}_0$, and $\hbar\omega_1 = g\mu_B H_1 = \hbar\omega_0$. The magnetic moment then precesses around the x' axis. The z component of $\vec{M}$ changes therefore its sign periodically at the resonance ($\vec{H}_E$ tends to $\vec{H}_1$ with increasing ω_1, see Fig. 11.6). At each change of sign of the z component, a photon energy $h\nu = g\mu_B H_1$ is absorbed or emitted by the magnetic moment. The absorption and emission of a photon by the system is a signature of the existence of a magnetic moment in the system.

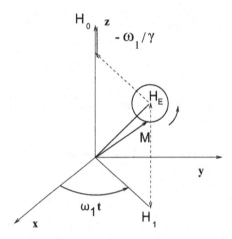

Fig. 11.6   A constant magnetic field $\vec{H}_0$ applied along the z direction and a rotating field $\vec{H}_1$ of frequency ω_1 applied in the xy plane: the magnetic moment $\vec{M}$ precesses around the axis of the effective field $\vec{H}_E$ with the frequency $\omega_0 - \omega_1$.

Problem 12. Nuclear Magnetic Resonance:

Solution: We use a quantum treatment here.

a) We consider first $H_1 = 0$. We show that the spin precesses around the z axis with frequency $\omega = \mu_B H_0/\hbar$ (Larmor frequency). The Hamiltonian reads $\mathcal{H} = -\mu_B \vec{H}_0 \cdot \vec{\sigma} = -\mu_B H_0 \sigma_z$. The time-dependent Schrödinger equation reads

$$i\hbar\frac{\partial \Psi(t)}{\partial t} = \mathcal{H}\Psi(t) = -\mu_B H_0 \sigma_z \Psi(t) \tag{11.37}$$

The solution is of the form $\Psi(t) = e^{i\omega\sigma_z t}\Psi(0)$. Replacing this into the above equation, we obtain $\omega = \mu_B H_0/\hbar$ and

$$e^{i\omega\sigma_z} = \begin{pmatrix} e^{i\omega t} & 0 \\ 0 & e^{i\omega t} \end{pmatrix}$$

When $t = 0$ (before the application of $\vec{H}_1$), the eigen-state is $\Psi(0) = a| \uparrow> +b| \downarrow>$ with $a = b = 1/\sqrt{2}$ (normalization coefficients) and the eigen-frequency is ω.

b) Now, we consider

$$\vec{H} = \begin{pmatrix} H_1 \cos \omega_r t \\ H_1 \sin \omega_r t \\ H_0 \end{pmatrix}. \tag{11.38}$$

The spin $\vec{\sigma}$ in the field $\vec{H}$ has the Hamiltonian $\mathcal{H} = -\mu_B \vec{H} \cdot \vec{\sigma}$. The time-dependent Schrödinger equation $\mathcal{H}\Psi(t) = i\hbar \, \partial\Psi(t)/\partial t$ gives

$$\frac{\partial\Psi(t)}{\partial t} = i \left[\omega_1 \sigma_x + \left(\omega_0 + \frac{\omega_r}{2} \right) \sigma_z \right] \Psi(t) \tag{11.39}$$

where $\omega_0 = \mu_B H_0/\hbar$ and $\omega_1 = \mu_B H_1/\hbar$.

Demonstration: The Schrödinger equation reads

$$i\hbar \frac{\partial\Psi(t)}{\partial t} = -\mu_B \vec{\sigma} \cdot \vec{H}\Psi(t)$$

$$\frac{\partial\Psi(t)}{\partial t} = i \left(\omega_1 \sigma_x \cos \omega_r t + \omega_1 \sigma_y \sin \omega_r t + \omega_0 \sigma_z \right) \Psi(t)$$

To eliminate the time dependence in the above expression, we use the transformation $\Psi(t) \to e^{-i\sigma_z \omega_r t/2}\Psi(t)$. The right-hand side becomes

$$-i\sigma_z \frac{\omega_r}{2} e^{-i\sigma_z \omega_r t/2}\Psi(t) + e^{-i\sigma_z \omega_r t/2} \frac{\partial\Psi(t)}{\partial t} = i(\omega_1 \sigma_x \cos \omega_r t$$

$$+ \omega_1 \sigma_y \sin \omega_r t + \omega_0 \sigma_z)e^{-i\sigma_z \omega_r t/2}\Psi(t)$$

$$\frac{\partial\Psi(t)}{\partial t} = ie^{i\sigma_z \omega_r t/2}[\omega_1 \sigma_x \cos \omega_r t + \omega_1 \sigma_y \sin \omega_r t$$

$$+ \left(\omega_0 + \frac{\omega_r}{2} \right) \sigma_z]e^{-i\sigma_z \omega_r t/2}\Psi(t) \quad (11.40)$$

We calculate each term of the right-hand side as follows

$$
e^{i\sigma_z \omega_r t/2} \sigma_x e^{-i\sigma_z \omega_r t/2} = \begin{pmatrix} e^{i\omega_r t/2} & 0 \\ 0 & e^{-i\omega_r t/2} \end{pmatrix} \begin{pmatrix} 0 & 1 \\ 1 & 0 \end{pmatrix}
$$

$$
\times \begin{pmatrix} e^{-i\omega_r t/2} & 0 \\ 0 & e^{i\omega_r t/2} \end{pmatrix}
$$

$$
= \begin{pmatrix} 0 & e^{i\omega_r t} \\ e^{-i\omega_r t} & 0 \end{pmatrix}
$$

$$
e^{i\sigma_z \omega_r t/2} \sigma_y e^{-i\sigma_z \omega_r t/2} = \begin{pmatrix} e^{i\omega_r t/2} & 0 \\ 0 & e^{-i\omega_r t/2} \end{pmatrix} \begin{pmatrix} 0 & -i \\ i & 0 \end{pmatrix}
$$

$$
\times \begin{pmatrix} e^{-i\omega_r t/2} & 0 \\ 0 & e^{i\omega_r t/2} \end{pmatrix}
$$

$$
= \begin{pmatrix} 0 & -ie^{i\omega_r t} \\ ie^{-i\omega_r t} & 0 \end{pmatrix}
$$

$$
e^{i\sigma_z \omega_r t/2} \sigma_z e^{-i\sigma_z \omega_r t/2} = \begin{pmatrix} e^{i\omega_r t/2} & 0 \\ 0 & e^{-i\omega_r t/2} \end{pmatrix} \begin{pmatrix} 1 & 0 \\ 0 & -1 \end{pmatrix}
$$

$$
\times \begin{pmatrix} e^{-i\omega_r t/2} & 0 \\ 0 & e^{i\omega_r t/2} \end{pmatrix}
$$

$$
= \sigma_z \tag{11.41}
$$

Using these expressions, the Schrödinger equation becomes

$$
\frac{\partial \Psi(t)}{\partial t}
$$

$$
= i[\omega_1 \begin{pmatrix} 0 & e^{i\omega_r t}\left(\cos \omega_r t - i \sin \omega_r t\right) \\ e^{-i\omega_r t}\left(\cos \omega_r t + i \sin \omega_r t\right) & 0 \end{pmatrix}
$$

$$
+ \left(\omega_0 + \frac{\omega_r}{2}\right)\sigma_z]\Psi(t) \tag{11.42}
$$

Replacing $\cos(\omega_r t) \pm i \sin(\omega_r t)$ by $e^{\pm i\omega_r t}$, we have

$$
\frac{\partial \Psi(t)}{\partial t} = i\left[\omega_1 \sigma_x + \left(\omega_0 + \frac{\omega_r}{2}\right)\sigma_z\right]\Psi(t) \tag{11.43}
$$

Resonance frequency:

From Eq. (11.39), we see that the precessing frequency is $(\omega_1, 0, \omega_0 + \omega_r/2)$. If we choose $\omega_r = -2\omega_0$, then the spin precesses around the x axis with frequency $2\omega_1$: the factor 2 comes from the fact that $\vec{\sigma}$ is twice its spin $\vec{S}$ [see (1.3)- (1.5)]. The z component of the spin changes its sign by this precession around the x axis, giving rise to a light emission. This is the NMR principle. It suffices to make a ω_r scanning to tune for resonance.

11.2 Solutions of problems of chapter 2

Problem 1. System of two electrons — Fermi hole:
Solution:

a) The Slater determinant wave function $\Psi(\vec{r}_1, \sigma_1, \vec{r}_2, \sigma_2)$ for a two-electron system is written as

$$\begin{vmatrix} \psi_{f_1}(q_1) & \psi_{f_1}(q_2) \\ \psi_{f_2}(q_1) & \psi_{f_2}(q_2) \end{vmatrix}$$

where $\psi_{f_i}(q_i) = \psi_{1s,\sigma_i}(\vec{r}_i, \zeta_i) = \varphi_{1s}(\vec{r}_i)S_{\sigma_i}(\zeta_i)$ [see Eq. (2.14)].
For two electrons in a He atom in the ground state for example, we have $\sigma_1 = +$ and $\sigma_2 = -$:
$S_{\sigma_1}(\uparrow) = 1$, $S_{\sigma_1}(\downarrow) = 0$, $S_{\sigma_2}(\uparrow) = 0$, $S_{\sigma_2}(\downarrow) = 1$.
We write $\Psi(\vec{r}_1, \sigma_1, \vec{r}_2, \sigma_2) = \varphi_{1s}(\vec{r}_1)S_{\sigma_1}(\uparrow)\varphi_{1s}(\vec{r}_2)S_{\sigma_2}(\downarrow) = \varphi_{1s}(\vec{r}_1)\varphi_{1s}(\vec{r}_2)$
The Hamiltonian is

$$\mathcal{H} = \frac{p_1^2}{2m} + \frac{p_2^2}{2m} + \frac{e^2}{|\vec{r}_1 - \vec{r}_2|} \tag{11.44}$$

The system energy is thus

$$E = \; < \varphi_{1s}(\vec{r}_1)\varphi_{1s}(\vec{r}_2)| \frac{p_1^2}{2m} + \frac{p_2^2}{2m} + \frac{e^2}{|\vec{r}_1 - \vec{r}_2|} |\varphi_{1s}(\vec{r}_1)\varphi_{1s}(\vec{r}_2) >$$

$$= E_1 + E_2 + < \varphi_{1s}(\vec{r}_1)\varphi_{1s}(\vec{r}_2)|\frac{e^2}{|\vec{r}_1 - \vec{r}_2|}|\varphi_{1s}(\vec{r}_1)\varphi_{1s}(\vec{r}_2) >$$

where $E_1 = E_2 =$ kinetic energy of each electron. The integral represents the direct interaction energy. The exchange interaction does not exist because the spins are antiparallel.

b) We take the Slater determinant with

$$\psi_{f_i}(q_j) = \frac{1}{\sqrt{2}} \frac{1}{\sqrt{\Omega}} e^{i\vec{k}_i \cdot \vec{r}_j} S_{\sigma_i}(\zeta_j)$$

where the normalization factor has been introduced.
We write for the case of two antiparallel spins: $\sigma_1 = +$, $\sigma_2 = -$.
Hence $S_{\sigma_1}(\uparrow) = 1$, $S_{\sigma_1}(\downarrow) = 0$, $S_{\sigma_2}(\downarrow) = 1$, $S_{\sigma_2}(\uparrow) = 0$. The corresponding wave function is

$$\Psi_{\vec{k}_1, \vec{k}_2}(\vec{r}_1, \uparrow, \vec{r}_2, \downarrow) = \frac{1}{2} \frac{1}{\Omega} e^{i\vec{k}_1 \cdot \vec{r}_1} e^{i\vec{k}_2 \cdot \vec{r}_2}$$

The probability of this state is

$$P(\uparrow,\downarrow) = \sum_{\vec{k}_1,\vec{k}_2} |\Psi_{\vec{k}_1,\vec{k}_2}(\vec{r}_1,\uparrow,\vec{r}_2,\downarrow)|^2 = \frac{1}{4}\frac{1}{\Omega^2}\sum_{\vec{k}_1,\vec{k}_2} 1 = \frac{1}{4}\frac{N^2}{\Omega^2}$$

(11.45)

We see that it is independent of the distance. The same thing is obtained for $P(\downarrow,\uparrow)$. The probability to find two antiparallel spins is thus $2 \times \frac{1}{4}\frac{N^2}{\Omega^2}$

In the case of two parallel spins, we have $\sigma_1 = +$, $\sigma_2 = +$, thus

$$\Psi_{\vec{k}_1,\vec{k}_2}(\vec{r}_1,\uparrow,\vec{r}_2,\uparrow) = \frac{1}{2}\frac{1}{\Omega}[e^{i\vec{k}_1\cdot\vec{r}_1}e^{i\vec{k}_2\cdot\vec{r}_2} - e^{i\vec{k}_2\cdot\vec{r}_1}e^{i\vec{k}_1\cdot\vec{r}_2}] \quad (11.46)$$

We have then

$$P(\uparrow,\uparrow) = \sum_{\vec{k}_1,\vec{k}_2} |\Psi_{\vec{k}_1,\vec{k}_2}(\vec{r}_1,\uparrow,\vec{r}_2,\uparrow)|^2$$

$$= \frac{1}{4}\frac{1}{\Omega^2}\sum_{\vec{k}_1,\vec{k}_2}(e^{-i\vec{k}_1\cdot\vec{r}_1}e^{-i\vec{k}_2\cdot\vec{r}_2} - e^{-i\vec{k}_2\cdot\vec{r}_1}e^{-i\vec{k}_1\cdot\vec{r}_2})$$

$$\times(e^{i\vec{k}_1\cdot\vec{r}_1}e^{i\vec{k}_2\cdot\vec{r}_2} - e^{i\vec{k}_2\cdot\vec{r}_1}e^{i\vec{k}_1\cdot\vec{r}_2})$$

$$= \frac{1}{4}\frac{1}{\Omega^2}\sum_{\vec{k}_1,\vec{k}_2} 2 - \frac{1}{4}\frac{1}{\Omega^2}\sum_{\vec{k}_1,\vec{k}_2}[e^{i\vec{k}_1\cdot(\vec{r}_2-\vec{r}_1)}e^{-i\vec{k}_2\cdot(\vec{r}_2-\vec{r}_1)}$$

$$+e^{-i\vec{k}_1\cdot(\vec{r}_2-\vec{r}_1)}e^{i\vec{k}_2\cdot(\vec{r}_2-\vec{r}_1)}]$$

$$= 2 \times \frac{1}{4}\frac{N^2}{\Omega^2} - \frac{1}{4}\frac{1}{\Omega^2}\times 2I^2$$

(11.47)

where

$$I \equiv \sum_{\vec{k}} e^{-i\vec{k}\cdot(\vec{r}_2-\vec{r}_1)}$$

$$= \frac{\Omega}{(2\pi)^3}\int d\vec{k}\, e^{-ikr\cos\theta} = \frac{\Omega}{(2\pi)^3}2\pi\int_0^{k_F} k^2 dk$$

$$\times \int_{-1}^{1} d(\cos\theta) e^{-ikr\cos\theta}$$

$$= \frac{\Omega}{(2\pi)^3}2\pi\int_0^{k_F} k^2 dk [e^{ikr} - e^{-ikr}]\frac{1}{ikr}$$

$$= \frac{\Omega}{(2\pi)^3}\frac{4\pi}{r}\int_0^{k_F} k\,dk \sin kr = \frac{\Omega}{(2\pi)^3}\frac{4\pi}{r}[-k_F\cos k_F r$$

$$+ \int_0^{k_F} dk\cos kr]$$

$$= \frac{\Omega}{(2\pi)^3}4\pi[\frac{-k_F\cos k_F r}{r} + \frac{\sin k_F r}{r^2}] \qquad (11.48)$$

Since $N = \sum_{\vec{k}} = \frac{\Omega}{(2\pi)^3}\int_0^{k_F} 4\pi k^2 dk = \frac{\Omega}{(2\pi)^3}[\frac{4\pi k_K^3}{3}]$, Eq. (11.47) becomes

$$P(\uparrow,\uparrow) = 2\times\frac{1}{4}\frac{N^2}{\Omega^2}\left[1 - \frac{1}{N^2}(4\pi)^2[\frac{-k_F\cos k_F r}{r} + \frac{\sin k_F r}{r^2}]\right]$$

$$= \frac{1}{2}\frac{1}{(2\pi)^6}[\frac{4\pi k_F^3}{3}]^2$$

$$\times\left[1 - [\frac{3}{4\pi}]^2(4\pi)^2[\frac{-k_F\cos k_F r + \sin k_F r}{k_F^3 r^3}]^2\right]$$

When $r \to 0$, $3[\frac{-k_F\cos k_F r+\sin k_F r}{k_F^3 r^3}] \to 1 - \frac{1}{10}k_F^3 r^2$ (expansion at the order of $(k_F r)^5$ of $\sin k_F r$ and of $\cos k_F r$). We see that $P(\uparrow,\uparrow) \to 0$ when $r \to 0$. This region of small r has a deficit of parallel spins. It explains the Pauli exclusion principle. It is called the Fermi hole. We show in Fig. 11.7 the function $\left[1 - 9[\frac{-k_F\cos k_F r+\sin k_F r}{k_F^3 r^3}]^2\right]$.

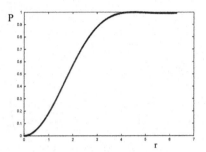

Fig. 11.7 Probability to find two parallel spins versus their relative distance, with $k_F = 1$.

Problem 2. Theorem of Koopmann:
   Solution:

$$E_{2N} \equiv\, < \Psi_{2N}|\mathcal{H}_{2N}|\Psi_{2N} >$$

$$= \sum_{i,\sigma} < \psi_{i,\sigma}(\vec{r}_i)|\frac{p_i^2}{2m} + V(\vec{r} - i)|\psi_{i,\sigma}(\vec{r}_i) >$$

$$+ \frac{1}{2} \sum_{i,j,\sigma,\sigma'} < \psi_{i,\sigma}(\vec{r}_i)\psi_{j,\sigma'}(\vec{r}_j)|\frac{e^2}{|\vec{r}_i - \vec{r}_j|}|\psi_{i,\sigma}(\vec{r}_i)\psi_{j,\sigma'}(\vec{r}_j) >$$

$$- \frac{1}{2} \sum_{i,j,\sigma,\sigma'} \delta(\sigma,\sigma')$$

$$\times\, < \psi_{i,\sigma}(\vec{r}_i)\psi_{j,\sigma'}(\vec{r}_j)|\frac{e^2}{|\vec{r}_i - \vec{r}_j|}|\psi_{j,\sigma'}(\vec{r}_i)\psi_{i,\sigma}(\vec{r}_j) >$$

Putting $\psi_{i,\sigma}(\vec{r}_i) \equiv \varphi_i(\vec{r})S_\sigma$ where S_σ is the spin part of the wave function, Eq. (2.14), we obtain

$$E_{2N} = 2 \sum_{i=1}^{N} T(i) + 2 \sum_{i,j=1}^{N} K(i,j) - \sum_{i,j=1}^{N} J(i,j) \qquad (11.49)$$

where the factors result from the spin sums: ($\sum_\sigma 1 = 2$, $\sum_{\sigma,\sigma'} 1 = 4$, $\sum_{\sigma,\sigma'} \delta(\sigma,\sigma')1 = 2$).
For a system of $2N - 1$ electrons, the sums on the orbitals i and j going from $i,j = 1$ to $i,j = N - 1$ (two electrons per orbital gives $2N - 2$ electrons). The last orbital is occupied only by one

electron. The same calculation gives

$$E_{2N-1} = 2 \sum_{i=1}^{N-1} T(i) + T(N) + 2 \sum_{i,j=1}^{N-1} K(i,j) + 2 \sum_{i}^{N-1} K(i,N)$$

$$- \sum_{i,j=1}^{N-1} J(i,j) - \sum_{i=1}^{N-1} J(i,N) \qquad (11.50)$$

The difference in energy is

$$E_{2N-1} - E_{2N} = -T(N) - 2 \sum_{i=1}^{N} K(i,N) + \sum_{i=1}^{N} J(i,N) \qquad (11.51)$$

On the other hand, the Hartree-Fock equation gives

$$\epsilon_i = T(i) + 2 \sum_{k=1}^{N} K(k,i) - \sum_{k=1}^{N} J(k,i) \qquad (11.52)$$

We take $i = N$, we obtain then
$$E_{2N-1} - E_{2N} = -\epsilon_N$$
The quantity $-\epsilon_N$ is thus the energy necessary to remove the electron in the state N from the system (theorem of Koopmann).

Problem 3. Screened Coulomb potential, Thomas-Fermi approximation:
Solution:

a) The Poisson equation is

$$\nabla^2 \varphi(\vec{r}) = \frac{e}{\epsilon_0}[n(\vec{r}) - n_0] \qquad (11.53)$$

where $n(\vec{r})$ is the density of electrons at $\vec{r}$, and n_0 the non perturbed electron density. We have in the non perturbed region

$$\epsilon_F^0 = \frac{\hbar^2}{2m}(3\pi^2 n_0)^{2/3} \quad \text{(Fermi level at } T = 0) \qquad (11.54)$$

b) In the perturbed region

$$\mu = \epsilon_F(\vec{r}) - e\varphi(\vec{r})$$

$$\simeq \frac{\hbar^2}{2m}[3\pi^2 n(\vec{r})]^{2/3} - e\varphi(\vec{r})$$

We suppose the perturbation is weak, we have $\mu \simeq \epsilon_F^0$ so that

$$\epsilon_F(\vec{r}) - \epsilon_F^0 \simeq e\varphi(\vec{r}) \qquad (11.55)$$

We expand $\epsilon_F(\vec{r})$ around ϵ_F^0:

$$\epsilon_F(\vec{r}) \simeq \epsilon_F^0 + (n - n_0)[\frac{d\epsilon_F(\vec{r})}{dn}]_{n=n_0}$$

$$\epsilon_F(\vec{r}) - \epsilon_F^0 = (n - n_0)[\frac{d\epsilon_F(\vec{r})}{dn}]_{n=n_0}$$

We have thus

$$(n - n_0)[\frac{d\epsilon_F(\vec{r})}{dn}]_{n=n_0} = e\varphi(\vec{r}) \qquad (11.56)$$

We calculate

$$[\frac{d\epsilon_F(\vec{r})}{dn}]_{n=n_0} = \frac{2}{3}\frac{\epsilon_F^0}{n_0} \qquad (11.57)$$

From (11.56), we have

$$(n - n_0) = \frac{3}{2}\frac{n_0}{\epsilon_F^0}e\varphi(\vec{r}) \qquad (11.58)$$

Equation (11.53) becomes

$$\nabla^2\varphi(\vec{r}) = \frac{3n_0e^2}{2\epsilon_0\epsilon_F^0}\varphi(\vec{r}) \equiv \lambda^2\varphi(\vec{r}) \qquad (11.59)$$

If $\varphi(\vec{r}) = \varphi(r)$ (spherical symmetry), then

$$\nabla^2\varphi(r) = (\frac{d^2}{dr^2} + \frac{2}{r}\frac{d}{dr})\varphi(r) = \lambda^2\varphi(r) \qquad (11.60)$$

We find a solution of the form $\varphi(r) = q\frac{e^{-\lambda r}}{r}$ by substituting this into the above equation. The larger λ is, the more strongly $\varphi(r)$ decreases with increasing r.

Remark: λ is proportional to $n_0^{1/2}$ so that the denser the gas is, the stronger the screening becomes. The above approximation is called "Thomas-Fermi approximation".

Problem 4. Paradox of the Hartree-Fock approximation:
Solution:

a) We show that the density of states $\rho(\epsilon_i)$ of an electron calculated using the Hartree-Fock result given by (2.48) is equal to 0 at $k_i = k_F$. The density of states (see Problem 3) can be written as

$$\rho(\epsilon) = \frac{d\mathcal{N}}{d\epsilon} = \frac{d\mathcal{N}(\epsilon)}{d\vec{k}} \cdot \frac{d\vec{k}}{d\epsilon} \qquad (11.61)$$

Using (2.48), we calculate $\frac{d\epsilon}{dk_i}$. We will see that this derivative diverges at $k_i = k_F$ due to the logarithmic term. The density of states is thus zero at k_F. This contradicts the definition of k_F which is the last *occupied* level.

We calculate the electron effective mass m^* defined by (2.49):

$$\frac{1}{m^*_{\alpha\beta}} = \frac{1}{\hbar^2} \frac{\partial^2 \epsilon_{\vec{k}_i}}{\partial k_\alpha \partial k_\beta} \tag{11.62}$$

where $\alpha, \beta = x, y, z$. Again here, the second derivative diverges at k_F using (2.48). So the effective mass is 0. This is not physically correct.

b) Introduce the screening factor $e^{-\mu r}$ to the Coulomb potential where μ is a screening constant [see Eq. (2.72)]. Using (2.44) without taking the limit $\mu \to 0$, we can calculate in the same manner as for (2.48) and we obtain

$$\epsilon_i = \frac{\hbar^2 k_i^2}{2m} - \frac{e^2}{2\pi} \left[\frac{k_F^2 + \mu^2 - k_i^2}{k_i} \ln \frac{(k_F + k_i)^2 + \mu^2}{(k_F - k_i)^2 + \mu^2} + 2k_F \right.$$
$$\left. + 2\mu (\arctan(\frac{k_i - k_F}{\mu}) - \arctan(\frac{k_i + k_F}{\mu})) \right] \tag{11.63}$$

The effective mass and the density of states do not diverge anymore at $k_i = k_F$. We have

$$\frac{1}{m^*} = \frac{1}{m} + \frac{1}{2\pi\hbar^2 k_F} \{ (1 + \frac{\mu^2}{2k_F^2}) \ln(1 + \frac{4k_F^2}{\mu^2}) - 2 \} \tag{11.64}$$

where m is the free electron mass (from the kinetic term). The second term is positive, therefore $m^* < m$.

Problem 5. Hydrogen molecule:

Solution:

a) The spatial Slater determinant wave function is written as

$$\Psi_\pm(\vec{r}_1, \zeta_1, \vec{r}_2, \zeta_2) = \frac{1}{\sqrt{2(1 \pm \alpha^2)}} [\varphi_A(\vec{r}_1)\varphi_B(\vec{r}_2) \pm \varphi_A(\vec{r}_2)\varphi_B(\vec{r}_1)] \tag{11.65}$$

where the positive and negative signs correspond respectively to symmetric and antisymmetric wave functions. When combined with spin wave functions the total wave function should be anti-symmetric with respect to the exchange of two particles. So, for

a symmetric spatial wave function, we should multiply by an antisymmetric spin function and vice-versa. For two spins, there are four spin functions:
- three triplets which are symmetric:

$|\uparrow>_1|\uparrow>_2$,

$|\uparrow>_1|\downarrow>_2 +|\downarrow>_1|\uparrow>_2,$

$|\downarrow>_1|\downarrow>_2$

and one singlet which is antisymmetric

$|\uparrow>_1|\downarrow>_2 -|\downarrow>_1|\uparrow>_2.$

The energies of these states are

$$
\begin{aligned}
E_\pm &= \frac{1}{2(1\pm\alpha^2)} < \varphi_A(\vec{r}_1)\varphi_B(\vec{r}_2)\\
&\pm\varphi_A(\vec{r}_2)\varphi_B(\vec{r}_1)|\mathcal{H}|\varphi_A(\vec{r}_1)\varphi_B(\vec{r}_2)\pm\varphi_A(\vec{r}_2)\varphi_B(\vec{r}_1) >\\
&= \frac{1}{1\pm\alpha^2}[< \varphi_A(\vec{r}_1)\varphi_B(\vec{r}_2)|\mathcal{H}|\varphi_A(\vec{r}_1)\varphi_B(\vec{r}_2)\\
&\pm < \varphi_A(\vec{r}_1)\varphi_B(\vec{r}_2)|\mathcal{H}|\varphi_A(\vec{r}_2)\varphi_B(\vec{r}_1) >]\\
&= \frac{1}{1\pm\alpha^2}[Q\pm J'] \qquad\qquad\qquad (11.66)
\end{aligned}
$$

We obtain

$$
\begin{aligned}
\Delta E &= E_+ - E_-\\
&= \frac{1}{1+\alpha^2}[Q+J'] - \frac{1}{1-\alpha^2}[Q-J'] = -2(J'-\alpha^2 Q)/(1-\alpha^4)
\end{aligned}
$$

b) The exchange interaction of two spins is

$$
\begin{aligned}
H &= -2J\vec{S}_1\cdot\vec{S}_2 = -J[(\vec{S}_1+\vec{S}_2)^2 - \vec{S}_1^2 - \vec{S}_2^2]\\
&= -J[S(S+1) - S_1(S_1+1) - S_2(S_2+1)]\\
&= -J[S(S+1) - 3/2] \qquad\qquad\qquad (11.67)
\end{aligned}
$$

because $S_1 = S_2 = 1/2$. In the triplet states $S = 1$ hence $E = 1/2$ while in the singlet state $S = 0$ which yields $E = -3/2$. The difference in energy is thus $2J$. Setting this equal to ΔE, we have

$$
J = (J'-\alpha^2 Q)/(1-\alpha^4)
$$

11.3 Solutions of problems of chapter 3

Problem 1. Free electron gas with second quantization:
Solution: In a free electron gas, only the kinetic term exists. In the second quantization, this term is given by (3.26). We have

$$
\begin{aligned}
\hat{T} &= \sum_{i,j}\langle i|H|j\rangle b_j^+ b_i = \sum_{i,j}\frac{1}{2m}\langle i|(-i\hbar\vec{\nabla})^2|j\rangle b_j^+ b_i \\
&= \sum_{\vec{k}_i,\vec{k}_j}\sum_{\sigma_i,\sigma_j} S_{\sigma_i}S_{\sigma_j}\frac{1}{\Omega}\frac{1}{2m}\int d\vec{r}\,e^{-i\vec{k}_i\cdot\vec{r}}(-i\hbar\vec{\nabla})^2 e^{-i\vec{k}_j\cdot\vec{r}}b_{\vec{k}_j,\sigma_j}^+ b_{\vec{k}_i,\sigma_i} \\
&= \frac{1}{\Omega}\sum_{\vec{k}_i,\vec{k}_j}\sum_{\sigma_i,\sigma_j}\frac{\hbar^2 k_j^2}{2m}\Omega\delta_{\sigma_i,\sigma_j}\delta_{\vec{k}_i,\vec{k}_j}b_{\vec{k}_j,\sigma_j}^+ b_{\vec{k}_i,\sigma_i} \\
&= \sum_{\vec{k}_i}\sum_{\sigma_i}\frac{\hbar^2 k_i^2}{2m}b_{\vec{k}_i,\sigma_i}^+ b_{\vec{k}_i,\sigma_i} \qquad\qquad (11.68)
\end{aligned}
$$

where we have used the orthogonality of individual wave functions. Let $|f\rangle$ be the state vector of the ground state. The kinetic energy at $T = 0$ is

$$
\begin{aligned}
E_0 &= \langle f|\hat{\mathcal{H}}_0|f\rangle = \sum_{\vec{k}_i,\sigma_i}\frac{\hbar^2 k_i^2}{2m}\langle f|b_{\vec{k}_i,\sigma_i}^+ b_{\vec{k}_i,\sigma_i}|f\rangle \\
&= \sum_{\vec{k}_i,\sigma_i}\frac{\hbar^2 k_i^2}{2m}\langle f|\hat{n}_{\vec{k}_i,\sigma_i}|f\rangle = \sum_{\vec{k}_i,\sigma_i}\frac{\hbar^2 k_i^2}{2m}n_{\vec{k}_i,\sigma_i}\langle f|f\rangle \\
&= \sum_{\vec{k}_i(k_i\leq k_F)}\sum_{\sigma_i}\frac{\hbar^2 k_i^2}{2m} \qquad\qquad (11.69)
\end{aligned}
$$

where we have replaced in the last line $n_{\vec{k}_i,\sigma_i}$ by 1 for $k_i \leq k_F$ and by 0 otherwise. Transforming the sum into an integral we obtain E_0 given by (2.50). In Rydberg, the average kinetic energy per electron at $T = 0$ is $2.21/r_s^2$ [see (2.54)].

Problem 2. Calculate the energy of an interacting electron gas at the first-order of perturbation with the second quantization:
Solution: At the first-order of perturbation, the energy is written

as

$$E_1 = \langle f|\hat{\mathcal{H}}_1|f\rangle = \frac{e^2}{2\Omega} \sum_{\vec{k},\vec{p},\vec{q}\neq 0} \sum_{\sigma_1,\sigma_2} \frac{4\pi}{q^2}$$

$$\langle f|b^+_{\vec{k}+\vec{q},\sigma_1} b^+_{\vec{p}-\vec{q},\sigma_2} b_{\vec{p},\sigma_2} b_{\vec{k},\sigma_1}|f\rangle \qquad (11.70)$$

where we have used the unscreened Coulomb potential ($\mu = 0$). For a non zero bracket, the orthogonality of the state vectors imposes that the operation of $b^+_{\vec{k}+\vec{q},\sigma_1} b^+_{\vec{p}-\vec{q},\sigma_2} b_{\vec{p},\sigma_2} b_{\vec{k},\sigma_1}$ on the ket $|f\rangle >$ should leave it unchanged. Since there are two creation operators and two annihilation operators, the condition imposed by the orthogonality is satisfied if the two particles created by the two creation operators are removed by the two annihilation operators. There are two possible ways:

(i) $(\vec{k} + \vec{q}, \sigma_1) = (\vec{k}, \sigma_1)$; $(\vec{p} - \vec{q}, \sigma_2) = (\vec{p}, \sigma_2)$

(ii) $(\vec{k} + \vec{q}, \sigma_1) = (\vec{p}, \sigma_2)$; $(\vec{p} - \vec{q}, \sigma_2) = (\vec{k}, \sigma_1)$

The first case is impossible because $\vec{q} \neq 0$. The second case gives

$$E_1 = \frac{e^2}{2\Omega} \sum_{\vec{k},\vec{p},\vec{q}\neq 0} \sum_{\sigma_1,\sigma_2} \frac{4\pi}{q^2}$$

$$\delta_{\vec{k}+\vec{q},\vec{p}}\delta_{\sigma_1,\sigma_2} \langle f|b^+_{\vec{k}+\vec{q},\sigma_1} b^+_{\vec{p}-\vec{q},\sigma_2} b_{\vec{k}+\vec{q},\sigma_1} b_{\vec{k},\sigma_1}|f\rangle$$

$$= -\frac{e^2}{2\Omega} \sum_{\vec{k},\vec{p},\vec{q}\neq 0} \sum_{\sigma_1,\sigma_2} \frac{4\pi}{q^2}$$

$$\delta_{\vec{k}+\vec{q},\vec{p}}\delta_{\sigma_1,\sigma_2} \langle f|b^+_{\vec{k}+\vec{q},\sigma_1} b_{\vec{k}+\vec{q},\sigma_1} b^+_{\vec{p}-\vec{q},\sigma_2} b_{\vec{k},\sigma_1}|f\rangle$$

$$= -\frac{e^2}{2\Omega} \sum_{\vec{k},\vec{p},\vec{q}\neq 0} \sum_{\sigma_1} \frac{4\pi}{q^2} n_{\vec{k}+\vec{q},\sigma_1} n_{\vec{k},\sigma_1} \langle f|f\rangle$$

$$= -\frac{e^2}{2\Omega} \sum_{\vec{k},\vec{q}\neq 0} \sum_{\sigma_1} \frac{4\pi}{q^2}\Theta(k_F - |\vec{k} + \vec{q}|)\Theta(k_F - |\vec{k}|) \quad (11.71)$$

where we have used the Heavyside function Θ to indicate that $n_{\vec{k}+\vec{q},\sigma_1}$ and $n_{\vec{k},\sigma_1}$ are equal to 1 if their wave vectors are inside the sphere of radius k_F (we are at $T=0$). They are zero otherwise. Transforming now the sums on $\vec{k}$ and $\vec{q}$ into integrals in the domains where $k_F - |\vec{k} + \vec{q}| > 0$ and $k_F - |\vec{k}| > 0$, we obtain the results of the Hartree-Fock approximation (2.55), (2.62) and (2.64), namely $E_1 = -\frac{0.916}{r_s}$.

Problem 3. Hubbard model: one-site case
   Solution:

a) We have $t = 0$, namely no hopping. The system is a collection of independent sites. For one site, Hamiltonian (3.71) becomes

$$\hat{\mathcal{H}} = U b_\uparrow^+ b_\uparrow b_\downarrow^+ b_\downarrow - \mu(b_\uparrow^+ b_\uparrow + b_\downarrow^+ b_\downarrow) \qquad (11.72)$$

There are four eigenstates $|0>$, $|\uparrow>$, $|\downarrow>$ and $|\uparrow\downarrow>$ with eigenenergies 0, $-\mu$, $-\mu$ and $U - 2\mu$, respectively. The corresponding grand partition function, the average occupation number and the average energy are (see Appendix A)

$$\mathcal{Z} = \sum_i \exp(-\beta \mathcal{H}_i) = 1 + 2\exp(\beta\mu) + \exp(2\beta\mu - \beta U)$$

$$<n> = \frac{1}{\beta}\frac{\partial \ln \mathcal{Z}}{\partial \mu} = 2\frac{\exp(\beta\mu) + \exp(2\beta\mu - \beta U)}{1 + 2\exp(\beta\mu) + \exp(2\beta\mu - \beta U)}$$

$$<H> = <E - \mu n> = -\frac{\partial \ln \mathcal{Z}}{\partial \beta}$$

$$= -\frac{2\mu\exp(\beta\mu) + (2\mu - U)\exp(2\beta\mu - \beta U)}{1 + 2\exp(\beta\mu) + \exp(2\beta\mu - \beta U)}$$

$$<E> = <H> + \mu<n> = \frac{U\exp(2\beta\mu - \beta U)}{1 + 2\exp(\beta\mu) + \exp(2\beta\mu - \beta U)}$$

The half-filling case $<n> = 1$ corresponds to $2[\exp(\beta\mu) + \exp(2\beta\mu - \beta U)] = [1 + 2\exp(\beta\mu) + \exp(2\beta\mu - \beta U)]$, namely $\exp(2\beta\mu - \beta U) = 1$ or $\mu = U/2$, independent of T. The curve of $<n>$ versus μ is shown in Fig. 11.8 where we see that $<n> = 1$ at $\mu = 2 = U/2$ for all T.

We see in this figure that (i) for $\mu < 2$ (less than half-filling), as T increases the density decreases (ii) for $\mu > 2$, as T increases, the density of spin increases beyond 1.

Remark: (i) The action of U comes into play when $<n>$ crosses the half-filling (> 1): At $T > 0.5$ for example, adding one electron at a lattice site with an electron already there costs an energy U. This is seen in the curve $<E>$ in Fig. 11.8 ($<E>$ becomes non zero) (ii) Imagine the situation at $T = 0$. The last filled energy level is the Fermi level: adding an electron to the system is to fill the next level. As long as the system is less than half-filled, this does not cost a U energy because one finds always an empty lattice to put the electron. But when the system reaches the half filling, the added electron comes on a site with another electron. This

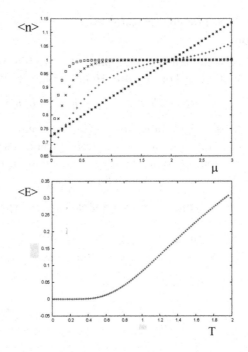

Fig. 11.8   Top: Density of spin versus chemical potential μ for $U = 4$ at temperatures $T = 2$ (stars), 0.5 (vertical crosses), 0.2 (oblique crosses) and 0.1 (squares). Bottom: Average energy $< E >$ versus T at half-filling ($\mu = U/2 = 2$).

costs U. The energy discontinuity from 0 to U at half filling is called "Mott gap". This gap is not necessarily the band gap but it plays the same role: a gap is an indication of the insulating phase.

b) When $U = 0$ we have

$$\hat{\mathcal{H}} = -t \sum_{i,i',\sigma} b^+_{i'\sigma} b_{i\sigma} - \mu \sum_n (b^+_{i\uparrow} b_{i\uparrow} + b^+_{i\downarrow} b_{i\downarrow}) \tag{11.73}$$

Using the Fourier transforms

$$b_{\ell\sigma} = \frac{1}{N^{1/2}} \sum_k b_{k\sigma} e^{ik\ell} \tag{11.74}$$

$$b^+_{\ell\sigma} = \frac{1}{N^{1/2}} \sum_k b^+_{k\sigma} e^{-ik\ell} \tag{11.75}$$

where N is the number of system sites, we obtain

$$\hat{\mathcal{H}} = -t \sum_{k,\sigma} b_{k\sigma}^+ b_{k\sigma} \sum_{\rho} \exp(ik\rho) - \mu \sum_k (b_{k\uparrow}^+ b_{k\uparrow} + b_{k\downarrow}^+ b_{k\downarrow})$$

$$= \sum_{k,\sigma} [-2t\cos(k) - \mu] b_{k\sigma}^+ b_{k\sigma} \qquad (11.76)$$

where $\rho = \ell - \ell' = \pm 1$ $(a = 1)$ is the vector connecting two nearest sites and where we have used the sum rules on k and n.

For a square lattice, the same calculation is performed with a Fourier transform in 2D. The structure factor $\sum_\rho \exp(ik\rho)$ becomes for the square lattice $\sum_{\vec{\rho}} \exp(i\vec{k} \cdot \vec{\rho}) = 2(\cos k_x + \cos k_y)$ [four nearest neighbors are $\vec{\rho} = (\pm 1, 0), (0, \pm 1)$]. The hopping energy is thus $\epsilon_{\vec{k}} = -2t(\cos k_x + \cos k_y)$.

Return to the 1D case. Using (11.76), the grand partition function (see Appendix A) is written as

$$\mathcal{Z} = \sum_{n_{k1,\sigma}, n_{k2,\sigma}, \cdots} \exp[-\beta \sum_{k,\sigma} (-2t\cos k - \mu) n_{k,\sigma}]$$

$$= \sum_{n_{k1,\sigma}} \exp[-\beta(-2t\cos k_1 - \mu) n_{k1,\sigma}]$$

$$\times \sum_{n_{k2,\sigma}} \exp[-\beta(-2t\cos k_2 - \mu) n_{k2,\sigma}] \cdots$$

$$= (1 + \exp[-\beta(-2t\cos k_1 - \mu)])(1 + \exp[-\beta(-2t\cos k_2 - \mu)]) \cdots$$

$$= \prod_k (1 + \exp[-\beta(-2t\cos k - \mu)]) \qquad (11.77)$$

where $n_{k,\sigma} = b_{k\sigma}^+ b_{k\sigma}$ and we have used for the n sum (second line) $n_{k_i,\sigma} = 0, 1$.

The average energy and average occupation number are given by

$$< n > = \frac{1}{\beta} \frac{\partial \ln \mathcal{Z}}{\partial \mu}$$

$$= \sum_k \frac{1}{1 + \exp[\beta(\epsilon_k - \mu)]}$$

$$< H > = < E - \mu n > = -\frac{\partial \ln \mathcal{Z}}{\partial \beta}$$

$$= \sum_k \frac{\epsilon_k - \mu}{1 + \exp[\beta(\epsilon_k - \mu)]}$$

$$< E > = < H > + \mu < n > = \sum_k \frac{\epsilon_k}{1 + \exp[\beta(\epsilon_k - \mu)]}$$

$$C = \frac{d < E >}{dT} = -\frac{\partial^2 \ln \mathcal{Z}}{\partial \beta^2} \frac{d\beta}{dT}$$

$$= \frac{1}{k_B T^2} \frac{\epsilon_k^2 exp[\beta(\epsilon_k - \mu)]}{(1 + \exp[\beta(\epsilon_k - \mu)])^2}$$

where $\epsilon_k = -2t \cos k$, C is the specific heat. The second equation is the Fermi-Dirac distribution. The plot of $< E >$ and C is shown in Fig. 11.9 with $t = 1$ at half filling ($\mu = U/2 = 0$).

Problem 4. Hubbard model on a two-site system
   Solution:

a) There are 16 configurations: (i) one configuration with no electron at two sites (ii) there are four configurations with one election $|0, \uparrow>$, $| \uparrow, 0 >$, $|0, \downarrow>$, $| \downarrow, 0 >$ (iii) six configurations with two electrons $| \downarrow, \uparrow>$, $|0, \downarrow\uparrow>$, $|0, \uparrow\downarrow>$, $| \uparrow, \downarrow>$, $| \uparrow, \uparrow>$, $| \downarrow, \downarrow>$ (the last two have no connected configurations) (iv) four configurations with three electrons (v) one configuration with four electrons.

b) There are two groups of connected configurations
   (i) $|\Phi_1 >= a| \uparrow, 0 > + b|0, \uparrow>$
   (ii) $|\Phi_2 >= c| \uparrow\downarrow, 0 > + d| \uparrow, \downarrow> + e| \downarrow, \uparrow> + f|0, \uparrow\downarrow>$
   where we did not consider up-down permutations. Writing the Schrödinger equation for $|\Phi >= |\Phi_1 > + |\Phi_2 >$, we have

$$\hat{\mathcal{H}}|\Phi >= E|\Phi > \tag{11.78}$$

Multiplying on the left of each side successively by $<\uparrow, 0|$, etc. and

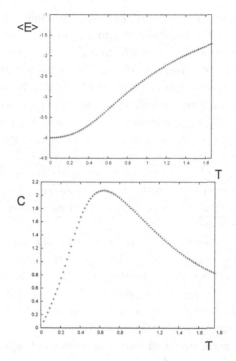

Fig. 11.9   Average energy $< E >$ (top) and specific heat C (bottom) versus T for $t = 1$ and half-filling $\mu = 0$ $(U = 0)$.

using the orthogonality of the state vectors

$$-t <\uparrow, 0|b_{n'\sigma}^{+} b_{n\sigma}| \uparrow, 0 > \, = -t <\uparrow, 0|0, \uparrow> = 0$$

$$-t < 0, \uparrow |b_{n'\sigma}^{+} b_{n\sigma}| \uparrow, 0 > \, = -t < 0, \uparrow |0, \uparrow> = -t$$

$$< 0, \uparrow\downarrow |U b_{n\uparrow}^{+} b_{n\uparrow} b_{n\downarrow}^{+} b_{n\downarrow}|0, \uparrow\downarrow> \, = U < 0, \uparrow\downarrow |(1)(1)|0, \uparrow\downarrow> = U$$

$$... = ...$$

$$E <\uparrow, 0|a| \uparrow, 0 > \, = Ea$$

$$... = ...$$

we arrive of a set of equations which can be written under the matrix form (3.73). We see that we can diagonalize it by writing

$$
\begin{pmatrix}
-E & -t & 0 & 0 & 0 & 0 \\
-t & -E & 0 & 0 & 0 & 0 \\
0 & 0 & U-E & -t & -t & 0 \\
0 & 0 & -t & -E & 0 & -t \\
0 & 0 & -t & 0 & -E & -t \\
0 & 0 & 0 & -t & -t & U-E
\end{pmatrix}
\begin{pmatrix}
a \\
b \\
c \\
d \\
e \\
f
\end{pmatrix}
= 0 \qquad (11.79)
$$

Non trivial solutions imply that the determinant det $|...|$ is zero. We decompose this determinant into a product of det $|2 \times 2||4 \times 4|=0$. The 2×2 block gives two solutions $E = \pm t$ with $a = b = 1/\sqrt{2}$ and $a = -b = 1/\sqrt{2}$, respectively. The 4×4 block can be diagonalized to give eigen-energies and statistical weights. We note that to make this block diagonal, it suffices to choose the following combinations $| \uparrow\downarrow, 0 > \pm |0, \uparrow\downarrow>$ and $c| \uparrow, \downarrow> \pm| \downarrow, \uparrow>$ and rewrite the Schrödinger equation for these four states, we obtain the diagonal 4×4 matrix. The eigen-energies are $E = U$, $E = 0$, $E = U/2 \pm \sqrt{(U/2)^2 + 4t^2}$. Note that the last two energies correspond to the spin singlet, while $E = 0$ corresponds to the $S_z = 0$ state of the spin triplet ($S_z = -1, 0, 1$). The four energies in increasing order are $E = U/2 - \sqrt{(U/2)^2 + 4t^2}, 0, U, U/2 + \sqrt{(U/2)^2 + 4t^2}$.

It is easy to see that the lowest energy is the singlet state with two antiparallel spins: they can hop back and forth between two sites. This is the reason why the Hubbard model yields an antiferromagnetic state at low temperatures. We can include the states with three spins in $|\Phi >$, their energies are $U - t$ and $U + t$. For the case of four spins, the energy is $2U$.

For a larger number of sites, the system of equations cannot be analytically solved. One has to use a numerical diagonalization of the matrix.

Problem 5. We show $[\hat{\mathcal{H}}, \hat{N}]=0$ where $\hat{N}$ is the occupation number field operator defined in (3.36) and $\hat{\mathcal{H}}$ the Hamiltonian in the second quantization (3.35).

Demonstration: We have

$$\hat{\mathcal{H}} = \sum_{\sigma} \int d\vec{r} \hat{\Psi}_\sigma^+(\vec{r}) H(\vec{r}) \hat{\Psi}_\sigma(\vec{r})$$

$$-\frac{1}{2} \sum_{\sigma\sigma'} \int \int d\vec{r}_1 d\vec{r}_2 \hat{\Psi}_\sigma^+(\vec{r}_1) \hat{\Psi}_{\sigma'}^+(\vec{r}_2) V(\vec{r}_1, \vec{r}_2) \hat{\Psi}_\sigma(\vec{r}_1) \hat{\Psi}_{\sigma'}(\vec{r}_2)$$

and

$$\hat{N} = \sum_{\beta} \int d\vec{r}_3 \hat{\Psi}_\beta^+(\vec{r}_3) \hat{\Psi}_\beta(\vec{r}_3) \qquad (11.80)$$

To calculate $[\hat{\mathcal{H}}, \hat{N}]$, we decompose the operators in the commutators as follows

$$[AB, C] = A[B, C] - [C, A]B \qquad (11.81)$$

$$[AB, C] = A[B, C]_+ - [A, C]_+ B \qquad (11.82)$$

for bosons and fermions, respectively. For $[\hat{\mathcal{H}}, \hat{N}]$, we should decompose several times. For example, with the kinetic term of $\hat{\mathcal{H}}$, we put
$A = \hat{\Psi}_\sigma^+(\vec{r})$, $B = H(\vec{r})\hat{\Psi}_\sigma(\vec{r})$, $C = \hat{\Psi}_\beta^+(\vec{r}_3)$, $D = \hat{\Psi}_\beta(\vec{r}_3)$. In the boson case, we have
$[AB, CD] = A[B, CD] - [CD, A]B = A[B, C]D - AC[D, B] - C[D, A]B + [A, C]DB$
We use now the commutation relations between field operators A, B, C, and D to get the desired results.

Problem 6. We show $\hat{\Psi}(\vec{r})\hat{N} = (\hat{N} + 1)\hat{\Psi}(\vec{r})$ for both boson and fermion cases.
Demonstration:

$$\hat{\Psi}(\vec{r})\hat{N} = \hat{\Psi}(\vec{r}) \int d\vec{r}' \hat{\Psi}^+(\vec{r}')\hat{\Psi}(\vec{r}')$$

$$= \int d\vec{r}' [\delta(\vec{r} - \vec{r}') \pm \hat{\Psi}^+(\vec{r}')\hat{\Psi}(\vec{r})]\hat{\Psi}(\vec{r}')$$

$$= \hat{\Psi}(\vec{r}) \pm \int d\vec{r}' \hat{\Psi}^+(\vec{r}')\hat{\Psi}(\vec{r})\hat{\Psi}(\vec{r}')$$

$$= \hat{\Psi}(\vec{r}) + \int d\vec{r}' \hat{\Psi}^+(\vec{r}')\hat{\Psi}(\vec{r}')\hat{\Psi}(\vec{r})$$

$$= (1 + \hat{N})\hat{\Psi}(\vec{r}) \qquad (11.83)$$

where the signs $\pm$ correspond to the boson and fermion cases, respectively. The last line is valid for two cases.

Problem 7. We show $\hat{\Psi}^+(\vec{r})|\text{vac}>$ ("vac" stands for vacuum) is a state in which there is a particle localized at $\vec{r}$.
Demonstration: Let $\rho(\vec{r}) = \hat{\Psi}^+(\vec{r})\hat{\Psi}(\vec{r})$ be the density operator.

We have

$$
\begin{aligned}
\rho(\vec{r})\hat{\Psi}^+(\vec{r}')|\text{vac}> &= \int d\vec{r}''\,\hat{\Psi}^+(\vec{r}'')\hat{\Psi}(\vec{r}'')\delta(\vec{r}-\vec{r}'')\hat{\Psi}^+(\vec{r}')|\text{vide}> \\
&= \int d\vec{r}''\,\hat{\Psi}^+(\vec{r}'')\delta(\vec{r}-\vec{r}'')[\delta(\vec{r}''-\vec{r}') \\
&\quad -\hat{\Psi}^+(\vec{r}')\hat{\Psi}(\vec{r}'')]|\text{vac}> \\
&= \int d\vec{r}''\,\hat{\Psi}^+(\vec{r}'')\delta(\vec{r}-\vec{r}'')\delta(\vec{r}''-\vec{r}')|\text{vac}> \\
&= \delta(\vec{r}-\vec{r}')\hat{\Psi}^+(\vec{r}')|\text{vac}>
\end{aligned}
\tag{11.84}
$$

The last equality means that $\delta(\vec{r}-\vec{r}')$ is the eigenvalue of $\rho(\vec{r})$ corresponding to the eigenvector $\Psi^+(\vec{r}')|\text{vac}>$. There is thus one particle at $\vec{r}$ in the state $\Psi^+(\vec{r})|\text{vac}>$.

Problem 8. Using the equation of motion $i\hbar\frac{d\hat{\Psi}(\vec{r})}{dt} = -[\hat{\mathcal{H}},\hat{\Psi}(\vec{r})]$ where $\hat{\mathcal{H}}$ is the Hamiltonian in the second quantization of a system of fermions, we show that we can obtain the Hartree-Fock equation by taking a first approximation of the right-hand side (linearization). **Demonstration:** We use the decomposition of the chains of operators shown in a previous exercise (Problem 5) we calculate $[\hat{\mathcal{H}},\hat{\Psi}(\vec{r})]$. We obtain

$$
i\hbar\frac{d\hat{\Psi}(\vec{r},t)}{dt} = -\frac{p^2}{2m}\hat{\Psi}(\vec{r},t) + \int d\vec{r}'\,\hat{\Psi}^+(\vec{r}',t)V(\vec{r},\vec{r}')\hat{\Psi}(\vec{r}',t)\hat{\Psi}(\vec{r},t)
\tag{11.85}
$$

where $\hat{\Psi}(\vec{r},t)$ and $\hat{\Psi}^+(\vec{r},t)$ are interaction representations of (3.29) and (3.30). We have

$$
\hat{\Psi}(\vec{r},t) = \sum_{\vec{k},\sigma} b_{\vec{k},\sigma}e^{-i\omega_{\vec{k}}t}\varphi_{\vec{k}}(\vec{r})
\tag{11.86}
$$

$$
\hat{\Psi}^+(\vec{r},t) = \sum_{\vec{k},\sigma} b_{\vec{k},\sigma}^+ e^{i\omega_{\vec{k}}t}\varphi_{\vec{k}}^+(\vec{r})
\tag{11.87}
$$

Replacing these expressions in (11.85), we obtain

$$\sum_{\vec{k},\sigma} \hbar\omega_{\vec{k}} b_{\vec{k},\sigma} e^{-i\omega_{\vec{k}}t} \varphi_{\vec{k}}(\vec{r}) = \sum_{\vec{k},\sigma} \frac{\hbar^2 k^2}{2m} b_{\vec{k},\sigma} e^{-i\omega_{\vec{k}}t} \varphi_{\vec{k}}(\vec{r})$$

$$+ \sum_{\vec{k},\vec{k}',\vec{k}'',\sigma,\sigma',\sigma''} \int d\vec{r}' \varphi_{\vec{k}'}^+(\vec{r}') V(\vec{r},\vec{r}') \varphi_{\vec{k}''}(\vec{r}') e^{i(\omega_{\vec{k}'}-\omega_{\vec{k}''})t}$$

$$\times b_{\vec{k}',\sigma'}^+ b_{\vec{k}'',\sigma''} b_{\vec{k},\sigma} e^{-i\omega_{\vec{k}}t} \varphi_{\vec{k}}(\vec{r}) \tag{11.88}$$

For a first approximation of the right-hand side of (11.85), we use the following decoupling called "random-phase approximation" (RPA) for the fermion case

$$b_{\vec{k}',\sigma'}^+ b_{\vec{k}'',\sigma''} b_{\vec{k},\sigma} \simeq <b_{\vec{k}',\sigma'}^+ b_{\vec{k}'',\sigma''}> b_{\vec{k},\sigma} - <b_{\vec{k}',\sigma'}^+ b_{\vec{k},\sigma}> b_{\vec{k}'',\sigma''} \tag{11.89}$$

where the negative sign of the right-hand side results from the permutation of $b_{\vec{k}'',\sigma''}$ and $b_{\vec{k},\sigma}$, and $< \ldots >$ denotes the average. This decoupling supposes that only terms of the type $<b_{\vec{k}',\sigma'}^+ b_{\vec{k}'',\sigma''}>$ are not zero. In the ground state, $<b_{\vec{k}',\sigma'}^+ b_{\vec{k}'',\sigma''}> = n_{\vec{k}',\sigma'} \delta(\vec{k}',\vec{k}'') \delta(\sigma',\sigma'') \Theta(k_F - k')$ etc.
Equation (11.88) becomes

$$\sum_{\vec{k},\sigma} \hbar\omega_{\vec{k}} \varphi_{\vec{k}}(\vec{r}) b_{\vec{k},\sigma} e^{-i\omega_{\vec{k}}t} = \sum_{\vec{k},\sigma} \frac{\hbar^2 k^2}{2m} b_{\vec{k},\sigma} e^{-i\omega_{\vec{k}}t} \varphi_{\vec{k}}(\vec{r})$$

$$+ \sum_{\vec{k},\sigma} \sum_{\vec{k}',\sigma'} \int d\vec{r}' \varphi_{\vec{k}'}^+(\vec{r}') V(\vec{r},\vec{r}') \varphi_{\vec{k}'}(\vec{r}') n_{\vec{k}',\sigma'} \varphi_{\vec{k}}(\vec{r}) b_{\vec{k},\sigma} e^{-i\omega_{\vec{k}}t}$$

$$- \sum_{\vec{k}'',\sigma''} \sum_{\vec{k}',\sigma'} \delta(\sigma',\sigma) \int d\vec{r}' \varphi_{\vec{k}'}^+(\vec{r}') V(\vec{r},\vec{r}') \varphi_{\vec{k}}(\vec{r}')$$

$$\times n_{\vec{k}',\sigma'} \varphi_{\vec{k}'}(\vec{r}) b_{\vec{k}'',\sigma''} e^{-i\omega_{\vec{k}''}t}$$

We change now the dummy variables $(\vec{k}'',\sigma'')$ into $(\vec{k},\sigma)$ in the last term and we remove the sums $\sum_{\vec{k},\sigma}$ on both sides, then we take off the factor $b_{\vec{k},\sigma} e^{-i\omega_{\vec{k}}t}$, we arrive at

$$\hbar\omega_{\vec{k}} \varphi_{\vec{k},\sigma}(\vec{r}) = \frac{\hbar^2 k^2}{2m} \varphi_{\vec{k},\sigma}(\vec{r})$$

$$+ \sum_{\vec{k}',\sigma'} \int d\vec{r}' \varphi_{\vec{k}',\sigma'}^+(\vec{r}') V(\vec{r},\vec{r}') \varphi_{\vec{k}',\sigma'}(\vec{r}') \varphi_{\vec{k},\sigma}(\vec{r})$$

$$- \sum_{\vec{k}',\sigma'} \delta(\sigma',\sigma) \int d\vec{r}' \varphi_{\vec{k}',\sigma'}^+(\vec{r}') V(\vec{r},\vec{r}') \varphi_{\vec{k},\sigma}(\vec{r}') \varphi_{\vec{k}',\sigma'}(\vec{r})$$

where we have replaced $n_{\vec{k}',\sigma'}$ by 1 (ground state) and we have transferred, for compactness, the spin indices σ, σ' of states $\vec{k}$ and $\vec{k}'$ on φ. This equation is the Hartree-Fock equation [see Eq. (2.28)].

Problem 9. Bardeen-Cooper-Schrieffer theory of supraconductivity:
  Solution: The reader can find the full theory on the superconductivity in any textbook on condensed matter physics, for example in Refs. [10, 26].

a) Using $i\frac{dc_{\vec{k}}}{dt} = [c_{\vec{k}}, \mathcal{H}]$ (taking $\hbar = 1$) with the decomposition of operator chains shown in Problem 5, we arrive at the two requested equations.

b) We replace the sums by $\Delta_{\vec{k}}$ and $\Delta_{\vec{k}}^*$ we obtain two coupled equations. The result is a kind of the Hartree-Fock approximation obtained from the linearization of the equation of motion (see Problem 8). A solution of the form $c_{\vec{k}} \propto \exp(-i\omega_{\vec{k}}t)$ leads to

$$\omega_{\vec{k}}c_{\vec{k}} = \epsilon_{\vec{k}}c_{\vec{k}} - c_{-\vec{k}}^+ \Delta_{\vec{k}} \tag{11.90}$$

$$\omega_{\vec{k}}c_{-\vec{k}}^+ = -\epsilon_{\vec{k}}c_{-\vec{k}}^+ - c_{\vec{k}}\Delta_{\vec{k}}^* \tag{11.91}$$

A non trivial solution imposes

$$\omega_{\vec{k}} = (\epsilon_{\vec{k}}^2 + \Delta^2)^{1/2} \tag{11.92}$$

c) We have

$$\begin{aligned}
[a_{\vec{k}}, a_{\vec{k}'}^+]_+ &= [u_{\vec{k}}c_{\vec{k}} - v_{\vec{k}}c_{-\vec{k}}^+, u_{\vec{k}'}c_{\vec{k}'}^+ - v_{\vec{k}'}c_{-\vec{k}'}]_+ \\
&= u_{\vec{k}}u_{\vec{k}'}[c_{\vec{k}}, c_{\vec{k}'}^+]_+ + v_{\vec{k}}v_{\vec{k}'}[c_{-\vec{k}}^+, c_{-\vec{k}'}]_+ \\
&= u_{\vec{k}}^2\delta(\vec{k},\vec{k}') + v_{\vec{k}}^2\delta(\vec{k},\vec{k}') \\
&= (u_{\vec{k}}^2 + v_{\vec{k}}^2)\delta(\vec{k},\vec{k}') = \delta(\vec{k},\vec{k}')
\end{aligned}$$

The other anticommutation relations are obtained in the same manner.
From (3.80)-(3.81), we have

$$c_{\vec{k}} = u_{\vec{k}}a_{\vec{k}} + v_{\vec{k}}a_{-\vec{k}}^+ \tag{11.93}$$

$$c_{\vec{k}}^+ = u_{\vec{k}}a_{\vec{k}}^+ + v_{\vec{k}}a_{-\vec{k}} \tag{11.94}$$

$$c_{-\vec{k}} = u_{\vec{k}}a_{-\vec{k}} - v_{\vec{k}}a_{\vec{k}}^+ \tag{11.95}$$

$$c_{-\vec{k}}^+ = u_{\vec{k}}a_{-\vec{k}}^+ + -v_{\vec{k}}a_{\vec{k}} \tag{11.96}$$

Replacing these relations in (11.90), we obtain

$$\omega_{\vec{k}} u_{\vec{k}} = \epsilon_{\vec{k}} u_{\vec{k}} + \Delta_{\vec{k}} v_{\vec{k}} \tag{11.97}$$

Putting this into square, we have

$$\omega_{\vec{k}}^2 u_{\vec{k}}^2 = \epsilon_{\vec{k}}^2 u_{\vec{k}}^2 + \Delta_{\vec{k}}^2 v_{\vec{k}}^2 + 2\epsilon_{\vec{k}} \Delta_{\vec{k}} u_{\vec{k}} v_{\vec{k}} = (\epsilon_{\vec{k}}^2 + \Delta_{\vec{k}}^2) u_{\vec{k}}^2 \tag{11.98}$$

where we have used (3.79) to obtain the last equality. Hence,

$$\Delta_{\vec{k}}^2 (u_{\vec{k}}^2 - v_{\vec{k}}^2) = 2\epsilon_{\vec{k}} \Delta_{\vec{k}} u_{\vec{k}} v_{\vec{k}} \tag{11.99}$$

Putting $u_{\vec{k}} = \cos(\theta_{\vec{k}}/2)$ and $v_{\vec{k}} = \sin(\theta_{\vec{k}}/2)$, we have

$$\Delta_{\vec{k}} \cos\theta_{\vec{k}} = \epsilon_{\vec{k}} \sin\theta_{\vec{k}} \tag{11.100}$$

Hence

$$\tan\theta_{\vec{k}} = \Delta_{\vec{k}}/\epsilon_{\vec{k}} \tag{11.101}$$

d) We show that $\phi^0 = \prod_{\vec{k}}(u_{\vec{k}} + v_{\vec{k}} c_{\vec{k}}^+ c_{-\vec{k}}^+)|\text{vac} >$ describes the ground state: The ground state in the superconducting regime contains pairs of electrons $(\vec{k}, -\vec{k})$ called Cooper's pairs near the Fermi level. If we consider only the subspace of these states, we see that ϕ^0 describes well these states thanks to operators $c_{\vec{k}}^+$ and $c_{-\vec{k}}^+$ which act on $|\text{vac} >$. The quantities u and v are for normalization purpose. However, their choice should respect the fermionic character of operators $c_{\vec{k}}^+$ and $c_{-\vec{k}}^+$ as we will see below.

$$< \phi^0|\phi^0 > = \prod_{\vec{k}} < \text{vac}|(u_{\vec{k}} + v_{\vec{k}} c_{-\vec{k}} c_{\vec{k}})(u_{\vec{k}} + v_{\vec{k}} c_{\vec{k}}^+ c_{-\vec{k}}^+)|\text{vac} >$$

$$= \prod_{\vec{k}} (u_{\vec{k}}^2 + v_{\vec{k}}^2) < \text{vac}|\text{vac} >= 1 \tag{11.102}$$

because

$$< \text{vac}|c_{-\vec{k}} c_{\vec{k}} c_{\vec{k}}^+ c_{-\vec{k}}^+|\text{vac} >= (1 - n_{\vec{k}})(1 - n_{-\vec{k}}) < \text{vac}|\text{vac} >$$
$$= < \text{vac}|\text{vac} >= 1$$

where we have used $n_{\vec{k}} = n_{-\vec{k}} = 0$ in the vacuum state, and $< \text{vac}|c_{\vec{k}} c_{-\vec{k}}|\text{vac} >= 0$, and $< \text{vac}|c_{\vec{k}}^+ c_{-\vec{k}}^+|\text{vac} >= 0$.

$$< \phi^0 |c_{\vec{k}'}^+ c_{\vec{k}'}| \phi^0 > = < \text{vac}|(u_{\vec{k}'} + v_{\vec{k}'} c_{-\vec{k}'} c_{\vec{k}'}) c_{\vec{k}'}^+ c_{\vec{k}'} (u_{\vec{k}'}$$
$$+ v_{\vec{k}'} c_{\vec{k}'}^+ c_{-\vec{k}'}^+) |\text{vac} > \prod_{\vec{k} \neq \vec{k}'} < \phi^0 | \phi^0 >$$
$$= < \text{vac}|[u_{\vec{k}'}^2 c_{\vec{k}'}^+ c_{\vec{k}'} + v_{\vec{k}'}^2 c_{-\vec{k}'} c_{\vec{k}'} c_{\vec{k}'}^+ c_{\vec{k}'} c_{\vec{k}'}^+ c_{-\vec{k}'}^+$$
$$+ u_{\vec{k}'} v_{\vec{k}'} c_{\vec{k}'}^+ c_{\vec{k}'} c_{\vec{k}'}^+ c_{-\vec{k}'}^+ + ...]|\text{vac} >$$
$$= [v_{\vec{k}'}^2] \tag{11.103}$$

The other terms in [...] are 0. In the same manner, we obtain

$$< \phi^0 |c_{\vec{k}'}^+ c_{-\vec{k}'}^+ c_{-\vec{k}''} c_{\vec{k}''}| \phi^0 > = u_{\vec{k}'} v_{\vec{k}'} u_{\vec{k}''} v_{\vec{k}''} \tag{11.104}$$

The energy of the ground state is thus

$$E_g = < \phi^0 |\mathcal{H}| \phi^0 > = 2 \sum_{\vec{k}} \epsilon_{\vec{k}} v_{\vec{k}'}^2 - V \sum_{\vec{k} \neq \vec{k}'} u_{\vec{k}} v_{\vec{k}} u_{\vec{k}'} v_{\vec{k}'}$$
$$= \sum_{\vec{k}} \epsilon_{\vec{k}} (1 - \cos \theta_{\vec{k}}) - \frac{V}{4} \sum_{\vec{k} \neq \vec{k}'} \sin \theta_{\vec{k}} \sin \theta_{\vec{k}'} \tag{11.105}$$

where the factor 2 in the first equality comes from the spin sum, and where we have used the definitions of $u_{\vec{k}}$ and $v_{\vec{k}}$. We obtain

$$E_g = - \sum_{\vec{k}} \epsilon_{\vec{k}} \cos \theta_{\vec{k}} - (\Delta^2/V) \tag{11.106}$$

where we have used the following considerations:
1) $\sum_{\vec{k}} \epsilon_{\vec{k}} = 0$ because $\epsilon_{\vec{k}}$ is symmetric with respect to the Fermi level, positive above and negative below,
2) We can show that $\sum_{\vec{k}} \sin \theta_{\vec{k}} = 2\Delta/V$.

Problem 10. Magnon-phonon interaction:
    Solution: The reader is recommended to read chapter 5 for more details on the theory of magnons. The theory of phonons can be found in any textbook on condensed matter physics, for instance Refs. [10, 26, 78].
    We consider the following Hamiltonian describing the interaction between magnon and phonon:

$$\mathcal{H} = \sum_{\vec{k}} \left[\omega_{\vec{k}}^m a_{\vec{k}}^+ a_{\vec{k}} + \omega_{\vec{k}}^p b_{\vec{k}}^+ b_{\vec{k}} + V_{\vec{k}}(a_{\vec{k}} b_{\vec{k}}^+ + a_{\vec{k}}^+ b_{\vec{k}}) \right] \qquad (11.107)$$

where $V_{\vec{k}}$ is the coupling constant, $\omega_{\vec{k}}^m$ and $\omega_{\vec{k}}^p$ are eigenfrequencies of magnon and phonon, respectively, a and a^+ denote annihilation and creation operators of magnon, while b and b^+ denote those of phonon.

a) We have

$$a_{\vec{k}} = \cos\theta_{\vec{k}} c_{\vec{k}} + \sin\theta_{\vec{k}} d_{\vec{k}} \qquad (11.108)$$

$$b_{\vec{k}} = \cos\theta_{\vec{k}} d_{\vec{k}} - \sin\theta_{\vec{k}} c_{\vec{k}} \qquad (11.109)$$

Multiplying the first equation by $\cos\theta_{\vec{k}}$ and the second by $\sin\theta_{\vec{k}}$, then taking the difference side by side, we have

$$a_{\vec{k}} \cos\theta_{\vec{k}} - b_{\vec{k}} \sin\theta_{\vec{k}} = c_{\vec{k}} \qquad (11.110)$$

where we used $\cos^2\theta_{\vec{k}} + \sin^2\theta_{\vec{k}} = 1$. Multiplying now (11.108) by $\sin\theta_{\vec{k}}$ and (11.109) by $\cos\theta_{\vec{k}}$ and adding two equations side by side we have

$$a_{\vec{k}} \sin\theta_{\vec{k}} + b_{\vec{k}} \cos\theta_{\vec{k}} = d_{\vec{k}} \qquad (11.111)$$

We write

$$
\begin{aligned}
[c_{\vec{k}}, c_{\vec{k}'}^+] &= [a_{\vec{k}} \cos\theta_{\vec{k}} - b_{\vec{k}} \sin\theta_{\vec{k}}, a_{\vec{k}'}^+ \cos\theta_{\vec{k}'} - b_{\vec{k}'}^+ \sin\theta_{\vec{k}'}] \\
&= [a_{\vec{k}} \cos\theta_{\vec{k}}, a_{\vec{k}'}^+ \cos\theta_{\vec{k}'}] - [a_{\vec{k}} \cos\theta_{\vec{k}}, b_{\vec{k}'}^+ \sin\theta_{\vec{k}'}] \\
&\quad -[b_{\vec{k}} \sin\theta_{\vec{k}}, a_{\vec{k}'}^+ \cos\theta_{\vec{k}'}] + [b_{\vec{k}} \sin\theta_{\vec{k}}, b_{\vec{k}'}^+ \sin\theta_{\vec{k}'}] \\
&= \cos\theta_{\vec{k}} \cos\theta_{\vec{k}'} [a_{\vec{k}}, a_{\vec{k}'}^+] - \cos\theta_{\vec{k}} \sin\theta_{\vec{k}'} [a_{\vec{k}}, b_{\vec{k}'}^+] \\
&\quad - \sin\theta_{\vec{k}} \cos\theta_{\vec{k}'} [b_{\vec{k}}, a_{\vec{k}'}^+] + \sin\theta_{\vec{k}} \sin\theta_{\vec{k}'} [b_{\vec{k}}, b_{\vec{k}'}^+] \\
&= \cos^2\theta_{\vec{k}} \delta_{k,k'} - 0 - 0 + \sin^2\theta_{\vec{k}} \delta_{k,k'} \\
&= \delta_{k,k'} \qquad (11.112)
\end{aligned}
$$

The other relations can be proved in the same manner.

b) Replacing the operators a and b in terms of c and d, we have

$$\mathcal{H} = \sum_{\vec{k}} \{\omega_{\vec{k}}^m [c_{\vec{k}}^+ c_{\vec{k}} \cos^2 \theta_{\vec{k}} + d_{\vec{k}}^+ d_{\vec{k}} \sin^2 \theta_{\vec{k}}$$

$$+ (c_{\vec{k}}^+ d_{\vec{k}} + d_{\vec{k}}^+ c_{\vec{k}}) \cos \theta_{\vec{k}} \sin \theta_{\vec{k}}]$$

$$+ \omega_{\vec{k}}^p [d_{\vec{k}}^+ d_{\vec{k}} \cos^2 \theta_{\vec{k}} + c_{\vec{k}}^+ c_{\vec{k}} \sin^2 \theta_{\vec{k}}$$

$$- (d_{\vec{k}}^+ c_{\vec{k}} + c_{\vec{k}}^+ d_{\vec{k}}) \cos \theta_{\vec{k}} \sin \theta_{\vec{k}}]$$

$$+ V_{\vec{k}} [c_{\vec{k}} d_{\vec{k}}^+ \cos^2 \theta_{\vec{k}} - d_{\vec{k}} c_{\vec{k}}^+ \sin^2 \theta_{\vec{k}}$$

$$+ (d_{\vec{k}} d_{\vec{k}}^+ - c_{\vec{k}} c_{\vec{k}}^+) \cos \theta_{\vec{k}} \sin \theta_{\vec{k}}$$

$$+ c_{\vec{k}}^+ d_{\vec{k}} \cos^2 \theta_{\vec{k}} - d_{\vec{k}}^+ c_{\vec{k}} \sin^2 \theta_{\vec{k}}$$

$$- (c_{\vec{k}}^+ c_{\vec{k}} - d_{\vec{k}}^+ d_{\vec{k}}) \cos \theta_{\vec{k}} \sin \theta_{\vec{k}}]\} \tag{11.113}$$

We can collect non diagonal terms:

$$\mathcal{H}_{\mathrm{nd}} = \sum_{\vec{k}} [(\omega_{\vec{k}}^m - \omega_{\vec{k}}^p)(c_{\vec{k}}^+ d_{\vec{k}} + d_{\vec{k}}^+ c_{\vec{k}}) \cos \theta_{\vec{k}} \sin \theta_{\vec{k}}$$

$$+ V_{\vec{k}} (c_{\vec{k}} d_{\vec{k}}^+ + c_{\vec{k}}^+ d_{\vec{k}}) \cos^2 \theta_{\vec{k}}$$

$$- V_{\vec{k}} (d_{\vec{k}} c_{\vec{k}}^+ + d_{\vec{k}}^+ c_{\vec{k}}) \sin^2 \theta_{\vec{k}}]$$

$$= \sum_{\vec{k}} [c_{\vec{k}}^+ d_{\vec{k}} + d_{\vec{k}}^+ c_{\vec{k}}] \{(\omega_{\vec{k}}^m - \omega_{\vec{k}}^p) \cos \theta_{\vec{k}} \sin \theta_{\vec{k}}$$

$$+ V_{\vec{k}} [\cos^2 \theta_{\vec{k}} - \sin^2 \theta_{\vec{k}}]\}$$

$$= \sum_{\vec{k}} \{(\omega_{\vec{k}}^m - \omega_{\vec{k}}^p) \sin(2\theta_{\vec{k}})/2 + V_{\vec{k}} \cos(2\theta_{\vec{k}})\} [c_{\vec{k}}^+ d_{\vec{k}} + d_{\vec{k}}^+ c_{\vec{k}}]$$

The non diagonal term is zero if the coefficient in the curly brackets is zero:

$$\{(\omega_{\vec{k}}^m - \omega_{\vec{k}}^p) \sin(2\theta_{\vec{k}})/2 + V_{\vec{k}} \cos(2\theta_{\vec{k}})\} = 0 \tag{11.114}$$

namely,

$$\tan(2\theta_{\vec{k}}) = \frac{2V_{\vec{k}}}{\omega_{\vec{k}}^p - \omega_{\vec{k}}^m} \tag{11.115}$$

We collect the diagonal terms of (11.113), putting $\omega_{\vec{k}}^m = \omega_{\vec{k}}^p = \omega$:

$$\mathcal{H}_{\mathrm{d}} = \sum_{\vec{k}} \{\omega [c_{\vec{k}}^+ c_{\vec{k}} + d_{\vec{k}}^+ d_{\vec{k}}] - V_{\vec{k}} [c_{\vec{k}}^+ c_{\vec{k}} + c_{\vec{k}} c_{\vec{k}}^+$$

$$- (d_{\vec{k}}^+ d_{\vec{k}} + d_{\vec{k}} d_{\vec{k}}^+)] \cos \theta_{\vec{k}} \sin \theta_{\vec{k}}$$

$$= \sum_{\vec{k}} \{\omega [c_{\vec{k}}^+ c_{\vec{k}} + d_{\vec{k}}^+ d_{\vec{k}}] - V_{\vec{k}} [2c_{\vec{k}}^+ c_{\vec{k}} + 1 - (2d_{\vec{k}}^+ d_{\vec{k}} + 1)] \frac{1}{2}\}$$

where we have used the fact that when $\omega_{\vec{k}}^m = \omega_{\vec{k}}^p$, we have $\tan(2\theta_{\vec{k}}) = \infty$, namely $2\theta_{\vec{k}} = \pi/2$ or $\theta_{\vec{k}} = \pi/4$, or $\cos\theta_{\vec{k}}\sin\theta_{\vec{k}} = 1/2$. We get

$$\mathcal{H} = \sum_{\vec{k}} \left[(\omega - V_{\vec{k}})c_{\vec{k}}^{+}c_{\vec{k}} + (\omega + V_{\vec{k}})d_{\vec{k}}^{+}d_{\vec{k}} \right] \qquad (11.116)$$

Comment: The magnon dispersion curve crosses the phonon dispersion curve when $\omega_{\vec{k}}^m = \omega_{\vec{k}}^p$. Due to their interaction, the degeneracy at $\omega \equiv \omega_{\vec{k}}^m = \omega_{\vec{k}}^p$ is removed: ω is split into two levels $\omega \pm V_{\vec{k}}$ as seen in the above equation.

11.4 Solutions of problems of chapter 4

Problem 1. Ising antiferromagnet: order parameter
Answer: We take the case of a chain of N Ising spins. The order parameter is defined by $M_s = \frac{1}{N}\sum_i(-1)^i S_i$ (staggered magnetization) where S_i is the spin at the lattice site i.

Problem 2. Potts model:
Answer:

a) Order parameter of q-state Potts model:

$$p = \left[q\frac{max(M_1, M_2, ..., M_q)}{N} - 1 \right] \frac{1}{q-1} \qquad (11.117)$$

where M_i is the number of spins in the state i $(i = 1, ..., q)$ and N, the number of spins in the system. In an ordered state, only one of the states is present, M_i is equal to N, so $p = 1$, and in the disordered state, all M_i are equal $(=N/q)$, so $p = 0$.

b) The ground state and its degeneracy when $J > 0$: When $J > 0$: in the ground state all lattice sites have the same Potts value.

c) If $J < 0$: the interaction of two different values has lower energy (zero). If $q = 2$ we see that the ground state is a configuration of alternating spins: this is the "antiferromagnetic ordering". The degeneracy is 2 (permutation of the two values of q).
For $q = 3$ $(i = 1, 2, 3)$, the ground state is constructed by choosing sequences of diagonal lines 1-2-3-1-2-3... or 1-3-2-1-3-2..., namely any sequence of diagonal lines with no adjacent similar numbers. An example is shown in Fig. 11.10. There are 3 ways for choosing

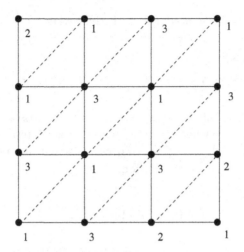

Fig. 11.10   An example of the construction of a ground-state configuration which has an order on each diagonal line (discontinued lines) but no order on the perpendicular diagonal lines.

a number for the first diagonal line, 2 ways for each of the following lines. Hence, the number of configurations in the ground state (degeneracy) is $3 \times (2)^{L-1} \times 2 \propto 2^L$, L being the number of diagonal lines (equal to the linear lattice size). The last factor 2 is to take into account the fact that there are two diagonal lines in the square lattice. This construction generates a semi-ordering: there is an ordering on each diagonal line but no ordering on the second diagonal line perpendicular to the first one.

There is another way to construct the ground state which is completely disordered: let us consider the square lattice defined on the xy plane. Each lattice site is defined by two indices (i, j). We fill the lattice sites line by line starting from $j = 0$, from $i = 0$ to $i = L$ (from left to right, bottom to top). The lattice sites on the first x line ($j = 0$) can be randomly filled with three values 1, 2, 3, never similar values at two adjacent sites. On the next line ($j = 1$), the lattice site at (i, j) has its neighbor at $(i, j - 1)$ (below, on the previous line): there are two ways to choose its value which should be different from that at $(i, j - 1)$. Once its value is chosen, we go next to its neighbor at $(i + 1, j)$: its value should be different from that at (i, j) and that at $(i + 1, j - 1)$ (see Fig. 11.11, left panel). If the value at (i, j) is equal to the value at $(i + 1, j - 1)$,

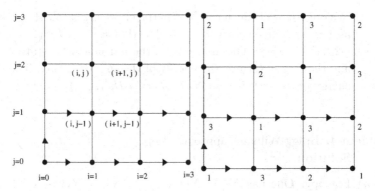

Fig. 11.11   Left: indexation of lattice sites for description in the text. Right: example of a random ground-state configuration constructed in the way described in the text.

then we have two possible values for $(i+1, j)$. If they are different, then we have only one possible value left (no choice) for $(i+1, j)$. We go to the next lattice site on the same line, namely the site at $(i+2, j)$, and we proceed in the same manner: we do not have any problem which can stop the pursuit of the construction of the line. We go to the next line at $j+1$ and continue the construction from left to right on the line. We see that the ground states obtained by such a way do not have a long-range order as seen in the example shown in Fig. 11.11 (right panel). Note that there are 2^N to place the numbers on the first line in the way described above. On each of the following lines, a number of the sites have only one choice as said above: only $\alpha_i N$ sites ($\alpha_i < 1$) have two choices. So, the random ground states have a degeneracy is of the order of $2^{N+\alpha_1 N+\alpha_2 N+\cdots} = 2^{aN^2}$ ($a < 1$). It is much larger than the degeneracy 2^N of the ordered ground states shown in Fig. 11.10.

  d) The Potts model is equivalent to the Ising model when $q = 2$: there are two states for each site of energy $-J$ and 0 for Potts model, and $\pm J$ for Ising model. There is only a shift of energy in the calculation. Physical results are identical.

Problem 3.   Domain walls:
  Solution: The interaction between two neighboring spins is $E = -2J\vec{S}_i \cdot \vec{S}_j = -2JS^2 \cos(\theta_i - \theta_j)$ where θ_i is the angle between the x axis and $\vec{S}_i$. In the ferromagnetic state, the angle $\phi = (\theta_i - \theta_j) = 0$.

In a domain wall which has N-spin thickness, ϕ is π/N. If N is large as it is the frequent case, ϕ is small so that $E \simeq -2JS^2(1 - \frac{1}{2}\phi^2) = -2JS^2 + JS^2\phi^2$. The first term is the ferromagnetic-state energy and the second term is the energy of the spin deviation. The total energy of the wall is $\Delta = NJS^2\phi^2 = NJS^2(\pi/N)^2 = JS^2\pi^2/N$.

Problem 4. Bragg-Williams approximation:
Solution:

a) Entropy: One has $N_+ = N(1 + X)/2$, $N_- = N(1 - X)/2$, with $0 \leq X \leq 1$. The number of configurations (microscopic states) is $W = N!/N_+!N_-!$. Using the Stirling formula (see Appendix A) $\ln n! \simeq n \ln n - n$ for large n, one has

$$S = k_B \ln W = k_B \ln[\frac{N!}{N_+!N_-!}]$$

$$= k_B[N \ln N - N - N_+ \ln N_+ + N_+ - N_- \ln N_- + N_-]$$

$$= k_B\{N \ln N - \frac{N(1 + X)}{2} \ln[\frac{N(1 + X)}{2}]$$

$$- \frac{N(1 - X)}{2} \ln[\frac{N(1 - X)}{2}]\}$$

$$= -k_B N \left\{ \frac{1 + X}{2} \ln[\frac{1 + X}{2}] + \frac{1 - X}{2} \ln[\frac{(1 - X)}{2}] \right\} \quad (11.118)$$

b) The probability to have an up spin at a lattice site: $p_+ = N_+/N$. The number of up-up spin pairs is thus

$$N_{up-up} = \frac{1}{2}zN_+p_+ = \frac{1}{8}zN(1 + X)^2$$

because there are z bonds around each site. The factor $1/2$ is to remove the double counting of each bond. In the same manner, the number of down-down spin pairs and the number of antiparallel spin pairs are

$$N_{down-down} = \frac{1}{2}zN_-p_- = \frac{1}{8}zN(1 - X)^2$$

$$N_{antip} = \frac{1}{2}z(N_+p_+ + N_-p_-) = \frac{1}{4}zN(1 - X^2)$$

c) The energy of the crystal is

$$E = -J(N_{up-up} + N_{down-down} - N_{antip}) = -\frac{1}{2}zJNX^2$$

d) The free energy

$$F = E - TS = -\frac{1}{2}zJNX^2 + k_BTN\{\frac{1+X}{2}\ln[\frac{1+X}{2}]$$
$$+\frac{1-X}{2}\ln[\frac{1-X}{2}]\} \tag{11.119}$$

The minimum of F corresponds to $\partial F/\partial X = 0$. This gives

$$0 = zNJX - \frac{1}{2}Nk_BT\ln\frac{1+X}{1-X}$$

$$\frac{2zJ}{k_BT}X = \ln\frac{1+X}{1-X}$$

$$\exp[\frac{2zJ}{k_BT}X] = \frac{1+X}{1-X}$$

$$X = \tanh[\frac{zJ}{k_BT}X] \tag{11.120}$$

This equation is equivalent to the mean-field equation (4.17) in the case where $S = \pm 1$.

e) Using an expansion of $\tanh[\frac{zJ}{k_BT}X]$ when $X \to 0$ up to third order in X and proceeding in the same manner as for (4.24), one obtains $T_c = zJ/k_B$. When $T > T_c$, one has $X = 0$. The entropy (11.118) is then $S = k_BN\ln 2$. This result can be directly obtained by the simple following argument: in the disordered phase, each spin is independent with two states ± 1. There are N sites so the total number of system spin configurations is just $W = 2^N$. The entropy is then $S = k_B\ln W = k_BN\ln 2$.

Remark: The Bragg-Williams approximation was initially used for binary alloys where an A atom is represented by an up spin, and a B atom by a down spin. The mixing of A and B atoms is favored if J is negative: this is equivalent to an antiferromagnetic ordering where an A atom (up spin) is surrounded by B atoms (down spins) at low temperatures.

Problem 5. Binary alloys by spin language:
    Solution:

a) Since $\epsilon > \phi$, when a spin is surrounded by neighbors of the other kind, its energy is lower. This yields a perfect antiferromagnetic ordering at $T = 0$.

b)   • One has $0 \leq N_{\uparrow,I} = N(1 + x)/4 \leq N/2$ hence $-1 \leq x \leq 1$. When $x = 0$, $N_{\uparrow,I} = N/4$: the system is in the disordered state. The number of $\uparrow$-spins occupying sites II is $N_{\uparrow,II} = N/2 - N_{\uparrow,I} = N(1 - x)/4$.

     For down spins (B atoms), one has $N_{\downarrow,II} + N_{\uparrow,II} = N/2$ hence $N_{\downarrow,II} = N(1 + x)/4$. One deduces $N_{\downarrow,I} = N/2 - N_{\downarrow,II} = N(1 - x)/4$.

     One considers the case $x > 0$ in the following.

  • The probability for a $\uparrow$-spin to be at a site of the type I is $P(\uparrow, I) = N_{\uparrow,I}/(N/2) = (1 + x)/2$, and that at a site of the type II is
$P(\uparrow, II) = N_{\uparrow,II}/(N/2) = (1 - x)/2$.
In the same way, for a $\downarrow$-spin one has
$P(\downarrow, I) = N_{\downarrow,I}/(N/2) = (1 - x)/2$ and $P(\downarrow, II) = N_{\downarrow,II}/(N/2) = (1 + x)/2$.

  • Let $N_{\uparrow,\uparrow}$, $N_{\downarrow,\downarrow}$, and $N_{\uparrow,\downarrow}$ be the numbers of $\uparrow\uparrow$, $\downarrow\downarrow$ and $\uparrow\downarrow$ pairs. The probability to have a $\uparrow\uparrow$ pair is
$P(\uparrow, \uparrow) = P(\uparrow, I)P(\uparrow, II) + P(\uparrow, II)P(\uparrow, I) = (1 - x^2)/2$ hence $N_{\uparrow,\uparrow} = NP(\uparrow, \uparrow) = N(1 - x^2)/2$. In the same way, one has:
$N_{\downarrow,\downarrow} = N(1 - x^2)/2$.
For a $\uparrow\downarrow$ pair, the probability is

$$P(\uparrow, \downarrow) = P(\uparrow, I)P(\downarrow, II) + P(\uparrow, II)P(\downarrow, I) + P(\downarrow, I)P(\uparrow, II)$$
$$+ P(\downarrow, II)P(\uparrow, I)$$
$$= (1 + x)^2/4 + (1 - x)^2/4 + (1 - x)^2/4 + (1 + x)^2/4$$
$$= 1 + x^2 \tag{11.121}$$

    Therefore $N_{\uparrow,\downarrow} = NP(\uparrow, \downarrow) = N(1 + x^2)$.

  • One has $E = [N_{\uparrow,\uparrow} + N_{\downarrow,\downarrow}]\epsilon + N_{\uparrow,\downarrow}\phi =$. Thus,

$$E = N(\epsilon + \phi) - N(\epsilon - \phi)x^2 \tag{11.122}$$

  • When E is given, x^2 is determined. One takes $x > 0$. Let $\Omega(E)$ be the number of microscopic states of energy equal to

E. To calculate $\Omega(E)$, one calculates the number of ways to choose $N(1+x)/4$ ↑-spins among $N/2$ sites of the type I and at the same time to choose $N(1+x)/4$ sites among $N/2$ sites of the type II to place them. One has

$$\Omega(E) = [C_{N/2}^{N(1+x)/4}]^2 = [\frac{(N/2)!}{[N(1+x)/4]![N(1-x)/4]!}]^2$$
(11.123)

Using the Stirling formula, one writes

$$\ln \Omega(E) = 2\{\ln(N/2)! - \ln[N(1+x)/4]! - \ln[N(1-x)/4]!\}$$
$$\simeq 2\{(N/2)\ln(N/2) - N/2 - [N(1+x)/4]$$
$$\times \ln[N(1+x)/4]$$
$$+N(1+x)/4 - [N(1-x)/4]$$
$$\times \ln[N(1-x)/4] + N(1-x)/4\}$$
$$= 2\{(N/2)\ln(N/2) - [N(1+x)/4]\ln[N(1+x)/4]$$
$$-[N(1-x)/4]\ln[N(1-x)/4]\}$$
(11.124)

The entropy is given by $S = k_B \ln \Omega$.

- The temperature T is calculated by (see Appendix A):

$$T^{-1} = \frac{\partial S}{\partial E} = k_B \frac{\partial \ln \Omega(E)}{\partial x}\frac{\partial x}{\partial E}$$
$$= \frac{k_B}{4(\epsilon - \phi)x}\{\ln[N(1+x)/4] - \ln[N(1-x)/4]\}$$
$$= \frac{k_B}{4(\epsilon - \phi)x}\ln\frac{1+x}{1-x}$$

One gets

$$\ln\frac{1+x}{1-x} = \frac{4(\epsilon - \phi)x}{k_B T}$$
$$\frac{1+x}{1-x} = \exp[\frac{4(\epsilon - \phi)x}{k_B T}]$$
$$x = \tanh[\frac{2(\epsilon - \phi)x}{k_B T}]$$
(11.125)

This equation is of the mean-field equation type [see (4.17)]. We have $x = 1$ at $T = 0$ and $x = 0$ at $T = \infty$. Between these limits, there exists a temperature below which x is not zero. An expansion at small x gives

$$x \simeq 2(\epsilon - \phi)x/(k_B T)$$
$$-[2(\epsilon - \phi)x/(k_B T)]^3/3$$
$$x[1 - 2(\epsilon - \phi)/(k_B T)] = -[2(\epsilon - \phi)/(k_B T)]^3 x^3$$
(11.126)

If $x \neq 0$, one can simplify the two sides to get

$$2(\epsilon - \phi)/(k_B T) - 1 = [2(\epsilon - \phi)/(k_B T)]^3 x^2 \qquad (11.127)$$

Since $2(\epsilon - \phi)/(k_B T) > 0$, the right-hand side is positive. This relation is satisfied if on the left-hand side one has $2(\epsilon - \phi)/(k_B T) - 1 > 0$ namely $T < 2(\epsilon - \phi)/k_B \equiv T_c$. T_c is the critical temperature.

If we return to the binary alloy language: we say we have an ordered binary alloy structure when $T < T_c$ and a disordered structure for $T > T_c$.

Problem 6. Critical temperature of ferrimagnet:

Solution: We make an expansion of (4.75) and (4.76) when $< S_A^z >$ and $< S_B^z >$ are small in the same manner as for (4.22). We have

$$< S_A^z > = a < S_B^z > - b < S_B^z >^3 + ... \qquad (11.128)$$

$$< S_B^z > = c < S_A^z > - d < S_A^z >^3 + ... \qquad (11.129)$$

where

$$a = \frac{S_A(S_A + 1)}{3} \frac{C J_1}{k_B T} \qquad (11.130)$$

$$b = \frac{[S_A^2 + (S_A + 1)^2] S_A(S_A + 1)}{90} (\frac{C J_1}{k_B T})^3 \qquad (11.131)$$

$$c = \frac{S_B(S_B + 1)}{3} \frac{C J_1}{k_B T} \qquad (11.132)$$

$$d = \frac{[S_B^2 + (S_B + 1)^2] S_B(S_B + 1)}{90} (\frac{C J_1}{k_B T})^3 \qquad (11.133)$$

Replacing (11.129) in (11.128), we write

$$< S_A^z > (ac - 1) = (ad + bc^3) < S_A^z >^3 \qquad (11.134)$$

For $< S_A^z > \neq 0$, we can simplify it on both sides. The remaining equation is

$$(ac - 1) = (ad + bc^3) < S_A^z >^2 \qquad (11.135)$$

The right-hand side of this equation is positive, therefore $ac - 1 > 0$. Replacing the coefficients a and c we obtain

$$k_B T < \frac{CJ_1}{3}\sqrt{S_A(S_A + 1)S_B(S_B + 1)} \equiv k_B T_N \qquad (11.136)$$

This means that non zero solutions of $< S_A^z >$ are found only below T_N. Note that replacing (11.128) in (11.129) gives the same solution.

Problem 7. Improvement of mean-field theory:
Solution:

a) We write

$$\vec{S}_1 \cdot \vec{S}_2 = S_1^z S_2^z + (S_1^+ S_2^- + S_1^- S_2^+)/2$$

The states of two spins $1/2$ are

$$\phi_1 = |1/2, 1/2\rangle, \quad \phi_2 = |1/2, -1/2\rangle,$$

$$\phi_3 = |-1/2, 1/2\rangle, \quad \phi_4 = |-1/2, -1/2\rangle.$$

To calculate

$$[-2J[S_1^z S_2^z + (S_1^+ S_2^- + S_1^- S_2^+)/2] - D[(S_1^z)^2 + (S_2^z)^2] - B(S_1^z + S_2^z)]|\phi_i\rangle$$

we use

$$S^{\pm}|jm\rangle = [j(j+1) - m(m \pm 1)]^{1/2}\hbar|j, m \pm 1\rangle \qquad (11.137)$$
$$(j = 1/2, m = \pm 1/2)$$
$$S^z|m\rangle = \hbar m|m\rangle \qquad (11.138)$$

We obtain a matrix 4×4. A simple diagonalization gives the following eigenvalues for the two-spin cluster:
$\epsilon_1 = -J/2 - D/2 - B$ ($\uparrow \uparrow$), $\epsilon_2 = 3J/2 - D/2$ ($\uparrow \downarrow - \downarrow \uparrow$), $\epsilon_3 = -J/2 - D/2$ ($\uparrow \downarrow + \downarrow \uparrow$), $\epsilon_4 = -J/2 - D/2 + B$($\downarrow \downarrow$) (we have taken $\hbar = 1$).

b) We put the cluster of two spins $\vec{S}_i$ and $\vec{S}_j$ in a lattice: it has $(Z-1)$ neighbors. We treat the interaction of the 4 cluster configurations found above in zero field $(B = 0)$ with these neighbors by the mean-field theory. The energy of the cluster in the crystal depends on the embedded cluster spin configurations, they are in increasing energies:

$\phi_1 = (\uparrow \uparrow) \rightarrow E_1 = -J/2 - 2J(Z-1) < S^z >$

$(\phi_2 + \phi_3)/2 = (\uparrow \downarrow + \downarrow \uparrow) \rightarrow E_2 = -J/2$

$\phi_4 = (\downarrow \downarrow) \rightarrow E_3 = -J/2 + 2J(Z-1) < S^z >$

$(\phi_2 - \phi_3)/2 = (\uparrow \downarrow - \downarrow \uparrow) \rightarrow E_4 = 3J/2$

We consider the cluster of two spins as a superspin with the z component $S^z = (S_i^z + S_j^z)/2$. We have

$$< S^z >= Tr\frac{1}{2}(S_i^z + S_j^z) \exp(-\beta E)/Tr \exp(-\beta E)$$

where

$$Tr \exp(-\beta E) = \exp(\beta J/2) \exp(\beta X) + \exp(\beta J/2)$$
$$+ \exp(\beta J/2) \exp(-\beta X)$$
$$+ \exp(-\beta 3J/2)$$
$$(X \equiv 2J(Z-1) < S^z >)$$

$$Tr\frac{1}{2}(S_i^z + S_j^z) \exp(-\beta E)] = \frac{1}{2} \exp(\beta J/2) \exp(\beta X) + 0$$
$$- \frac{1}{2} \exp(\beta J/2) \exp(-\beta X) + 0$$
$$= \exp(\beta J/2) \sinh \beta X$$

Hence,

$$< S^z >= \frac{\sinh \beta X}{2 \left[\cosh \beta X + \exp(-\beta J) \cosh \beta J\right]} \qquad (11.139)$$

We see that $< S^z >= 0$ is a solution of this equation. An expansion around $< S^z >= 0$ gives

$$2 < S^z > [\frac{-3 + 2\beta(Z-1)J - e^{-2\beta J}}{2}] = \beta^2 4(Z-1)^2 J^2 < S^z >^3 \qquad (11.140)$$

The solution $< S^z > \neq 0$ is possible if
$$-3 + 2\beta(Z-1)J - e^{-2\beta J} > 0.$$

T_c is obtained by solving $-3 + 2\beta_c(Z-1)J - e^{-2\beta_c J} = 0$ where $\beta_c = (k_B T_c)^{-1}$. We obtain

$$e^{-2J/k_B T_c} + 3 - 2(Z-1)J/k_B T_c = 0 \qquad (11.141)$$

Problem 8. Interaction between next-nearest neighbors in mean-field treatment:

Solution:

a) All spins are parallel at $T = 0$. All interactions are fully satisfied.

b) The hypothesis of the mean-field theory: all neighboring spins of a spin are replaced by an average value which is used to calculate the value of the spin under consideration.

c) The energy of a spin at $T = 0$ is $E = -Z_1 J_1 - Z_2 J_2$ where Z_1 and Z_2 are the numbers of nearest neighbors and of next-nearest neighbors, respectively. For a body-centered cubic lattice, $Z_1 = 8$, $Z_2 = 6$. E is the energy which maintains the spin ordering: the lower it is, the higher the temperature is needed to destroy the ordering. Thus, the stronger J_2 is, the higher the transition temperature becomes.

d) The same calculation as that in the chapter by replacing CJ with $Z_1 J_1 + Z_2 J_2$ in Eqs. (4.12)-(4.14) and in the following equations to obtain the final mean-field equation.

e) The critical temperature is obtained by replacing CJ in Eq. (4.24) by $Z_1 J_1 + Z_2 J_2$.

f) Now we suppose $J_2 < 0$. When $|J_2| \gg J_1$, it is obvious that the J_2 interaction imposes the antiferromagnetic ordering to make the overall energy negative. The spins on the cube corners form an antiferromagnetic sublattice, the centered spins form another antiferromagnetic sublattice, independent of the first one. Since each spin has 4 up neighbors and 4 down neighbors (make a figure to convince yourself), its interaction energy with nearest neighbors is zero, independent of J_1. The energy of such a spin configuration is thus $E_{Antif} = Z_2 J_2 = -Z_2 |J_2|$.

If $|J_2| \ll J_1$ then the ferromagnetic state is more favorable. Its energy is $E_{Ferro} = -Z_1 J_1 - Z_2 J_2 = -Z_1 J_1 + Z_2 |J_2|$.

The critical value of $|J_2|$ below which the ferromagnetic state is stable is determined by solving $E_{Ferro} < E_{Antif}$. We have $|J_2^c| = \frac{Z_1 J_1}{2 Z_2}$ or $J_2^c = -\frac{Z_1 J_1}{2 Z_2}$. When $|J_2| < |J_2^c|$ (or $J_2 > J_2^c$) we have the ferromagnetic ordering. Otherwise, we have the antiferromagnetic one.

Problem 9. Improved mean-field theory: Bethe's approximation

Solution: We consider the spin σ_0 surrounded by its nearest neighbors σ_i. The Hamiltonian of this "cluster" embedded in a crystal is given by

$$\mathcal{H} = -J \sum_{i=1}^{z} \sigma_0 \sigma_i - \mu_B B \sigma_0 - \mu_B(B + H) \sum_{i=1}^{z} \sigma_i \qquad (11.142)$$

where z is the coordination number, B an applied magnetic field, and H the molecular field acting on the neighboring spins of σ_0 from spins outside of the cluster. H is thus given by $H = zJ\overline{\sigma}/\mu_B$. The mean-field equation will be obtained at the end by setting $\overline{\sigma_0} = \overline{\sigma_i}$. We have

$$Z = \sum_{\sigma_0 = \pm 1} \sum_{\sigma_1 = \pm 1} \cdots \sum_{\sigma_z = \pm 1} e^{\beta[J \sum_{i=1}^{z} \sigma_0 \sigma_i + \mu_B B \sigma_0 + \mu_B(B+H) \sum_{i=1}^{z} \sigma_i]}$$

$$= \sum_{\sigma_0 = \pm 1} \sum_{\sigma_1 = \pm 1} \cdots \sum_{\sigma_z = \pm 1} e^{a\sigma_0 \sum_{i=1}^{z} \sigma_i + b\sigma_0 + (b+c) \sum_{i=1}^{z} \sigma_i} \qquad (11.143)$$

where $a = J/k_B T$, $b = \mu_B B/k_B T$ and $c = \mu_B H/k_B T = zJ\overline{\sigma}/k_B T$. Summing on $\sigma_0 = \pm 1$ and factorizing the other sums, we have

$$Z = Z_+ + Z_- \quad \text{where}$$

$$Z_\pm = \sum_{\sigma_1 = \pm 1} \cdots \sum_{\sigma_z = \pm 1} e^{\pm a \sum_{i=1}^{z} \sigma_i \pm b + (b+c) \sum_{i=1}^{z} \sigma_i]}$$

$$= e^{\pm b}[2\cosh(\pm a + b + c)]^z \qquad (11.144)$$

The averaged $\overline{\sigma_0}$ and $\overline{\sigma_i}(i = 1, ..., z)$ are given by

$$\overline{\sigma_0} = \frac{Z_+ - Z_-}{Z} \qquad (11.145)$$

$$\overline{\sigma_i} = \frac{1}{z} \sum_{i=1}^{z} \overline{\sigma_i} = \frac{1}{z} \frac{\partial Z/\partial c}{Z}$$

$$= [Z_+ \tanh(a + b + c) + Z_- \tanh(-a + b + c)]/Z \qquad (11.146)$$

Setting $\overline{\sigma_0} = \overline{\sigma_i} = \overline{\sigma}$, we have

$$Z_+[1 - \tanh(a + b + c)] = Z_-[1 + \tanh(-a + b + c)] \qquad (11.147)$$

Replacing $Z_\pm$ by (11.144), we obtain

$$\left[\frac{\cosh(a + b + c)}{\cosh(-a + b + c)} \right]^{z-1} = e^{2c} \qquad (11.148)$$

This equation allows us to determine self-consistently c, namely $\bar{\sigma}$. In zero field $B = 0$, namely $b = 0$, we have

$$\frac{\cosh(a + c)}{\cosh(-a + c)} = e^{2c/(z-1)}$$

$$\frac{c}{z - 1} = \frac{1}{2} \ln \frac{\cosh(a + c)}{\cosh(-a + c)} \qquad (11.149)$$

We see that $c = 0$ is a solution of the last equation. However there is a nonzero solution by making an expansion at small c (small $\bar{\sigma}$) to the third order:

$$\frac{c}{z - 1} \simeq \frac{1}{2} \ln \frac{\cosh a + c \sinh a + (1/2)c^2 \cosh a + \ldots}{\cosh a - c \sinh a + (1/2)c^2 \cosh a + \ldots}$$

$$= [c - \frac{1}{3} \frac{c^3}{\cosh^2 a} + \ldots] \tanh a$$

$$\frac{1}{z - 1} = [1 - \frac{1}{3} \frac{c^2}{\cosh^2 a} + \ldots] \tanh a \qquad (11.150)$$

Therefore

$$c^2 = 3 \frac{\cosh^3 a}{\sinh a} \left[\tanh a - \frac{1}{z - 1} + \ldots \right] \qquad (11.151)$$

This is satisfied if $[\tanh a - \frac{1}{z-1}] > 0$ since the left-hand side is positive. This means that the nonzero solution exists if $T < T_c$ where T_c is given by

$$\tanh(J/k_B T_c) = \frac{1}{z - 1} \qquad (11.152)$$

It is interesting to note that for $z = 1$ there is no solution for T_c and for $z = 2$ (1D) we have $T_c = 0$. This corresponds more to reality (see Problems 1, 2 and 3 of section 7.7).

Problem 10. We repeat Problem 7 in the case of an antiferromagnet:

Solution: With $J < 0$: the changes with respect to the ferromagnetic case are

i) in question a): no change

ii) in question b): we have the inverse order of energies $E_4 < E_3 < E_2 < E_1$, because $J < 0$. The remaining calculation is exactly the same. The four configurations embedded in the crystal give the following energies:

ϕ_1: $E_1 = -J/2$

$(\phi_2 + \phi_3)/2$: $E_2 = -J/2 - 2J(Z - 1) < S^z >$

ϕ_4: $E_3 = -J/2$

$(\phi_2 - \phi_3)/2$: $E_4 = 3J/2 + 2J(Z-1) < S^z >$

When putting these energies in the calculation of the average $<(S_i^z + S_j^z)/2>$, be careful to use the energy of the corresponding crystal-field spin configuration in the argument of the exponential and to use the correct sign of each neighboring spin. Make a draw to help. We will have

$$e^{+2J/k_B T_c} + 3 + 2(Z-1)J/k_B T_c = 0 \qquad (11.153)$$

It is the same as the ferromagnetic result, bearing in mind that $J < 0$ here.

Problem 11. The critical field H_c:

Answer:

a) in simple cubic lattice with antiferromagnetic interaction J between first neighbors: $H_c = 6|J|$.

b) in a square lattice with first- and second-neighbor interactions: $H_c = 4|J_1| + 4J_2$

11.5   Solutions of problems of chapter 5

Problem 1. Demonstration of (5.63)-(5.64):

Demonstration: In (5.62), by replacing $\epsilon_{\vec{k}} \simeq 2JS(ka)^2$ and using (5.58), we have

$$\sum_{\vec{k}} \epsilon_{\vec{k}} < n_{\vec{k}} > \simeq \frac{N}{(2\pi)^2} \int_0^\infty 2JS(ka)^2 \sum_{l=1}^{l=\infty} e^{-l\beta 2JS(ka)^2} k^2 dk$$

$$(11.154)$$

Putting $x = l\beta 2JSk^2$ $(a = 1)$ and integrating, we obtain (5.63) and then (5.64).

Problem 2. Chain of Heisenberg spins:

Solution:

a) $\omega = 2J_1 SZ(1 - \cos(ka)) + 2J_2 SZ(1 - \cos(2ka))$ (Z=2, number of neighbors)

b)If $J_2 < 0$, $\omega = 2J_1 SZ(1 - \cos(ka)) - 2|J_2|SZ(1 - \cos(2ka))$. We plot ω versus k. We see that ω is strongly affected by J_2 when $k \to 0$. Analytically, we take the derivative of ω with respect to k, we have

$d\omega/dk = 2J_1 SZa \sin(ka) - 4a|J_2|SZ \sin(2ka) = 2SZa[J_1 \sin(ka) - 4|J_2| \sin(ka) \cos(ka)] = 2SZa \sin(ka)[J_1 - 4|J_2| \cos(ka)]$

This derivative is zero at $k = 0$ (uniform mode) and at $\cos(ka) = \frac{J_1}{4|J_2|} = -\frac{J_1}{4J_2}$ ($J_2 < 0$). This second case is called "soft mode" because the slope (stiffness) of ω is zero at this value of k. We have a helimagnetic ordering for $J_2 < -J_1/4$ (see sections 5.4 and 7.5).

Problem 3. Heisenberg spin systems in two dimensions:

Solution:

a)$\omega = 2JSZ(1 - \gamma_k)$ where $Z = 4$ (number of nearest neighbors in a square lattice), $\gamma_k = (\cos(k_x a) + \cos(k_y a))/2$. $\omega \to 2JS(ka)^2$ when $\vec{k} \to 0$.

b) $< S^z > = 1/2 - A \int_{ZB} \frac{2\pi k dk}{\exp(\beta\omega)-1}$ [A: constant, see (5.55)-(5.55)]. The most important contribution to the integral comes from the small k region where $\omega \to 2JS(ka)^2$. We have $< S^z > \simeq 1/2 - A \int_{ZB} \frac{2\pi k dk}{1+\beta JS(ka)^2-1} \simeq 1/2 - A \int_{ZB} \frac{2\pi k dk}{\beta JS(ka)^2}$. This integral diverges at $k = 0$, hence $< S^z >$ is not defined if $T \neq 0$. There is no long-range order for $T \neq 0$ in 2D (see the rigorous theorem of Mermin-Wagner in Ref. [95]) .

Note: In 3D, we replace in the integral $2\pi k dk$ by $4\pi k^2 dk$. The integral does not diverge at $k = 0$. The long-range ordering exists at $T \neq 0$ in 3D.

Problem 4. Demonstration of Eqs. (5.143)-(5.145):

Demonstration: We have

$$a_{\vec{k}} = \alpha_{\vec{k}} \cosh \theta_k - \alpha^+_{-\vec{k}} \sinh \theta_k \qquad (11.155)$$

$$a^+_{\vec{k}} = \alpha^+_{\vec{k}} \cosh \theta_k - \alpha_{-\vec{k}} \sinh \theta_k \qquad (11.156)$$

where we can show that $\alpha_{\vec{k}}$ and $\alpha^+_{\vec{k}}$ obey the boson commutation relations (see similar demonstration in Problem 9). Replacing these expressions in the Hamiltonian (5.137), we have

$$
\begin{aligned}
\mathcal{H} &= -NSJ(\vec{Q}) + \frac{S}{2}\sum_{\vec{k}}[A(\vec{k},\vec{Q})(a_{\vec{k}}a_{\vec{k}}^+ + a_{\vec{k}}^+ a_{\vec{k}}) \\
&\quad + B(\vec{k},\vec{Q})(a_{\vec{k}}a_{-\vec{k}} + a_{\vec{k}}^+ a_{-\vec{k}}^+)] \\
&= -NSJ(\vec{Q}) + \frac{S}{2}\sum_{\vec{k}}\{\sqrt{A(\vec{k},\vec{Q})^2 - B(\vec{k},\vec{Q})^2}[\alpha_{\vec{k}}\alpha_{\vec{k}}^+ + \alpha_{\vec{k}}^+\alpha_{\vec{k}}] \\
&\quad + [B(\vec{k},\vec{Q})\cosh 2\theta_k - A(\vec{k},\vec{Q})\sinh 2\theta_k][\alpha_{\vec{k}}\alpha_{-\vec{k}} + \alpha_{\vec{k}}^+\alpha_{-\vec{k}}^+]\}
\end{aligned}
$$

The Hamiltonian is diagonal if the second term in the curly brackets $\{...\}$ is zero, namely

$$
\tanh(2\theta_k) = \frac{B(\vec{k},\vec{Q})}{A(\vec{k},\vec{Q})} \tag{11.157}
$$

Omitting the constant term, we have

$$
\mathcal{H} = \frac{S}{2}\sum_{\vec{k}}\hbar\omega_{\vec{k}}[\alpha_{\vec{k}}^+\alpha_{\vec{k}} + \alpha_{\vec{k}}\alpha_{\vec{k}}^+] \tag{11.158}
$$

where the energy of the magnon of mode $\vec{k}$ is

$$
\hbar\omega_{\vec{k}} = \sqrt{A(\vec{k},\vec{Q})^2 - B(\vec{k},\vec{Q})^2} \tag{11.159}
$$

Problem 5. 'Union-Jack' lattice:

    Solution: We write the energy expression for each kind of configuration. Then, we compare two by two to determine the frontier between them. The result is shown in Fig. 11.12.

Problem 6. Ground-state spin configuration of a triangular lattice with XY spins:

    Solution: In the case of the triangular plaquette, suppose that spin $\mathbf{S}_i$ $(i = 1, 2, 3)$ of magnitude S makes an angle θ_i with the **Ox** axis. Writing E and minimizing it with respect to the angles θ_i,

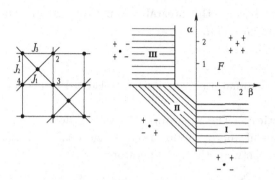

Fig. 11.12 Left: Union-Jack lattice: diagonal, vertical and horizontal bonds denote the interactions J_1, J_2 and J_3, respectively. Right: Phase diagram of the ground state shown in the plane ($\alpha = J_2/J_1$, $\beta = J_3/J_1$). Heavy lines separate different phases and spin configuration of each phase is indicated (up, down and free spins are denoted by +, - and o, respectively). The three kinds of partially disordered phases and the ferromagnetic phase are denoted by I, II , III and F, respectively.

one has

$$E = J(\mathbf{S}_1 \cdot \mathbf{S}_2 + \mathbf{S}_2 \cdot \mathbf{S}_3 + \mathbf{S}_3 \cdot \mathbf{S}_1)$$
$$= JS^2 \left[\cos(\theta_1 - \theta_2) + \cos(\theta_2 - \theta_3) + \cos(\theta_3 - \theta_1) \right],$$

$$\frac{\partial E}{\partial \theta_1} = -JS^2 \left[\sin(\theta_1 - \theta_2) - \sin(\theta_3 - \theta_1) \right] = 0,$$

$$\frac{\partial E}{\partial \theta_2} = -JS^2 \left[\sin(\theta_2 - \theta_3) - \sin(\theta_1 - \theta_2) \right] = 0,$$

$$\frac{\partial E}{\partial \theta_3} = -JS^2 \left[\sin(\theta_3 - \theta_1) - \sin(\theta_2 - \theta_3) \right] = 0.$$

A solution of the last three equations is $\theta_1 - \theta_2 = \theta_2 - \theta_3 = \theta_3 - \theta_1 = 2\pi/3$. One can also write

$$E = J(\mathbf{S}_1 \cdot \mathbf{S}_2 + \mathbf{S}_2 \cdot \mathbf{S}_3 + \mathbf{S}_3 \cdot \mathbf{S}_1) = -\frac{3}{2}JS^2 + \frac{J}{2}(\mathbf{S}_1 + \mathbf{S}_2 + \mathbf{S}_3)^2.$$

The minimum of E corresponds to $\mathbf{S}_1 + \mathbf{S}_2 + \mathbf{S}_3 = 0$ which yields the 120° structure. This is also true for Heisenberg spins.

Problem 7. Uniaxial anisotropy:
Answer:
a) to follow the method of the chapter.

b) Yes, because the integral does not diverge any more at $k = 0$ in the presence of d.

Problem 8. Commutation relations of Holstein-Primakoff operators:

Solution: The operators a^+ and a defined in the Holstein-Primakoff approximation respect rigorously the commutation relations between the spin operators:

$$[S_l^+, S_m^-] = 2S_l^z \delta_{lm} \quad \text{and} \quad [S_l^z, S_m^\pm] = \pm S_l^\pm \delta_{lm}$$

Demonstration: Replacing the spin operators by the Holdstein-Primakoff operators, one has

$$[S_l^+, S_m^-] = 2S[f_l(S)a_l, a_m^+ f_m(S)]$$
$$= 2S[f_l(S)a_l a_m^+ f_m(S) - a_m^+ f_m(S) f_l(S) a_l]$$

If $l = m$, one has

$$[S_l^+, S_l^-] = 2S[f(S)aa^+ f(S) - a^+ f(S)f(S)a]$$
$$= 2S[(1 - \frac{a^+a}{2S})^{1/2}(a^+a + 1)(1 - \frac{a^+a}{2S})^{1/2}$$
$$-a^+(1 - \frac{a^+a}{2S})a]$$
$$= 2S[(1 - \frac{a^+a}{2S})(a^+a + 1) - a^+a + \frac{a^+a^+aa}{2S}]$$
$$= 2S[1 - \frac{a^+a}{S}]$$
$$= 2[S - a^+a] = 2S_l^z \tag{11.160}$$

If $l \neq m$, one obtains in the same manner $[S_l^+, S_m^-] = 0$. For the second relation, when $l = m$ one has

$$[S_l^z, S_l^+] = \sqrt{2S}[(S - a^+a)fa - fa(S - a^+a)]$$
$$= -\sqrt{2S}[faS - faa^+a - Sfa + a^+afa]$$
$$= -\sqrt{2S}[-faa^+a + fa^+aa] = -\sqrt{2S}[-faa^+a$$
$$+f(aa^+ - 1)a]$$
$$= \sqrt{2S}fa = S_l^+ \tag{11.161}$$

If $l \neq m$, one obtains $[S_l^z, S_m^\pm] = 0$. Similarly, one has $[S_l^z, S_m^-] = -S_l^- \delta_{lm}$.

Remark: One has used $a^+af = fa^+a$ in the above demonstration of (11.160) because
$$a^+af = a^+a(1 - \tfrac{a^+a}{2S})^{1/2} = (a^+aa^+a - \tfrac{a^+aa^+aa^+a}{2S})^{1/2} = (1 - \tfrac{a^+a}{2S})^{1/2}a^+a = fa^+a.$$

Problem 9. Operators defined in (5.86)-(5.89) obey the commutation relations:

Demonstration: We have

$$
\begin{aligned}
[\alpha_{\vec{k}}, \alpha_{\vec{k}'}^+] &= [a_{\vec{k}}\cosh\theta_k + b_{\vec{k}}^+\sinh\theta_k, a_{\vec{k}'}^+\cosh\theta_k' + b_{\vec{k}'}\sinh\theta_k'] \\
&= \cosh\theta_k\cosh\theta_k'[a_{\vec{k}}, a_{\vec{k}'}^+] + \cosh\theta_k\sinh\theta_k'[a_{\vec{k}}, b_{\vec{k}'}^+] \\
&\quad + \sinh\theta_k\cosh\theta_k'[b_{\vec{k}}^+, a_{\vec{k}'}^+] + \sinh\theta_k\sinh\theta_k'[b_{\vec{k}}^+, b_{\vec{k}'}] \\
&= \cosh\theta_k\cosh\theta_k'\delta(k,k') + 0 + 0 - \sinh\theta_k\sinh\theta_k'\delta(k,k') \\
&= [\cosh^2\theta_k - \sinh^2\theta_k]\delta(k,k') = \delta(k,k') \qquad (11.162)
\end{aligned}
$$

The same demonstration is done for the other relations.

Problem 10. Magnon soft mode:

Demonstration: The magnon spectrum (5.125) becomes unstable when the interaction between next-nearest neighbors defined in ϵ, Eq. (5.119), is larger than a critical constant.

The spectrum becomes unstable when one of its frequencies tends to zero: this mode is termed as "soft mode". Numerically, we plot (5.125) versus $\vec{k}$ for various values of ϵ and determine its critical value. Analytically, we see that interaction J_2 affects modes near $k_x = k_y = k_z = \pi/a$. To increase J_2 makes the frequencies of these modes decrease. The first mode to become zero occurs at
$$\epsilon = \epsilon_c = \frac{2}{3}\frac{1-|\alpha|}{1+|\alpha|}$$

11.6 Solutions of problems of chapter 6

Problem 1. Proofs of (6.13):

Solution:

We consider the following integral in the complex plane

$$\int_C \frac{e^{-iz(t-t')}}{z+i\epsilon}dz \tag{11.163}$$

where C is an integral contour chosen in the plane. This integral has a pole at $z_0 = -i\epsilon$ on the imaginary axis in the lower half plane. If $t - t' > 0$, we choose the contour in the lower half plane including the pole z_0: $C = -R \to +R + C'$ where $\pm R$ are on the real axis and C' is the half circle going from R to $-R$ in the lower half plane. Using the theorem of residues we write

$$\int_C \frac{e^{-iz(t-t')}}{z+i\epsilon}dz = \int_{-R}^{R} \frac{e^{-ix(t-t')}}{x+i\epsilon}dx + \int_{C'} \frac{e^{-iz(t-t')}}{z+i\epsilon}dz$$
$$= -2\pi i \text{Residue} \tag{11.164}$$

where the minus sign come from the sense of the contour and the only pole lying inside the contour is z_0:

$$\text{Residue} = \lim_{z\to z_0} \frac{e^{-iz(t-t')}}{z+i\epsilon} \times (z - z_0)$$
$$= e^{-iz_0(t-t')} = e^{-\epsilon(t-t')}$$
$$= 1 \quad (\lim \quad \epsilon \to 0)$$

Taking the limit $R \to \infty$, we can show that the integral $\int_{C'\to\infty} \frac{e^{-iz(t-t')}}{z+i\epsilon}dz$ goes to zero. Therefore,

$$\lim_{\epsilon\to 0+} \int_{-\infty}^{\infty} \frac{e^{-ix(t-t')}}{x+i\epsilon}dx = -2\pi i \quad \text{for} \quad t - t' > 0 \tag{11.165}$$

Now, if $t - t' < 0$, we choose the contour C in the upper half plane. The right-hand side of (11.164) is zero because there is no pole inside C. We have thus

$$\lim_{\epsilon\to 0+} \int_{-\infty}^{\infty} \frac{e^{-ix(t-t')}}{x+i\epsilon}dx = 0 \quad \text{for} \quad t - t' < 0 \tag{11.166}$$

Combining (11.165) and (11.166) and using the Heavyside function $\Theta(t - t') = 1$ if $t - t' > 0$, $= 0$ if $t - t' < 0$, we obtain the formula (6.13):

$$\Theta(t - t') = \lim_{\epsilon\to 0+} \frac{i}{2\pi} \int_{-\infty}^{\infty} \frac{e^{-ix(t-t')}}{x+i\epsilon}dx \tag{11.167}$$

Problem 2. Demonstration of Eq. (6.22): We have (6.21)

$$G_{lm}(t-t') = <<S_l^+(t); S_m^-(t') >>$$
$$= -i\theta(t-t') < \left[S_l^+(t), S_m^-(t')\right] > \quad (11.168)$$

Noting that $\theta(t) = \int_{-\infty}^t \delta(t)dt$ so that $d\theta(t)/dt = \delta(t)$, we write

$$i\frac{dG_{lm}(t-t')}{dt} = \frac{d\theta(t-t')}{dt} < \left[S_l^+(t), S_m^-(t')\right] >$$

$$+\theta(t-t')\frac{d < \left[S_l^+(t), S_m^-(t')\right] >}{dt}$$

$$= \delta_{tt'} 2 < S_l^z > \delta_{lm} - << \left[\mathcal{H}, S_l^+\right](t); S_m^-(t') >>$$

$$= 2 < S_l^z > \delta_{lm}\delta_{tt'} - << \left[\mathcal{H}, S_l^+\right](t); S_m^-(t') >>$$

$$(11.169)$$

where we have used in the second equality a commutation relation for the first term and the equation of motion $id\hat{O}/dt = -[\mathcal{H}, \hat{O}]$ for the second term. We calculate now the commutator $A = [\mathcal{H}, S_l^+]$. Using $\mathcal{H}$ of (5.43) without the factor 2 of J and without the applied field term, we have

$$A = -\frac{1}{2}J \sum_{<l',m'>} \{[S_{l'}^z S_{m'}^z + \frac{1}{2}(S_{l'}^+ S_{m'}^- + S_{l'}^- S_{m'}^+), S_l^+]\}$$

$$= -\frac{1}{2}J \sum_{<l',m'>} \{[S_{l'}^z S_{m'}^z, S_l^+] + \frac{1}{2}[S_{l'}^+ S_{m'}^-, S_l^+]$$

$$+\frac{1}{2}[S_{l'}^- S_{m'}^+, S_l^+]\}$$

$$= -\frac{1}{2}J \sum_{<l',m'>} \{[S_{l'}^z, S_l^+]S_{m'}^z + S_{l'}^z[S_{m'}^z, S_l^+] + \frac{1}{2}[S_{l'}^+, S_l^+]S_{m'}^-$$

$$+\frac{1}{2}S_{l'}^+[S_{m'}^-, S_l^+] + \frac{1}{2}[S_{l'}^-, S_l^+]S_{m'}^+ + \frac{1}{2}S_{l'}^-[S_{m'}^+, S_l^+]\}$$

$$= -\frac{1}{2}J \sum_{<l',m'>} \{S_l^+ S_{m'}^z \delta_{ll'} + S_{l'}^z S_l^+ \delta_{m'l} + 0 - S_{l'}^+ S_l^z \delta_{lm'}$$

$$-S_l^z \delta_{ll'} S_{m'}^+ + 0\}$$

$$= -\frac{1}{2}J[\sum_{m'} S_l^+ S_{m'}^z + \sum_{l'} S_{l'}^z S_l^+ - \sum_{l'} S_{l'}^+ S_l^z - \sum_{m'} S_l^z S_{m'}^+]$$

$$= -J \sum_{l'} [S_{l'}^z S_l^+ - S_l^z S_{l'}^+] \quad (11.170)$$

where the pre-factor $\frac{1}{2}$ was added to remove the double counting due to the double sum and where we have used $[AB, C] = [A, C] B + A [B, C]$ and

$$[S_l^+, S_m^-] = 2S_l^z \delta_{lm}$$
$$[S_m^z, S_l^\pm] = \pm S_l^\pm \delta_{lm}$$

Note that in the last line of Eq. (11.170) we gathered the sums by changing the dummy variables and permuted operators such as $S_l^+ S_{l'}^z = S_{l'}^z S_l^+$ because the indices l and l' indicate two different neighboring sites [in the Hamiltonian, l' and m' are neighboring sites, one of these sites is equal to l because of the delta functions in the 4th equality of Eq. (11.170)]. Inserting (11.170) into (11.169) and putting $l' = l + \vec{\rho}$ where $\vec{\rho}$ are the vectors connecting l to its neighbors, we obtain Eq. (6.22):

$$i\hbar \frac{dG_{lm}(t)}{dt} = 2 < S_l^z > \delta_{lm}\delta(t) - << [\mathcal{H}, S_l^+] (t); S_m^- >>$$
$$= 2 < S_l^z > \delta_{lm}\delta(t)$$
$$- J \sum_{\vec{\rho}} << S_l^z(t)S_{l+\vec{\rho}}^+(t) - S_l^+(t)S_{l+\vec{\rho}}^z(t); S_m^- >>$$

Problem 3. Helimagnet by Green's function method:
   Solution: We consider a crystal of simple cubic lattice with Heisenberg spins of amplitude $1/2$. The interaction J_1 between nearest neighbors is ferromagnetic. Suppose that along the y axis there exists an antiferromagnetic interaction J_2 between next nearest neighbors, in addition to J_1.

a) In the xz plane the only interaction is J_1, so the spins in the plane are parallel. Along the y axis, the competition between the ferromagnetic J_1 and the antiferromagnetic J_2 can give rise to a helical structure (see section 5.4). Let θ be the helical angle between two nearest neighboring spins in the y direction. The energy of a spin is written as

$$E = -4J_1 - 2J_1 \cos \theta + 2|J_2| \cos(2\theta) \qquad (11.171)$$

where the first term is the energy in the xz plane, the second and third terms are the energy in the y direction. Minimizing E with respect to θ we have

$$\frac{dE}{d\theta} = 0 = 2J_1 \sin\theta - 4|J_2|\sin(2\theta)$$

$$0 = 2J_1 \sin\theta - 8|J_2|\sin\theta\cos\theta$$

$$= 2\sin\theta(J_1 - 4|J_2|\cos\theta) \qquad (11.172)$$

The solutions are

i) $\sin\theta = 0$, namely $\theta = 0$ (solution 1), $\theta = \pi$ (solution 2)

ii) $\cos\theta = \frac{J_1}{4|J_2|}$ (solution 3) if $-1 \le \frac{J_1}{4|J_2|} \le 1$, namely $\frac{|J_2|}{J_1} \ge \frac{1}{4} \equiv \alpha_c$. This solution corresponds to the helimagnetic configuration. We can compare the energies of the three solutions

$$E_1 = -4J_1 - 2J_1 + 2|J_2|$$

$$E_2 = -4J_1 + 2J_1 + 2|J_2|$$

$$E_3 = -4J_1 - 2\frac{J_1^2}{4|J_2|} + 2|J_2|(2\frac{J_1^2}{16|J_2|^2} - 1)$$

We see that E_2 is a maximum, and

$$E_1 < E_3 \text{ when } \frac{|J_2|}{J_1} < \alpha_c$$

$$E_3 < E_1 \text{ when } \frac{|J_2|}{J_1} > \alpha_c$$

b) Let θ be the helical angle between two nearest neighboring spins in the y direction. The Hamiltonian in terms of θ is given by (5.136) (without the anisotropy term):

$$\mathcal{H} = -\frac{1}{4}\sum_{(i,j)} J(\vec{R}_{ij})\{(S_i^+ S_j^- + S_i^- S_j^+)[1 + \cos(\vec{\theta}\cdot\vec{R}_{ij})]$$

$$-(S_i^+ S_j^+ + S_i^- S_j^-)[1 - \cos(\vec{\theta}\cdot\vec{R}_{ij})] + 4S_i^z S_j^z \cos(\vec{\theta}\cdot\vec{R}_{ij})$$

$$+2[(S_i^+ + S_i^-)S_j^z - S_i^z(S_j^+ + S_j^-)]\sin(\vec{\theta}\cdot\vec{R}_{ij})\} \qquad (11.173)$$

where $\vec{R}_{ij}$ is the distance vector between the two spins i and j, $\vec{\theta}$ is the vector of magnitude θ perpendicular to the angle plane (xz). In the present problem, we have

i) $\cos(\vec{\theta} \cdot \vec{R}_{ij}) = 0$ for nearest neighbors i and j belonging to the xz plane,

ii) $\cos(\vec{\theta} \cdot \vec{R}_{ij}) = \cos \theta$ for nearest neighbors i and j on the y axis (lattice constant=1),

iii) $\cos(\vec{\theta} \cdot \vec{R}_{ij}) = \cos(2\theta)$ for next nearest neighbors i and j on the y axis (distance $R_{ij} = 2$).

c) We define two Green's functions by (6.68)-(6.69) which lead to, using the RPA decoupling scheme,

$$i\hbar \frac{dG_{m\ell}(t - t')}{dt} = 2 < S_k^z > \delta_{m\ell}\delta(t)$$

$$-\frac{1}{2}\sum_{m'} J_{m,m'}[< S_k^z > (\cos \theta_{m,m'} - 1)F_{m'\ell}(t - t')$$

$$+ < S_k^z > (\cos \theta_{m,m'} + 1)G_{m'\ell}(t - t')$$

$$-2 < S_{m'}^z > \cos \theta_{m,m'} G_{m\ell}(t - t')] \tag{11.174}$$

$$i\hbar \frac{dF_{m\ell}(t - t')}{dt} = \frac{1}{2}\sum_{m'} J_{m,m'}[< S_m^z > (\cos \theta_{m,m'} - 1)G_{m'\ell}(t - t')$$

$$+ < S_m^z > (\cos \theta_{m,m'} + 1)F_{m'\ell}(t - t')$$

$$-2 < S_{m'}^z > \cos \theta_{m,m'} F_{m\ell}(t - t')] \tag{11.175}$$

Using the Fourier transforms and summing on neighbors with their corresponding angles mentioned above, we obtain

$$\mathbf{M}(\omega)\,\mathbf{g} = \mathbf{u}, \tag{11.176}$$

where

$$\mathbf{g} = \begin{pmatrix} g_{n,n'} \\ f_{n,n'} \end{pmatrix}, \qquad \mathbf{u} = \begin{pmatrix} 2\langle S^z \rangle \delta_{n,n'} \\ 0 \end{pmatrix}, \tag{11.177}$$

and

$$\mathbf{M}(\omega) = \begin{pmatrix} \omega + A & B \\ -B & \omega - A \end{pmatrix}, \tag{11.178}$$

where

$$A = J_1 < S^z > \Big[Z(\gamma_{\vec{k}} - 1) + (\cos \theta + 1)\cos k_y$$

$$-2\cos \theta + \frac{|J_2|}{J_1}[\cos(2\theta) + 1]\cos(2k_y) - 2\frac{|J_2|}{J_1}\cos(2\theta)\Big]$$

$$B = J_1 < S^z > (\cos \theta - 1)\cos k_y$$

$$-|J_2| < S^z > [\cos(2\theta) - 1]\cos(2k_y) \tag{11.179}$$

in which, $Z = 4$ is the number of nearest neighbors in the xz plane, $\theta = \arccos(J_1/4|J_2|)$, and $\gamma_{\vec{k}} = [2\cos(k_x) + 2\cos(k_z)]/Z$. Non trivial solutions of (11.176) impose that

$$\det \begin{vmatrix} \omega + A & B \\ -B & \omega - A \end{vmatrix} = 0$$

We have

$$0 = (\omega + A)(\omega - A) + B^2 \tag{11.180}$$

$$\omega = \pm\sqrt{A^2 - B^2} \tag{11.181}$$

We show in Fig. 11.13 the spin-wave spectrum (11.181) as a function of the wave-vector in the helical direction k_y, for $\theta = \pi/3$ ($|J_2|/J_1 = 0.5$) at $k_x = k_y = 0$. We see that $\omega = 0$ at θ.

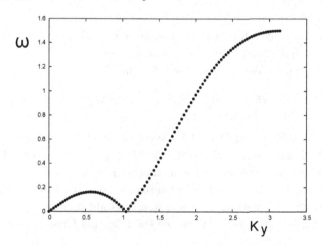

Fig. 11.13 Spin-wave spectrum versus the wave-vector k_y in the simple cubic lattice with a helical structure in the y axis, in the case $\theta = \pi/3$ (namely $|J_2|/J_1 = 0.5$), and $k_x = k_z = 0$.

d) Note that
i) when $\cos\theta = 1$ (ferromagnet), taking $J_2 = 0$, we have

$$A = J_1 < S^z > \left[4(2\cos k_x + 2\cos k_z)/4 - 4 + 2\cos k_y - 2\right]$$

$$= J_1 < S^z > \left[2\cos k_x + 2\cos k_z + 2\cos k_y - 6\right]$$

$$= J_1 < S^z > Z'(\gamma'_{\vec{k}} - 1)$$

The expression (11.181) is reduced to the dispersion relation (6.32) for ferromagnets $\omega = Z'J_1 \langle S^z \rangle (1 - \gamma'_{\vec{k}})$ where $Z' = 6$ and $\gamma'_{\vec{k}} = [2\cos(k_x) + 2\cos k_y + 2\cos(k_z)]/Z'$ [we have in this case $B = 0$ in (11.181)].

ii) when $\cos\theta = -1$ (collinear antiferromagnets), for $J_2 = 0$, we have, by returning to (11.174)-(11.175) to calculate A and B ($J_1 < 0$),

$$A = |J_1| < S^z > Z'$$
$$B = |J_1| < S^z > Z'\gamma'_{\vec{k}} \tag{11.182}$$

Eq. (11.181) then gives the dispersion relation (6.58) of antiferromagnets: $\omega = \pm Z'|J_1| \langle S^z \rangle \sqrt{1 - (\gamma'_{\vec{k}})^2}$.

Problem 4. Green's function method for a system of Ising spins $\pm 1/2$ in an applied magnetic field in one dimension:

Solution: We consider the following Hamiltonian

$$\mathcal{H} = -2J \sum_n S_n^z S_{n+1}^z - g\mu_B H \sum_n S_n^z \tag{11.183}$$

where $S_n^z = \pm 1/2$ is the z component of the spin given by the Pauli matrices. $2S_n^z = \sigma_n^z = \pm 1$ [Eqs. (1.3)-(1.5)]. The Ising model corresponds to the assumption that only the z component is taken into account. For simplicity, we often take $S_z = \pm 1$, the results are however identical to a constant factor.

Consider the site i. We define the following Green's function

$$G_i(t) \equiv <<S_i^+(t); S_i^- >> \tag{11.184}$$

The equation of motion of $G_i(t)$ is ($\hbar = 1$)

$$i\frac{dG_i(t)}{dt} = 2 < S_i^z > \delta(t) + 2J << S_i^+(t)[S_{i+1}^z(t) + S_{i-1}^z(t)]; S_i^- >>$$
$$+ g\mu_B H << S_n^+(t); S_n^- >> \tag{11.185}$$

This equation generates the following Green's function

$$G_{i+1}(t) = << S_i^+(t)[S_{i+1}^z(t) + S_{i-1}^z(t)]; S_i^- >> \tag{11.186}$$

Equation (11.185) becomes

$$i\frac{dG_i(t)}{dt} = 2 < S_i^z > \delta(t) + 2G_{i+1}(t) + g\mu_B H G_i(t) \qquad (11.187)$$

The equation of motion of $G_{i+1}(t)$ generates the following Green's function

$$G_{i+2}(t) \equiv << S_i^+(t)S_{i-1}^z(t)S_{i+1}^z(t); S_i^- >> \qquad (11.188)$$

We write the equation of motion of $G_{i+2}(t)$ but we shall neglect the Green's functions of higher orders. We have

$$i\frac{dG_i(t)}{dt} = 2 < S_i^z > \delta(t) + 2G_{i+1}(t) + g\mu_B H G_i(t)$$

$$i\frac{dG_{i+1}(t)}{dt} = 2 < S_i^z(S_{i-1}^z + S_{i+1}^z > \delta(t) + JG_i(t)$$
$$+4JG_{i+2}(t) + g\mu_B H G_{i+1}(t)$$

$$i\frac{dG_{i+2}(t)}{dt} = 2 < S_{i-1}^z S_i^z S_{i+1}^z > \delta(t) + \frac{J}{2}G_i(t) + g\mu_B H G_{i+2}(t)$$

The Fourier transforms $G_k(t) = \int_{-\infty}^{\infty} G_k(E)e^{-iEt}dE$ $(k = i, i + 1, i + 2)$ yield, putting $E' = E - g\mu_B H$,

$$E'G_i(E') - 2G_{i+1}(E') = \frac{1}{\pi}x_1 \qquad (11.189)$$

$$-JG_i(E') + E'G_{i+1}(E') - 4JG_{i+2}(E') = \frac{2}{\pi}x_2 \qquad (11.190)$$

$$-\frac{J}{2}G_{i+1}(E') + E'G_{i+2}(E') = \frac{1}{\pi}x_3 \qquad (11.191)$$

where

$$x_1 = < S_i^z >, \quad 2x_2 = < S_i^z(S_{i-1}^z + S_{i+1}^z) >, \quad x_3 = < S_i^z S_{i-1}^z S_{i+1}^z > .$$

The solutions of these equations are

$$G_i(E') = \frac{1}{\pi}\left[\frac{x_1/2 - 2x_3}{E'} + \frac{x_1/4 + x_2 + x_3}{E' - 2J} + \frac{x_1/4 - x_2 + x_3}{E' + 2J}\right]$$

$$G_{i+1}(E') = \frac{1}{\pi}\left[\frac{x_1/4 + x_2 + x_3}{E' - 2J} + \frac{-x_1/4 + x_2 - x_3}{E' + 2J}\right]$$

$$G_{i+2}(E') = \frac{1}{\pi}[\frac{-x_1/8 + x_3/2}{E'} + \frac{x_1/16 + x_2/4 + x_3/4}{E' - 2J}$$
$$+ \frac{x_1/16 - x_2/4 + x_3/4}{E' + 2J}]$$

The spectral theorem [see Eq. (6.39)] gives

$$< S_i^- S_i^+ > = \frac{x_1 - 4x_3}{e^{\beta g \mu_B H} - 1} + \frac{x_1/2 + 2x_2 + 2x_3}{e^{\beta(g \mu_B H + 2J)} - 1}$$

$$+ \frac{x_1/2 - 2x_2 + 2x_3}{e^{\beta(g \mu_B H - 2J)} - 1}$$

$$< S_i^- S_i^+ (S_{i-1}^z + S_{i+1}^z) > = \frac{x_1/2 + 2x_2 + 2x_3}{e^{\beta(g \mu_B H + 2J)} - 1}$$

$$+ \frac{-x_1/2 + 2x_2 - 2x_3}{e^{\beta(g \mu_B H - 2J)} - 1}$$

$$< S_i^- S_i^+ S_{i-1}^z S_{i+1}^z > = \frac{-x_1/4 + x_3}{e^{\beta g \mu_B H} - 1} + \frac{x_1/8 + x_2/2 + x_3/2}{e^{\beta(g \mu_B H + 2J)} - 1}$$

$$+ \frac{x_1/8 - x_2/2 + x_3/2}{e^{\beta(g \mu_B H - 2J)} - 1}$$

We expand these expressions at small values of H. At the first order in H, we have

$$x_1 - 4x_3 = \beta g \mu_B H [\frac{1}{2} - 2x_2 \coth \beta J] \qquad (11.192)$$

$$\left[-1 + \frac{1}{2} \coth \beta J \right] x_1 + 2x_3 \coth \beta J$$

$$= \frac{\beta g \mu_B H}{\sinh^2 \beta J} x_2 \qquad (11.193)$$

$$x_1 - 4x_3 = 2\beta g \mu_B H [-x_2' + x_2 \coth \beta J] \qquad (11.194)$$

where $x_2' = < S_{i-1}^z S_{i+1}^z >$.

We have three equations for four unknowns x_1, x_2, x_3 and x_2'. It is not possible to solve them. However, when $H = 0$, we have $x_1 = x_3 = 0$, namely there is no ordering at $T \neq 0$ in one dimension (remember $x_1 = < S_i^z >$), as we will show by the renormalization group in Problem 2 of section 7.7.

To go further, we calculate the correlation function between the spin at the site 0 and the spin at the site n: we consider the following Green's functions

$$G_{0,n} = << S_0^+(t) S_n^z(t); S_0^- >> \qquad (11.195)$$

$$G_{1,n} = << S_0^+(t)[S_{-1}^z(t) + S_1^z(t)] S_n^z(t); S_0^- >> \qquad (11.196)$$

$$G_{2,n} = << S_0^+(t) S_{-1}^z(t) S_1^z(t) S_n^z(t); S_0^- >> \qquad (11.197)$$

Following the same method as above, we obtain

$$2x_{1,n} - 8x_{3,n} = \beta g \mu_B H [x_1 - 4x_{2,n} \coth \beta J]$$

$$< S_1^z S_n^z > + < S_{-1}^z S_n^z > = (x_{1,n} + 4x_{3,n}) \coth \beta J - 2 \frac{\beta g \mu_B H}{\sinh^2 \beta J} x_{2,n}$$

$$4x_{3,n} - x_{1,n} = 2\beta g \mu_B H [< S_{-1}^z S_1^z S_n^z > -x_{2,n} \coth \beta J]$$

where $x_{1,n} =< S_0^z S_n^z >$, $2x_{2,n} =< S_0^z (S_{-1}^z + S_1^z) >$, $x_{3,n} =< S_0^z S_{-1}^z S_1^z S_n^z >$

When $H = 0$, we have $x_{1,n} = 4x_{3,n}$. Hence

$$< S_{-1}^z S_n^z > + < S_1^z S_n^z > = 2 < S_0^z S_n^z > \coth \beta J \qquad (11.198)$$

We take a solution of the form $< S_m^z S_n^z >= AX^{n-m}$ where A is a constant: when $m = n$, we have $< S_n^z S_n^z >= A = 1/4$. Setting $n = 0$, $m = -1$, $m = 1$ and $m = 0$ in (11.198) we have

$$X + \frac{1}{X} = 2 \coth \beta J \qquad (11.199)$$

The solution of this equation is $X = A \tanh \frac{\beta J}{2}$, namely

$$< S_0^z S_n^z >= \frac{1}{4} [\tanh \frac{\beta J}{2}]^n \qquad (11.200)$$

We see that, for $n = 0$, we have $< S_0^z S_0^z >= 1/4$ as expected for a spin $1/2$. For $n = 1$, we have

$$x_2 =< S_0^z S_1^z >= \frac{1}{4} \tanh \frac{\beta J}{2} \qquad (11.201)$$

Substituting this in (11.192) and (11.193) we obtain

$$x_1 =< S_0^z >= \beta g \mu_B H e^{\beta J} \qquad (11.202)$$

The susceptibility is thus

$$\chi = N g \mu_B < S_0^z > /H = \beta (g \mu_B)^2 e^{\beta J} = \frac{(g \mu_B)^2 e^{\beta J}}{k_B T} \qquad (11.203)$$

This result is also obtained by the transfer matrix method [see (11.249)]. Note that the argument of the exponential is βJ because of the factor 2 in the Hamiltonian and spins $1/2$: to find the argument of the exponential of χ in Problem 3 [Eq. (11.249)] of section 7.7 we divide J by 2 and multiply by 4 (inverse of the square of spin magnitude).

Problem 5. Effect of next-nearest neighbor interaction:

Guide: We follow the same calculation in section 6.3 in adding the interaction J_2 in the equations (6.22), (6.26)-(6.33). We obtain, instead of (6.29),

$$\hbar\omega G_{\vec{k}}(\omega) = \frac{<S^z>}{\pi} + J<S^z> ZG_{\vec{k}}(\omega)[1 - \frac{1}{Z}\sum_{\vec{\rho}} e^{i\vec{k}\cdot\vec{\rho}}]$$

$$+ J_2 <S^z> Z_2 G_{\vec{k}}(\omega)[1 - \frac{1}{Z_2}\sum_{\vec{\rho}_2} e^{i\vec{k}\cdot\vec{\rho}_2}] \quad (11.204)$$

where Z is the number of nearest neighbors and Z_2 that of next-nearest neighbors. We have

$$G_{\vec{k}}(\omega) = \frac{<S^z>}{\pi}$$

$$\times \frac{1}{\hbar\omega - ZJ<S^z>(1-\gamma(\vec{k})) - Z_2 J_2 <S^z>(1-\gamma_2(\vec{k}))}$$

$$= \frac{<S^z>}{\pi} \frac{1}{\hbar\omega - \epsilon_{\vec{k}}} \quad (11.205)$$

where

$$\gamma(\vec{k}) = \frac{1}{Z}\sum_{\vec{\rho}} e^{i\vec{k}\cdot\vec{\rho}} \quad (11.206)$$

$$\gamma_2(\vec{k}) = \frac{1}{Z_2}\sum_{\vec{\rho}_2} e^{i\vec{k}\cdot\vec{\rho}_2} \quad (11.207)$$

and

$$\epsilon_{\vec{k}} = ZJ<S^z>(1-\gamma(\vec{k})) + Z_2 J_2 <S^z>(1-\gamma_2(\vec{k})) \quad (11.208)$$

Problem 6. Magnon spectrum in Heisenberg triangular antiferromagnet: Green's function method

Solution: Using the ground state of a triangular lattice determined in Problem 6 of section 5.6 for the angle $\theta_{k,k'}$ between two neighboring spins, the Fourier transformations of Eqs. (6.70)-(6.71) can be written as

$$\mathbf{M}(\omega)\mathbf{g} = \mathbf{u}, \quad (11.209)$$

where

$$\mathbf{g} = \begin{pmatrix} g_{n,n'} \\ f_{n,n'} \end{pmatrix}, \qquad \mathbf{u} = \begin{pmatrix} 2\langle S^z\rangle \delta_{n,n'} \\ 0 \end{pmatrix}, \quad (11.210)$$

and

$$\mathbf{M}(\omega) = \begin{pmatrix} A^+ & B \\ -B & A^- \end{pmatrix}, \qquad (11.211)$$

where

$$A^{\pm} = \omega \pm \left[\frac{1}{2} J \langle S^z \rangle \left(Z \gamma_{\vec{k}} \right) (\cos\theta + 1) \right.$$

$$\left. - J \langle S^z \rangle Z \cos\theta \right]$$

$$B = \frac{1}{2} J \langle S^z \rangle (\cos\theta - 1) \left(Z \gamma_{\vec{k}} \right), \qquad (11.212)$$

in which, $Z = 6$ is the number of nearest neighbors, $\theta = 2\pi/3$ the angle between two neighboring spins, and $\gamma_{\vec{k}} = \left[2\cos(k_x a) + 4\cos(k_x a/2)\cos(k_y a\sqrt{3}/2) \right]/Z$.

Note that we can also follow the method outlined in subsection 9.7.3 for a thin film. For only one layer ($n = 1$), Eqs. (9.75)-(9.77) give the same equations as those displayed above.

Non trivial solutions of (11.209) impose

$$\det \begin{vmatrix} A^+ & B \\ -B & A^- \end{vmatrix} = 0$$

We have

$$0 = A^+ A^- + B^2$$

$$\omega = \pm \sqrt{ \left[\frac{1}{2} J \langle S^z \rangle \left(Z \gamma_{\vec{k}} \right) (\cos\theta + 1) - J \langle S^z \rangle Z \cos\theta \right]^2 - B^2 }$$

$$= \pm Z J \langle S^z \rangle \sqrt{ \left[\frac{1}{2} \gamma_{\vec{k}} (\cos\theta + 1) - \cos\theta \right]^2 - \left[\frac{1}{2} (\cos\theta - 1) \gamma_{\vec{k}} \right]^2 }$$

$$= \pm Z J \langle S^z \rangle \sqrt{ \cos\theta (1 - \gamma_{\vec{k}})(\cos\theta - \gamma_{\vec{k}}) } \qquad (11.213)$$

We show in Fig. 11.14 the spin-wave frequency ω versus k_y in the first Brillouin zone for $k_x = 0$.

Note that when $\cos\theta = 1$, the expression (11.213) is reduced to $\omega = ZJ\langle S^z \rangle (1 - \gamma_{\vec{k}})$ which is the dispersion relation (6.32) for ferromagnets. When $\cos\theta = -1$, Eq. (11.213) becomes $\omega = ZJ\langle S^z \rangle \sqrt{1 - \gamma_{\vec{k}}^2}$ which is precisely the dispersion relation (6.58) of collinear antiferromagnets.

Using the formula (5.109) for $T = 0$, we have

$$M \simeq \frac{N}{2}(S - \Delta S) \qquad (11.214)$$

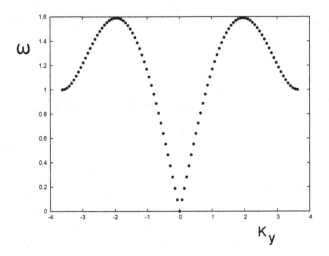

Fig. 11.14   Spin-wave spectrum versus the wave-vector k_y in the case of a triangular antiferromagnet ($S = 1/2$, $k_x = 0$, $J = 1$).

where $\Delta S = \sum_{\vec{k}} \sinh^2 \Theta_k$ is independent of T. Θ_k is given by (5.95)

$$\tanh(2\Theta_k) = \gamma_{\vec{k}} \qquad (11.215)$$

Writing $\sinh \Theta_k$ as a function of $\tanh(2\Theta_k)$, one can calculate numerically the sum in ΔS using the wave vectors in the first Brillouin zone. We have the result $\Delta S \simeq 0.21$, namely the spin length at $T = 0$ is reduced from $S = 1/2$ to 0.29 (compared to 0.303 for a square antiferromagnet). This strong zero-point contraction is due to the frustration of the triangular antiferromagnet. Other more precise methods taking into account higher-order fluctuations should yield a smaller value for the spin length at $T = 0$.

Remark: The integral should be performed inside the first Brillouin zone which is a hexagon.

Problem 7. Quantum gas by the Green's method:
Solution:
We define the following Green's function:

$$G_{\alpha,\beta}(\vec{r}t, \vec{r}'t') = -i < \varphi_0 | \hat{T}[\hat{\Psi}_{H\alpha}(\vec{r}t) \hat{\Psi}^+_{H\beta}(\vec{r}'t')]|\varphi_0 >$$

where α and β are the spins.

a) For free fermions, $\hat{\Psi}_{H\alpha}(\vec{r}t) = \hat{\Psi}_{I\alpha}(\vec{r}t)$ (interaction representation). We write

$$\hat{\Psi}_{I\alpha}(\vec{r}t) = \sum_{\vec{k}} c_{\vec{k},\alpha}(t)e^{i\vec{k}\cdot\vec{r}} = \sum_{\vec{k}} c_{\vec{k},\alpha}e^{-i\omega_{\vec{k}}t} \, e^{i\vec{k}\cdot\vec{r}} \qquad (11.216)$$

where $c_{\vec{k},\alpha}$ is time-independent and $\omega_{\vec{k}} = \hbar k^2/2m$.
At $T = 0$ we have for $t - t' > 0$

$$G^0_{\alpha,\beta}(\vec{r}t, \vec{r}'t') = -i < \varphi_0|\hat{\Psi}_{I\alpha}(\vec{r}t)\hat{\Psi}^+_{I\beta}(\vec{r}'t')|\varphi_0 >$$

$$= -i < \varphi_0| \sum_{\vec{k},\vec{k}'} c_{\vec{k},\alpha}c^+_{\vec{k}',\beta}e^{-i\,(\omega_{\vec{k}}t-\omega_{\vec{k}'}t')}e^{i\,(\vec{k}\cdot\vec{r}-\vec{k}'\cdot\vec{r}')}$$

$$\times|\varphi_0 >$$

$$= -i \sum_{\vec{k},\vec{k}'} e^{-i\,(\omega_{\vec{k}}t-\omega_{\vec{k}'}t')}e^{i\,(\vec{k}\cdot\vec{r}-\vec{k}'\cdot\vec{r}')} < \varphi_0|c_{\vec{k},\alpha}c^+_{\vec{k}',\beta}$$

$$\times|\varphi_0 >$$

$$= -i \sum_{\vec{k},\vec{k}'} e^{-i\,(\omega_{\vec{k}}t-\omega_{\vec{k}'}t')}e^{i\,(\vec{k}\cdot\vec{r}-\vec{k}'\cdot\vec{r}')}$$

$$\times\Theta(k - k_F)\delta_{\vec{k},\vec{k}'}\delta_{\alpha,\beta}$$

$$= -i\delta_{\alpha,\beta} \sum_{\vec{k}} e^{-i\,\omega_{\vec{k}}(t-t')}e^{i\,\vec{k}\cdot(\vec{r}-\vec{r}')}\Theta(k - k_F)$$

$$(11.217)$$

where $|\varphi_0 >$ is the ground state. Note that we have used the fact that $c^+_{\vec{k}',\beta}|\varphi_0 >$ is not zero only if $k' > k_F$ because all state with $k' \leq k_F$ are occupied in the ground state. That is why we have $\Theta(k - k_F)$ in the result. For $t - t' < 0$, in the same manner we have $G^0_{\alpha,\beta}(\vec{r}t, \vec{r}'t') \propto < \varphi_0|c^+_{\vec{k}',\beta}c_{\vec{k},\alpha}|\varphi_0 >= n_{\vec{k}}\Theta(k_F - k)\delta_{\vec{k},\vec{k}'}$ due to the action of $c_{\vec{k},\alpha}$ on $|\varphi_0 >$ (nothing outside the Fermi sphere to destroy). Note that for $t - t' \to 0^+$ we have from (11.217)

$$G^0_{\alpha,\beta}(\vec{r}t, \vec{r}'t') = -i\delta_{\alpha,\beta} \sum_{\vec{k}} e^{i\,\vec{k}\cdot(\vec{r}-\vec{r}')} = -i\delta_{\alpha,\beta}\delta_{\vec{r},\vec{r}'}$$

Now, combining the two cases $t - t' \gtrless 0$ and changing the sum into an integral we have

$$iG^0_{\alpha,\beta}(\vec{r}t, \vec{r}'t') = \delta_{\alpha\beta}(\frac{1}{2\pi})^3 \int d\vec{k}e^{i\vec{k}\cdot(\vec{r}-\vec{r}')}e^{-i\omega_k(t-t')}$$

$$\times [\Theta(t - t')\Theta(k - k_F) - \Theta(t' - t)\Theta(k_F - k)]$$

b) For $t - t' > 0$, the time and spatial Fourier transform is written as

$$G^0_{\alpha,\beta}(\vec{k}, \omega) = \delta_{\alpha\beta} \int_0^\infty d(t - t') e^{i\omega(t-t')}$$

$$\times \int_{-\infty}^\infty d(\vec{r} - \vec{r}') G^0_{\alpha,\beta}(\vec{r}t, \vec{r}'t') e^{-i\vec{k}\cdot(\vec{r}-\vec{r}')}$$

$$= \delta_{\alpha\beta} \int_0^\infty d(t - t') e^{i(\omega - \omega_{\vec{k}})(t-t')}$$

$$\times \int_{-\infty}^\infty d(\vec{r} - \vec{r}') \sum_{\vec{k}'} e^{-i(\vec{k}-\vec{k}')\cdot(\vec{r}-\vec{r}')} \Theta(k - k_F)$$

$$= \delta_{\alpha\beta} \int_0^\infty d(t - t') e^{i(\omega - \omega_{\vec{k}})(t-t')} \sum_{\vec{k}'} \delta_{\vec{k},\vec{k}'} \Theta(k - k_F)$$

$$= \delta_{\alpha\beta} \int_0^\infty d(t - t') e^{i(\omega - \omega_{\vec{k}})(t-t')} \Theta(k - k_F)$$

$$= \lim_{\eta \to 0} \delta_{\alpha\beta} \int_0^\infty d(t - t') e^{i(\omega - \omega_{\vec{k}} + i\eta)(t-t')} \Theta(k - k_F)$$

$$= \delta_{\alpha\beta} \frac{\Theta(k - k_F)}{\omega - \omega_{\vec{k}} + i\eta} \tag{11.218}$$

For $t - t' < 0$, we proceed in the same manner we obtain the second term of the following result:

$$G^0_{\alpha,\beta}(\vec{k}, \omega) = \delta_{\alpha\beta} \left[\frac{\theta(k - k_F)}{\omega - \omega_{\vec{k}} + i\eta} + \frac{\theta(k_F - k)}{\omega - \omega_{\vec{k}} - i\eta} \right]$$

where η is an infinitesimal positive constant.

c) We consider both bosons and fermions in this question. We establish some relations to observables:

In the second quantization, a one-particle operator $\hat{J}$ such as the kinetic-energy operator is written as (see chapter 3)

$$\hat{J}(\vec{r}t) = \sum_{\alpha,\beta} \hat{\Psi}^+_{H\beta}(\vec{r}t) J_{\beta\alpha}(\vec{r}t) \hat{\Psi}_{H\alpha}(\vec{r}t) \tag{11.219}$$

The expectation value in the ground state is

$$< \hat{J}(\vec{r}t) > = \frac{< \varphi_0 | \hat{J}(\vec{r}t) | \varphi_0 >}{< \varphi_0 | \varphi_0 >}$$

$$= \lim_{\vec{r}' \to \vec{r}} \sum_{\alpha,\beta} J_{\beta\alpha}(\vec{r}t) \frac{< \varphi_0 | \hat{\Psi}_{H\beta}^{+}(\vec{r}'t') \hat{\Psi}_{H\alpha}(\vec{r}t) | \varphi_0 >}{< \varphi_0 | \varphi_0 >}$$

$$= \pm \lim_{t' \to t^+} \lim_{\vec{r}' \to \vec{r}} \sum_{\alpha,\beta} J_{\beta\alpha}(\vec{r}t) G_{\alpha,\beta}(\vec{r}t, \vec{r}'t')$$

$$= \pm \lim_{t' \to t^+} \lim_{\vec{r}' \to \vec{r}} \text{Tr}[J(\vec{r}t) G(\vec{r}t, \vec{r}'t')] \tag{11.220}$$

Consider the case where $\hat{J}$ is the density operator. We have $< \hat{n}(\vec{r}) > = \pm i \text{Tr} G(\vec{r}t, \vec{r}t^+)$ from the definition of $\hat{n}(\vec{r})$ (see chapter 3). For the total density $N = \int d\vec{r} < \hat{n}(\vec{r}) >$ we have, using the above Fourier transform,

$$N = \pm i \frac{\Omega}{(2\pi)^4} \lim_{\eta \to 0^+} \int d\vec{k} \int_{-\infty}^{\infty} d\omega e^{i\omega\eta} \text{Tr} G(\vec{k}, \omega)$$

For the kinetic energy, we have

$$< \hat{T} > = \pm i \int d\vec{r} \lim_{\vec{r}' \to \vec{r}} [-\frac{\hbar^2 \nabla^2}{2m} \text{Tr} G(\vec{r}t, \vec{r}'t^+)] \tag{11.221}$$

The total energy $E = < \hat{T} > + < V >$ can be calculated as follows. We write the Heisenberg equation of motion

$$i \frac{\partial}{\partial t} \hat{\Psi}_{H\alpha}(\vec{r}t) = e^{i\hat{\mathcal{H}}t/\hbar} [\hat{\Psi}_{\alpha}(\vec{r}t), \hat{\mathcal{H}}] e^{-i\hat{\mathcal{H}}t/\hbar} \tag{11.222}$$

where $\hat{\mathcal{H}}$ is given by Eq. (3.35). We calculate the commutator $[\hat{\Psi}_{\alpha}(\vec{r}t), \hat{\mathcal{H}}]$ using the relations (11.81)-(11.82) given in the solution of Problem 5 of chapter 3. We then obtain (see details p. 68 of Ref. [50]):

$$E = \pm \frac{i}{2} \frac{\Omega}{(2\pi)^4} \lim_{\eta \to 0^+} \int d\vec{k} \int_{-\infty}^{\infty} d\omega e^{i\omega\eta} (\frac{\hbar^2 k^2}{2m} + \hbar\omega) \text{Tr} G(\vec{k}, \omega).$$

11.7   Solutions of problems of chapter 7

Problem 1. Solution for an Ising chain:

We write the Hamiltonian as

$$\mathcal{H} = -J \sum_{n=1}^{N} \sigma_n \sigma_{n+1} \tag{11.223}$$

with $\sigma_{N+1} = \sigma_1$. The partition function is written as

$$Z = \sum_{\sigma_1=\pm 1} \sum_{\sigma_2=\pm 1} \cdots \sum_{\sigma_N=\pm 1} e^{\beta \sum_{n=1}^{N} \sigma_n \sigma_{n+1}}$$

$$= \sum_{\sigma_1=\pm 1} \sum_{\sigma_2=\pm 1} \cdots \sum_{\sigma_N=\pm 1} \prod_{n=1}^{N} e^{\beta \sigma_n \sigma_{n+1}} \tag{11.224}$$

where $\beta = J/k_B T$. Since $\sigma_n \sigma_{n+1} = \pm 1$, we have the following identity (by verification)

$$e^{\beta \sigma_n \sigma_{n+1}} = \cosh \beta + \sigma_n \sigma_{n+1} \sinh \beta \tag{11.225}$$

Equation (11.224) becomes

$$Z = \sum_{\sigma_1=\pm 1} \sum_{\sigma_2=\pm 1} \cdots \sum_{\sigma_N=\pm 1} [\cosh \beta + \sigma_1 \sigma_2 \sinh \beta]$$
$$\times [\cosh \beta + \sigma_2 \sigma_3 \sinh \beta]...$$
$$= \sum_{\sigma_1=\pm 1} \sum_{\sigma_2=\pm 1} \cdots \sum_{\sigma_N=\pm 1} [(\cosh \beta)^N + (\cosh \beta)^{N-1} \sinh \beta (\sigma_2 \sigma_3)$$
$$+... + \cosh \beta (\sinh \beta)^{N-1} (\sigma_3 \sigma_4)(\sigma_4 \sigma_5)...$$
$$...(\sigma_{N+1} \sigma_1) + (\sinh \beta)^N (\sigma_1 \sigma_2)(\sigma_2 \sigma_3)...(\sigma_{N+1} \sigma_1)] \tag{11.226}$$

Except the first and the last terms, all other terms of the sum in the square brackets [...] are zero because in each term there is one σ which appears once in the factor giving rise, when summed up, two terms of opposite signs. The first term in (11.226) does not depend on σ, it gives $2^N (\cosh \beta)^N$. The last term yields $2^N (\sinh \beta)^N$ because each σ appears twice in its factor. We have

$$Z = 2^N [(\cosh \beta)^N + (\sinh \beta)^N] \tag{11.227}$$

For $T \neq 0$ ($\beta \neq \infty$), we have $\cosh \beta > \sinh \beta$. With $N \gg 1$, we can neglect $(\sinh \beta)^N$ compared to $(\cosh \beta)^N$. Thus,

$$Z = 2^N (\cosh \beta)^N [1 + (\frac{\sinh \beta}{\cosh \beta})^N] = 2^N (\cosh \beta)^N \tag{11.228}$$

The free energy is $F = -k_B T \ln Z = -N k_B T \ln[2 \cosh(J/k_B T)]$. The average energy is [see (A.10)]:

$$\overline{E} = -\frac{\partial \ln Z}{\partial \beta} = -N J \tanh(\beta J) \tag{11.229}$$

The heat capacity is

$$C = dE/dT = Nk_B[\frac{k_BT}{J}\cosh\frac{J}{k_BT}]^{-2} \qquad (11.230)$$

The energy and the heat capacity per spin are shown in Fig. 11.15. We see that C has a maximum but no divergence. Thus, there is no phase transition at finite T.

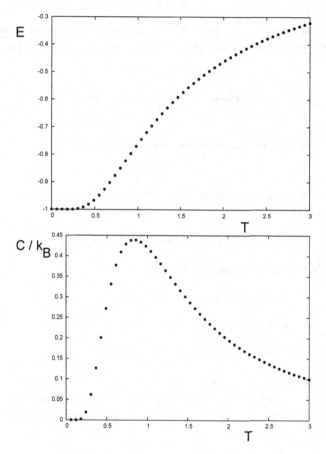

Fig. 11.15 Energy E (top) and heat capacity C (bottom) per spin versus temperature T. $J/k_B = 1$ has been used.

Problem 2. Renormalization group applied to an Ising chain:

Solution: We consider the following Hamiltonian

$$\mathcal{H} = -K \sum_n \sigma_n \sigma_{n+1} \qquad (11.231)$$

where $K = J/k_B T$, J being a ferromagnetic interaction between nearest neighbors. The partition function is given by

$$Z = Tr \exp(-\mathcal{H}) \qquad (11.232)$$

To study this spin chain, we use the decimation method. We divide the system into three-spin blocks as shown in Fig. 11.16.

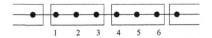

Fig. 11.16   Blocks of three spins used for the decimation.

We write for the two blocks in the figure the corresponding factors in Z

$$\exp(K\sigma_2\sigma_3)\exp(K\sigma_3\sigma_4)\exp(K\sigma_4\sigma_5) \qquad (11.233)$$

Using the following equality for the case $\sigma_n = \pm 1$

$$\exp(K\sigma_2\sigma_3) = \cosh K(1 + x\sigma_2\sigma_3) \qquad (11.234)$$

where $x = \tanh K$, we rewrite (11.233) as

$$\cosh^3 K(1 + x\sigma_2\sigma_3)(1 + x\sigma_3\sigma_4)(1 + x\sigma_4\sigma_5) \qquad (11.235)$$

Writing explicitly each term of the product and making the sum on $\sigma_3 = \pm 1$ and $\sigma_4 = \pm 1$ (decimation of spins at block borders), we see that the odd terms of these variables give zero contributions. There remains

$$2^2 \cosh^3 K(1 + x^3\sigma_2\sigma_5)$$

We can rewrite

$$2^2 \cosh^3 K(1 + x^3\sigma_2\sigma_5) = \exp[K'\sigma_2\sigma_5 + C] \qquad (11.236)$$

where C is a constant. If we forget C, the right-hand side is similar to the initial Hamiltonian with a new interaction K' between the remaining spins σ_2 and σ_5. To calculate K' we write

$$\exp[K'\sigma_2\sigma_5 + C] = \exp(C)\exp(K'\sigma_2\sigma_5)$$
$$= \exp(C)\cosh K'(1 + x'\sigma_2\sigma_5)$$

By identifying this with the left-hand side of (11.236), we obtain $x' = \tanh K' = x^3$ where $K' = \tanh^{-1}[\tanh^3 K]$ and

$$\exp(C)\cosh K' = 2^2 \cosh^3 K$$
$$\exp(C) = \frac{2^2 \cosh^3 K}{\cosh K'}$$
$$C = -\ln[\frac{\cosh^3 K}{\cosh K'}] - 2\ln 2 \qquad (11.237)$$

We renumber the spins after the first decimation as follows: $\sigma_1' = \sigma_2$, $\sigma_2' = \sigma_5$, ... (every three old spins). The new Hamiltonian is thus

$$\mathcal{H}' = -K' \sum_n \sigma_n'\sigma_{n+1}' - C\frac{N}{3} \qquad (11.238)$$

where $N/3$ is the number of three-spin blocks (N: initial number of spins).
The equation of the renormalization group is thus

$$K' = \tanh^{-1}[\tanh^3 K] \qquad (11.239)$$

We see that $K' = K$ if $K = 0$ and $K = \infty$.
At high T, $K \to 0^+$, $\tanh^3 K < 1$, hence $K' \to 0$ after successive decimations. The fixed point at $T = \infty$ ($K = 0$) is thus stable.
At low T, $K \to \infty$, $\tanh^3 K \to 1^-$, hence $K' \to 0$ after successive decimations, namely a "run away" flow. The fixed point at $T = 0$ ($K = \infty$) is thus unstable.
The flow diagram is shown in Fig. 11.17.

Fig. 11.17   Flow diagram of a chain of Ising spins.

Any point between $K = 0$ and $K = \infty$ moves to $K = 0$ after successive decimations. The nature of any point between these limits is therefore the same as that of $K = 0$ $(T = \infty)$, namely it belongs to the paramagnetic phase.

Problem 3. Transfer matrix method applied to an Ising chain:

Solution: Let N be the total number of spins. The Hamiltonian is given by

$$\mathcal{H}_0 = -J \sum_{n=1}^{N-1} \sigma_n \sigma_{n+1} - J\sigma_N \sigma_1 \tag{11.240}$$

where the last term expresses the periodic boundary condition. We have $\sigma_i = \pm 1$. We can define new variables $\alpha_n = \sigma_n \sigma_{n+1}$. α_n takes the values ± 1 as σ_i. We can rewrite $\mathcal{H}$ as

$$\mathcal{H}_0 = -J \sum_{n=1}^{N-1} \alpha_n - J\alpha_N \tag{11.241}$$

The partition function is then

$$Z = Tr \exp[\beta J \sum_{n=1}^{N} \alpha_n] = Tr \prod_{n=1}^{N} \exp(\beta J \alpha_n)$$

$$= \prod_{n=1}^{N} [\exp(\beta J) + \exp(-\beta J)] = [2 \cosh \beta J]^N \tag{11.242}$$

This result is the same as Eq. (11.228). The average energy is calculated by $E = -\partial ln Z / \partial \beta$ [see (A.10)]:

$$\overline{E} = -\frac{\partial \ln Z}{\partial \beta} = -NJ \tanh(\beta J) \tag{11.243}$$

We obtain the paramagnetic heat capacity

$$C_V = dE/dT = Nk_B [\frac{k_B T}{J} \cosh \frac{J}{k_B T}]^{-2} \tag{11.244}$$

In an applied magnetic field H, we proceed as follows:

$$\mathcal{H} = \mathcal{H}_0 - H \sum_{n=1}^{N} \sigma_n \tag{11.245}$$

where $\mathcal{H}_0$ is given by (11.240). The partition function is $Z = \prod_{n=1}^{N} V_n$ where

$$V_n = \exp[\beta(J\sigma_n \sigma_{n+1} + H\sigma_n)] \tag{11.246}$$

$$V_N = \exp[\beta(J\sigma_N \sigma_1 + H\sigma_N)] \tag{11.247}$$

The matrix elements V_n, of dimension 2x2, depend on σ_n and σ_{n+1}. We have

$$V_n(1,1) = \exp[\beta(J + H)] \quad (\sigma_n = 1, \sigma_{n+1} = 1)$$
$$V_n(1,2) = \exp[\beta(-J + H)] \quad (\sigma_n = 1, \sigma_{n+1} = -1)$$
$$V_n(2,1) = \exp[\beta(-J - H)] \quad (\sigma_n = -1, \sigma_{n+1} = 1)$$
$$V_n(2,2) = \exp[\beta(J - H)] \quad (\sigma_n = -1, \sigma_{n+1} = -1)$$

The matrix V_n is called "transfer matrix". We note that all V_n ($n = 1, N$) have the same elements, say V. We thus have $Z = TrV^N$. Let z_1 and z_2 be the eigenvalues of V obtained by diagonalizing V:

$$z_1 = \exp(\beta J)\cosh(\beta H) + \sqrt{\exp(2\beta J)\cosh^2(\beta H) - 2\sinh(2\beta J)}$$

$$z_2 = \exp(\beta J)\cosh(\beta H) - \sqrt{\exp(2\beta J)\cosh^2(\beta H) - 2\sinh(2\beta J)}$$

We obtain then

$$Z = z_1^N + z_2^N = z_1^N(1 + \exp[-N\ln(z_1/z_2)]) \tag{11.248}$$

where z_1 denotes the larger eigenvalue. When $N \to \infty$, we have $Z = z_1^N$.

The susceptibility is calculated by $\chi = (dM/dH)_{H \to 0}$ where $M = -\partial F/\partial H$ with $F = -k_B T \ln Z$. We obtain

$$\chi \simeq \frac{1}{T}\exp[2J/k_B T] \tag{11.249}$$

This result shows that there is no phase transition in 1D (absence of anomaly of χ with varying T as seen in Fig. 11.18.

Problem 4. Low- and high-temperature expansions of the Ising model on the square lattice:

Solutions: We have

$$\mathcal{H} = -J \sum_{<i,j>} \sigma_i \sigma_j \tag{11.250}$$

where the sum is performed over nearest neighbors and $\sigma_{i(j)} = \pm 1$.

a) The partition function

$$Z = \sum_{\sigma_1 = \pm 1, \ldots} \exp(K \sum_{<i,j>} \sigma_i \sigma_j)$$

$$= \sum_{\sigma_1 = \pm 1, \ldots} \prod_{<ij>} \exp(K\sigma_i \sigma_j) \tag{11.251}$$

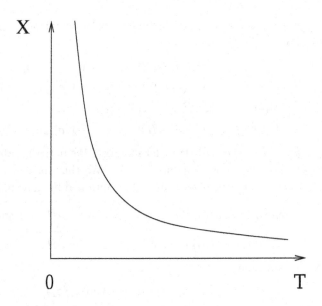

Fig. 11.18 χ versus T.

where $K = J/(k_B T)$. In the ground state (GS), we have $E_0 = -2JN = -N_b J$ and $Z = 2\exp(N_b K)$ where the factor 2 comes from the GS degeneracy (reversing all spins), $N_b = 2N$ is the total number of links.

Let us construct the dual lattice by drawing the links (broken lines) perpendicular to the real links (solid lines) as shown in Fig. 11.19. The case of the square lattice is special: its dual lattice formed by the broken lines is also a square lattice. This is not the case in general: for example the dual lattice of the triangular lattice is the honeycomb lattice.

b) Low-temperature expansion:

• one reversed spin: there are 4 broken links around it, degeneracy=N (the number of choices of a spin among N spins), the energy is increased from $-4J$ to $+4J$ so that $E = E_0 + 8J$ (see Fig. 11.20),

• two reversed neighbors: 6 broken links, degeneracy=N_b (the number of choices of a link among N_b links), the energy is increased from $-6J$ to $+6J$: $E = E_0 + 12J$,

• three reversed neighboring spins: 8 broken links for both configurations (trimer with two links on a line, or trimer with two

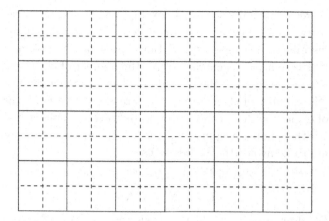

Fig. 11.19   Real square lattice (solid lines) and its dual lattice (broken lines).

perpendicular links), degeneracy=$6N$ (there are N choices of the central site and 6 choices to form a trimer with the central site: one straight trimer along x axis and the second straight trimer along y axis, and four perpendicular trimers), $E = E_0 + 16J$.

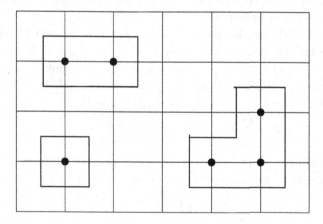

Fig. 11.20   Graphs (heavy lines) crossing broken links for one, two three reversed spin clusters. The reversed spins are shown by black circles, other spins are not shown.

Calculation of Z with the first excited states of energies $E_0 + 8J$, $E_0 + 12J$, $E_0 + 16J$: to do that, we have to find all excited states for each level. The first two levels concern one and two reversed neighboring spins as found above. However, the level $E_0 + 16J$

corresponds not only to the excited trimer shown above but also to two other cases: a cluster of four sites forming a square, two disconnected reversed spins. Both cases have 8 broken links. The first case has a degeneracy of N (the number of choices of the first site of the square), the second case has a degeneracy of $N(N-5)$ (the number of choices of the first reversed spin is N, the number of choices of the second disconnected reversed spin is $N-5$ where 5 is the number of spins concerned by the first reversed spins which are to be avoided). We finally have

$$Z = 2e^{2NK}[1 + Ne^{-8K} + 2Ne^{-12K}$$
$$+[6N + N + N(N-5)]e^{-16K} + ...] \qquad (11.252)$$

c) We draw a path P which encircles each cluster of reversed spins: this path crosses the broken links around the cluster. For such a closed path, we can verify that it crosses an even number of broken links as follows. Imagine a rectangular path, the number of broken links is even. Now, including an additional site anywhere around the path will add two additional broken links. Excluding a site inside the path will reduce the number of broken links by 2. Such a construction shows clearly that any closed path crosses an even number of broken links.

Let $\ell(P)$ be the number of broken links crossed by the path. Since the variation of energy when breaking a link is $\Delta E = J - (-J) = 2J$. A path crossing $\ell(P)$ broken links corresponds to $\Delta E = +2\ell(P) J$, so that the system energy is $E(P) = E_0 + \Delta E = E_0 + 2\ell(P) J$. The partition functions is thus

$$Z = \sum_P e^{-E(P)/k_B T} = 2e^{N_b K} \sum_P e^{-2K\ell(P)} \qquad (11.253)$$

d) High-temperature expansion: Using Eq. (11.225) or Eq. (11.234) we write the partition function as

$$Z = \sum_{\sigma_1 = \pm 1,...} \prod_{<ij>} \exp(K\sigma_i \sigma_j)$$

$$= \sum_{\sigma_1 = \pm 1,...} \prod_{<ij>} (\cosh K + \sigma_i \sigma_j \sinh K)$$

$$= (\cosh K)^{N_b} \sum_{\sigma_1 = \pm 1,...} \prod_{<ij>} (1 + \sigma_i \sigma_j \tanh K) \quad (11.254)$$

Since there are N_b links, there are N_b factors in the product. We draw a link between two nearest sites in each factor (this link is in the real space). We expand the product, we see that if a given spin appears an odd number of times in a term of the resulting polynomial, then when summing on its values, this term gives two opposite values yielding a zero contribution. Each nonzero term contains an even number of times of each spin, so that the spin factor in front of $\tanh K$ is equal to 1. The power of $\tanh K$ for this term is nothing but the number of links of spins in front of it. Consider for example the term $x = \sigma_1\sigma_2\ \sigma_2\sigma_3\ \sigma_3\sigma_{10}\ \sigma_{10}\sigma_9\ \sigma_9\sigma_8\ \sigma_8\sigma_1 \tanh^6 K$ shown in Fig. 11.21. This term has 6 links each of which connects two spins. Each spin appears twice because there are two links emanating from it. Hence, $x = \tanh^6 K$. The corresponding graph is shown by the lower left graph in Fig. 11.21. Therefore, we can replace the polynomial by the sum

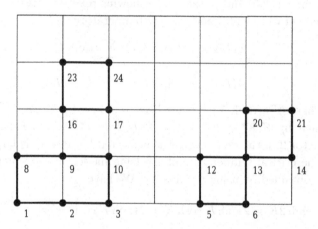

Fig. 11.21 Graphs linking nearest sites: the lower left graph represents the term $\sigma_1\sigma_2\ \sigma_2\sigma_3\ \sigma_3\sigma_{10}\ \sigma_{10}\sigma_9\ \sigma_9\sigma_8\ \sigma_8\sigma_1\ \tanh^6 K = \tanh^6 K$, the right one represents the term $\sigma_5\sigma_6\ \sigma_6\sigma_{13}\ \sigma_{13}\sigma_{14}\ \sigma_{14}\sigma_{21}\ \sigma_{21}\sigma_{20}\ \sigma_{20}\sigma_{13}\ \sigma_{13}\sigma_{12}\ \sigma_{12}\sigma_5\ \tanh^8 K = \tanh^8 K$. The upper left graph represents the term $\sigma_{16}\sigma_{17}\ \sigma_{17}\sigma_{24}\ \sigma_{24}\sigma_{23}\ \sigma_{23}\sigma_{16}\ \tanh^4 K = \tanh^4 K$. The spins crossed by the graphs are shown by black circles, other spins are not shown.

$$Z = 2^N(\cosh K)^{N_b} \sum_P (\tanh K)^{\ell(P)} \tag{11.255}$$

where 2^N comes from the sum $\sum_{\sigma_1 = \pm 1, \dots}$ and $\ell(P)$ is the number of links in the closed graph P.

e) Duality:

The partition functions Z in (11.253) and (11.255) have the same structure: since the prefactors are non singular, the summations over the closed paths determine the singularity of Z. Note that $\ell(P)$ in (11.253) corresponds to the path drawn in the dual lattice (see Fig. 11.20). While, $\ell(P)$ in (11.255) corresponds to the path drawn in the real lattice (see Fig. 11.21). Nevertheless, these two kinds of path have the same structure with an even number of links in each path. Therefore, we can connect the two Z by fixing

$$e^{-2K^*} = \tanh K \tag{11.256}$$

$$\text{Hence,} \quad K^* = -\frac{1}{2}\ln\tanh K \tag{11.257}$$

where K^* corresponds to the low-T phase and K to the high-T phase. The above relation (11.257) is called the "duality" condition which connects the low- and the high-T phases. We deduce from (11.255) and (11.253) the following relation between the high-temperature $Z(K)$ and the low-temperature $Z(K^*)$:

$$Z(K) = 2^N (\cosh K)^{N_b} \sum_P (\tanh K)^{\ell(P)}$$

$$Z(K) = 2^N (\cosh K)^{N_b} e^{-N_b K^*} Z(K^*) \tag{11.258}$$

where $\sum_P (\tanh K)^{\ell(P)}$ has been replaced by $\sum_P e^{-2K^*\ell(P)}$ and then by $Z(K^*)/(2e^{N_b K^*}) = Z(K^*)/(e^{N_b K^*})$ from (11.253) (the factor 2 in the denominator is neglected because N is large).

f) The critical temperature of the Ising model on the square lattice is obtained by using the duality. We have

$$\sinh 2K = 2\sinh K \cosh K = 2\tanh K \cosh^2 K = \frac{2\tanh K}{1 - \tanh^2 K}$$

$$= \frac{2e^{-2K^*}}{1 - e^{-4K^*}} = \frac{2}{e^{2K^*} - e^{-2K^*}}$$

$$= \frac{1}{\sinh 2K^*} \tag{11.259}$$

from which

$$\sinh 2K \ \sinh 2K^* = 1 \tag{11.260}$$

This relation is symmetric with respect to K and K^*: when K^* increases, K decreases, and vice-versa. If the system undergoes a single phase transition, it should undergo at the same point of K and K^*, namely $K_c^* = K_c$. To satisfy (11.260), we should have

$$\sinh 2K_c = \sinh 2K_c^* = 1 \tag{11.261}$$

Thus,

$$\frac{e^{2K_c} - e^{-2K_c}}{2} = 1$$

$$e^{2K_c} - 2e^{-2K_c} - 2 = 0$$

$$e^{4K_c} - 1 - 2e^{2K_c} = 0$$

$$X^2 - 2X - 1 = 0 \quad (X = e^{2K_c})$$

$$X = 1 + \sqrt{2} \quad \text{(positive solution)}$$

$$K_c = \frac{\ln(1 + \sqrt{2})}{2} \tag{11.262}$$

Therefore, $k_B T_c / J = 1/K_c \simeq 2.27$ which is the exact Onsager's solution.

Problem 5. Critical temperatures of the triangular lattice and the honeycomb lattice by duality:

Solutions: We have

$$\mathcal{H} = -J \sum_{<i,j>} \sigma_i \, \sigma_j \tag{11.263}$$

where the sum is performed over nearest neighbors on the triangular lattice and $\sigma_{i(j)} = \pm 1$.

The dual lattice by construction is a honeycomb lattice shown by the broken lines in Fig. 11.22.

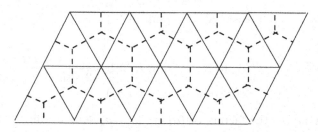

Fig. 11.22 Triangular lattice (solid lines) and its dual honeycomb lattice (broken lines).

First, we follow the same method as for the square lattice in Problem 4 above: writing the partition functions of the triangular lattice using low-temperature expansion with graphs on links of the dual lattice and high-temperature expansion with graphs on links of the real lattice, we obtain a relation similar to Eq. (11.258)

$$Z_t(K) = 2^N (\cosh K)^{N_b} e^{-N_b K} Z_h(K^*) \tag{11.264}$$

where $Z_t(K)$ denotes the low-temperature partition function of the triangular lattice, $Z_h(K^*)$ the high-temperature dual (honeycomb) lattice and $N_b = 3N$ is the total number of links.

Now, we can relate the two partition functions in another relation as follows. For convenience, the dual lattice is shifted as shown in Fig. 11.23. We consider the plaquette defined by three sites 1,2 and 3 with a site at the center. We calculate the following quantity

$$A_p(K^*) = \sum_{\sigma_0 = \pm 1} \exp[K^* \sigma_0 (\sigma_1 + \sigma_2 + \sigma_3)]$$

$$= \sum_{\sigma_0 = \pm 1} \cosh^3 K^* (1 + \sigma_0\,\sigma_1 \tanh K^*)(1 + \sigma_0\,\sigma_2 \tanh K^*)$$
$$\times (1 + \sigma_0\,\sigma_3 \tanh K^*)$$
$$= 2 \cosh^3 K^* [1 + \tanh^2 K^* (\sigma_1\,\sigma_2 + \sigma_2\,\sigma_3 + \sigma_3\,\sigma_1)]$$
$$(11.265)$$

where we used the remark before Eq. (11.255) to expand the product of the first equality.

Now, we consider the plaquette defined by three sites 1, 2 and 3 of the triangular lattice. We calculate the following quantity

$$B_p(K^+) = \exp[K^+ (\sigma_1\,\sigma_2 + \sigma_2\,\sigma_3 + \sigma_3\,\sigma_1)]$$
$$= \cosh^3 K^+ (1 + \sigma_1\,\sigma_2 \tanh K^+)(1 + \sigma_2\,\sigma_3 \tanh K^+)$$
$$(1 + \sigma_3\,\sigma_1 \tanh K^+)$$
$$= \cosh^3 K^+ [1 + \tanh^3 K^+ + (\tanh K^+ + \tanh^2 K^+)$$
$$\times (\sigma_1\,\sigma_2 + \sigma_2\,\sigma_3 + \sigma_3\,\sigma_1)]$$
$$= (\cosh^3 K^+ + \sinh^3 K^+)[1 + \frac{\sinh 2K^+}{2 \cosh 2K^+ - \sinh 2K^+}$$
$$\times (\sigma_1\,\sigma_2 + \sigma_2\,\sigma_3 + \sigma_3\,\sigma_1)]\qquad(11.266)$$

If we set the factors of $(\sigma_1\,\sigma_2 + \sigma_2\,\sigma_3 + \sigma_3\,\sigma_1)$ in (11.265) and (11.266) to be equal, then we have

$$\tanh^2 K^* = \frac{\sinh 2K^+}{2 \cosh 2K^+ - \sinh 2K^+}\qquad(11.267)$$

so that

$$\sum_{\sigma_0 = \pm 1} \exp[K^* \sigma_0 (\sigma_1 + \sigma_2 + \sigma_3)] = \frac{2 \cosh^3 K^*}{\cosh^3 K^+ + \sinh^3 K^+}$$
$$\times \exp[K^+ (\sigma_1\,\sigma_2 + \sigma_2\,\sigma_3 + \sigma_3\,\sigma_1)]\qquad(11.268)$$

This relation connects the two dual lattices. To find the full partition functions, it suffices to sum over all spins of the plaquette and to take the product over all plaquettes on each side of the above equation, we then have

$$Z_h(K^*) = (\frac{2\cosh^3 K^*}{\cosh^3 K^+ + \sinh^3 K^+})^N Z_t(K^+) \qquad (11.269)$$

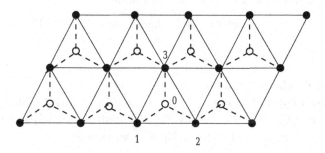

Fig. 11.23 Triangular lattice (solid lines) and its dual honeycomb lattice (broken lines) shifted for convenience: the four sites are numbered from 0 to 3 for calculating the partition function.

Replacing $Z_h(K^*)$ given by (11.269) in (11.264), we have

$$Z_t(K) = 2^N (\cosh K)^{3N} e^{-3NK^*} (\frac{2\cosh^3 K^*}{\cosh^3 K^+ + \sinh^3 K^+})^N Z_t(K^+)$$
$$(11.270)$$

Using (11.267) to eliminate K^* in the above equation, and after some algebra, we get

$$\frac{Z_t(K)}{(\sinh 2K)^{N/2}} = \frac{Z_t(K^+)}{(\sinh 2K^+)^{N/2}} \qquad (11.271)$$

Let us calculate the critical temperature of the triangular lattice. We note that by graph constructions for low- and high-temperatures, we obtained Eq. (11.256) which is very general, independent of the lattice structure: it was established using the square lattice, but all arguments leading to it are also valid for the triangular lattice. Using Eq. (11.256) to eliminate K^* in Eq. (11.267), and after some algebra, we obtain

$$(e^{4K} - 1)(e^{4K^+} - 1) = 4 \qquad (11.272)$$

As before in the case of the square lattice, this relation shows that if K increases, K^+ decreases, and vice-versa. The transition should occur at the same critical temperature $K_c = K_c^+$. The solution of Eq. (11.272) at K_c is thus

$$e^{4K_c} - 1 = e^{4K_c^+} - 1 = 2, \quad \text{hence} \quad k_B T_c / J = 3.640$$

To find the critical temperature of the honeycomb lattice, we follow the same method as above: we obtain

$$\cosh 2K_c = 2, \quad \text{hence} \quad k_B T_c / J = 1.518.$$

Problem 6. Ground state of Villain's model:

Solution: The energy of a plaquette of the 2D Villain's model with XY spins defined in Fig. 7.7 with $\mathbf{S}_1$ and $\mathbf{S}_2$ linked by the antiferromagnetic interaction η, is written as

$$H_p = \eta \mathbf{S}_1 \cdot \mathbf{S}_2 - \mathbf{S}_2 \cdot \mathbf{S}_3 - \mathbf{S}_3 \cdot \mathbf{S}_4 - \mathbf{S}_4 \cdot \mathbf{S}_1 \tag{11.273}$$

where $(\mathbf{S}_i)^2 = 1$. The variational method gives

$$\delta[H_p - \frac{1}{2} \sum_{i=1}^{4} \lambda_i (\mathbf{S}_i)^2] = 0 \tag{11.274}$$

By symmetry, $\lambda_1 = \lambda_2 \equiv \lambda$, $\lambda_3 = \lambda_4 \equiv \mu$. We have

$$\lambda \mathbf{S}_1 - \eta \mathbf{S}_2 + \mathbf{S}_4 = 0 \tag{11.275}$$

$$-\eta \mathbf{S}_1 + \lambda \mathbf{S}_2 + \mathbf{S}_3 = 0 \tag{11.276}$$

$$\mathbf{S}_2 + \mu \mathbf{S}_3 + \mathbf{S}_4 = 0 \tag{11.277}$$

$$\mathbf{S}_1 + \mathbf{S}_3 + \mu \mathbf{S}_4 = 0 \tag{11.278}$$

Hence,

$$(\lambda - \mu)(\mathbf{S}_1 + \mathbf{S}_2) + (\mathbf{S}_3 + \mathbf{S}_4) = 0 \tag{11.279}$$

$$(\mathbf{S}_1 + \mathbf{S}_2) + (\mu + 1)(\mathbf{S}_3 + \mathbf{S}_4) = 0 \tag{11.280}$$

We deduce

$$\mu = -\left[\frac{1 + \eta}{\eta}\right]^{1/2} \tag{11.281}$$

$$\lambda = \eta \mu = -[\eta(1 + \eta)]^{1/2} \tag{11.282}$$

To calculate the angle between two spins, for instance $\mathbf{S}_1$ and $\mathbf{S}_4$, we write

$$(\lambda \mathbf{S}_1 + \mathbf{S}_4)^2 = (-\eta \mathbf{S}_2)^2$$

Hence

$$\mathbf{S}_1 \cdot \mathbf{S}_4 = \cos\theta_{14} = \frac{1}{2\lambda}(\eta^2 - \lambda^2 - 1) = \frac{1}{2}[\frac{\eta+1}{\eta}]^{1/2}$$

We find in the same manner,

$$\cos\theta_{23} = \cos\theta_{34} = \cos\theta_{41} = \frac{1}{2}[\frac{\eta+1}{\eta}]^{1/2}$$

We have

$$\theta_{14} = \theta_{23} = \theta_{34} = \theta_{41} \equiv \theta.$$

Note that $|\theta_{12}| = 3|\theta|$. These solutions exist if $|\cos\theta| \leq 1$, namely $\eta > \eta_c = 1/3$. When $\eta = 1$, we have $\theta = \pi/4$, $\theta_{12} = 3\pi/4$.

Problem 7. Proofs of Eq. (7.85):
Solution: Using Eqs. (7.72), (7.80) and (7.84), we rewrite Eq. (7.83) as

$$G(\vec{r}) = \, < \exp[i[\theta(\vec{r}) - \theta(0)]] >$$

$$= \frac{\int D[\theta] \exp[i[\theta(\vec{r}) - \theta(0)]] \exp\{-\beta\frac{J}{2}\int d\vec{r}[\vec{\nabla}\theta(\vec{r})]^2\}}{\int D[\theta] \exp\{-\beta\frac{J}{2}\int d\vec{r}[\vec{\nabla}\theta(\vec{r})]^2\}}$$

$$= \frac{\int D[\theta] \exp\left\{-\int \frac{d\vec{k}}{(2\pi)^d}\left[\beta\frac{J}{2}k^2\theta(\vec{k})\theta(-\vec{k}) - i\theta(\vec{k})[e^{i\vec{k}\cdot\vec{r}} - 1]\right]\right\}}{\int D[\theta] \exp\{-\beta\frac{J}{2}\int \frac{d\vec{k}}{(2\pi)^d}k^2\theta(\vec{k})\theta(-\vec{k})\}}$$

To integrate this integral, let us separate the real and imaginary parts as follows: we put $\theta(\vec{k}) = R_{\vec{k}} + iI_{\vec{k}}$. Since $\theta(\vec{r})$ is real, we have $\theta(-\vec{k}) = \theta^*(\vec{k})$ so that $R_{\vec{k}}$ is even and $I_{\vec{k}}$ odd in $\vec{k}$. Then we can write the argument of the exponential of the numerator as

$$\int \frac{d\vec{k}}{(2\pi)^d} \beta \frac{J}{2} k^2 \Big[\Big(R_{\vec{k}} - i\frac{\cos(\vec{k}\cdot\vec{r})-1}{\beta J k^2} \Big)^2$$

$$+ \Big(I_{\vec{k}} + i\frac{\sin(\vec{k}\cdot\vec{r})}{\beta J k^2} \Big)^2 \Big] + \int \frac{d\vec{k}}{(2\pi)^d} \frac{1-\cos(\vec{k}\cdot\vec{r})}{\beta J k^2} \qquad (11.283)$$

where we have used $[\cos(\vec{k}\cdot\vec{r})-1]^2 + \sin^2(\vec{k}\cdot\vec{r}) = 2[1-\cos(\vec{k}\cdot\vec{r})]$. The integrals over $R_{\vec{k}}$ and $I_{\vec{k}}$ are gaussian, they give the same value as the gaussian integrals of the denominator although they do not have the same gaussian centers as the denominator. Hence, these gaussian integrals cancel out. Only the last term in (11.283) remains. We have then

$$G(\vec{r}) = \exp[-\frac{k_B T}{J} \int \frac{d\vec{k}}{(2\pi)^d} \frac{1-\cos(\vec{k}\cdot\vec{r})}{k^2}] = \exp[-g(\vec{r})]$$

This is Eq. (7.85).

Problem 8. Critical line of an antiferromagnet in an applied magnetic field:

Solution: An antiferromagnet under an applied magnetic field can have a phase transition at a finite temperature T_c (see discussion in subsection 4.4.3). The critical line $T_c(H)$ belongs to the universality class of the zero field Ising model. The leading singularity is from the following reduced free energy

$$f_s \simeq a\tau^2 \ln(\frac{1}{|\tau|}) \qquad (11.284)$$

where $\tau(t, H)$ is the scaling field measuring the distance to the critical line [76]. For small H, we have

$$\tau \simeq t + uH^2 \qquad (11.285)$$

where u is a coefficient to be determined. The zero-field susceptibility is known. We use it to compute u as follows:

$$K^2\chi_s = \frac{\partial^2 f_s}{\partial H^2}\Big|_{H=0} = \frac{\partial^2 f_s}{\partial \tau^2}(\frac{\partial \tau}{\partial H})^2 + \frac{\partial f_s}{\partial \tau}\frac{\partial^2 \tau}{\partial H^2}$$

$$= 2a\tau \ln(\frac{1}{|\tau|}) \times 2u \quad \text{(the first term=0 at } H = 0\text{)}$$

$$= 4aut \ln(\frac{1}{|t|}), \quad \text{therefore}$$

$$\chi_s = Dt \ln(\frac{1}{|t|}), \quad \text{where} \quad D = \frac{4au}{K_c^2}$$

With $D = 0.1935951863$, $a = [\ln(1 + \sqrt{2})]^2/\pi$, $[T_c^0]^{-1} = K_c = \ln(1 + \sqrt{2})/2$, we obtain $u = 0.0380123259$. Setting $\tau = 0$, we get the critical line $T_c(H) = T_c^0(1 - uH^2)$ for small H. This relation has been schematically shown in Fig. 4.10 of subsection 4.4.3.

11.8   Solutions of problems of chapter 8

Problem 1. Program for Ising spins:

   Solution:

a) Heat capacity and the magnetic susceptibility C_v and χ are

   CV=N*N*(U**2-E2M)/T**2

   CHI=N*N*(AM**2-A2M)/T

   where E2M is the average of the square of the energy par spin and A2M the average of the square of the magnetization.

b) The case of a simple cubic lattice:

   DO I=1,N

   ...

   DO J=1,N

   ...

   DO K=1,N

   KP=K+1-(K/N)*N

   KM=K-1+(1/K)*N

   E=-S(I,J,K)*(...+S(I,J,KP)+S(I,J,KM))

   ENDDO

   ENDDO

   ENDDO

c) Case of a centered-cubic lattice:
Two sublattices:

```
DO I=1,N
...
DO J=1,N
...
DO K=1,N
KP=K+1-(K/N)*N
KM=K-1+(1/K)*N
DO M=1,2 ! two sublattices (corner spins and centered spins)
IF(M.EQ.1)THEN
E=-
S(I,J,K,1)*(S(I,J,K,2)+S(IM,J,K,2)+S(I,JM,K,2)+S(IM,JM,K,2)+
S(I,J,KM,2)+S(IM,J,KM,2)+S(I,JM,KM,2)+S(IM,JM,KM,2))
ELSE
E=-
S(I,J,K,2)*(S(I,J,K,1)+S(IP,J,K,1)+S(I,JP,K,1)+S(IP,JP,K,1)+
S(I,J,KP,1)+S(IP,J,KP,1)+S(I,JP,KP,1)+S(IP,JP,KP,1))
ENDIF
ENDDO
ENDDO
ENDDO
...
```

Problem 2. Heisenberg spin model:

Solution: For the square lattice:

```
E=-SX(I,J)*(SX(IP,J)+SX(IM,J)+SX(I,JP)+SX(I,JM))        -
SY(I,J)*(SY(IP,J)+SY(IM,J)+SY(I,JP)+SY(I,JM))-
SZ(I,J)*(SZ(IP,J)+SZ(IM,J)+SZ(I,JP)+SZ(I,JM))
```

To generate a random Heisenberg spin:

```
PI=ACOS(-1.)
SZ(I,J)=2*RAND()-1
PHI=2*PI*RAND()
SX(I,J)=SQRT(1-Z(I,J)**2)*COS(PHI)
SY(I,J)=SQRT(1-Z(I,J)**2)*SIN(PHI)
```

The remaining part is the same as in the sample program given in Appendix B.

Problem 3. Energy histogram H(E) for the program shown in Appendix B:

We consider a cubic sample with periodic boundary conditions described in Appendix B. During the averaging (subroutine MC2), at each Monte Carlo step we have the instantaneous energy E. This energy is used to establish the energy histogram in the do loop 490 as follows:

DO 490 IJ = 1, INTV
CALL IMC(AJ1,E, T, S)
E1 = E1 + E/(2*NN)
IF(E.LT.EMINN.OR.E.GT.EMAXN)GO TO 1225
IBIN=(E-EMINN+0.5*D)/D
WRITE(6,*)'IBIN=',IBIN
H(IBIN)=H(IBIN)+1
1225 CONTINUE
490 CONTINUE

Remark:

1) We have to choose EMINN (energy minimum) and EMAXN (energy maximum) in such a way that E falls into the interval [EMINN,EMAXN]. The number of subintervals NBIN in the interval [EMINN,EMAXN] can be chosen equal to the total number of spins. The width of each subinterval is D=(EMAXN-EMINN)/NBIN.

2) We can also make histograms of the instantaneous magnetization M, M**2, M**4 which will be used to calculate the susceptibility, cumulants, etc.

Problem 4. Steepest-descent method:

Solution:

- we generate a random spin configuration on a lattice,
- we calculate the local field acting on a spin from its neighbors according to the Hamiltonian,
- we align the spin under consideration in the direction of its local field (this is to minimize its interaction energy),
- then we take another spin and repeat the previous steps until all spins are considered: we say we did a sweeping,
- we make another sweeping: at each sweeping we record the

system energy. We repeat sweeping iterations many times,

- we stop the iteration process when the system energy converges to a fixed value or a value within a desirable precision,
- we analyze the snapshot of the final spin configuration to determine its nature (remark: except in spin-glass or alike, the final spin configuration corresponds to the ground state)

Below is the program which realizes the above steps with the Ising model on a square lattice with nearest-neighbor interaction J_1 and next-nearest neighbor interaction J_2. We can use it to determine the phase diagram at temperature $T = 0$ in the space (J_1, J_2).

```
C PROGRAM
PARAMETER (N=30)
REAL*4 J1,J2
J1=1.
J2=-0.5
DO I=1,N
DO J=1,N
S(I,J)=SIGN(1.,0.5-RAND())
ENDDO
ENDDO
C ITERATION
DO IT=1,1000
E=0.
DO I=1,N
IP=I+1-I/N*N
IM=I-1+1/I*N
DO J=1,N
JP=J+1-J/N*N
JM=J-1+1/J*N
H= J1*(S(IP,J)+S(IM,J)+S(I,JP)+S(I,JM))
&+J2*(S(IP,JP)+S(IM,JP)+S(IM,JM)+S(IP,JM))
E=E-S(I,J)*H
S(I,J)=H/ABS(H)
ENDDO
ENDDO
ENERGY(IT)=E
WRITE(6,*)IT,ENERGY(IT)
ENDDO
```

WRITE(6,*)((S(I,J), I=1,N),J=1,N)
STOP
END

Look at the energy ENERGY(IT) as a function of time (IT). You will see if this converges to a fixed value. Look at the printed final configuration to have an idea about its symmetry. Change J2 and repeat the calculation.

11.9 Solutions of problems of chapter 9

Problem 1. Surface magnon:

Solution: In the ferromagnetic case, we just write the equation of motion for S_m^+ . We use next the Fourier transform in the xy plane. We obtain then, for $n > 2$,

$$(E - E_n)U_n = \left[4\gamma_1(\vec{k}_\parallel)(U_{n-1} + U_{n+1}) + \epsilon(U_{n-2} + U_{n+2})\right] \tag{11.286}$$

and, for the first two layers,

$$(E - E_1)U_1 = \left[4\gamma_1(\vec{k}_\parallel)U_2 + \epsilon U_3\right] \tag{11.287}$$

$$(E - E_2)U_2 = \left[4\gamma_1(\vec{k}_\parallel)(U_1 + U_3) + \epsilon U_4\right] \tag{11.288}$$

where $E = \frac{\hbar\omega}{J_1 S}$, $\epsilon = \frac{J_2}{J_1}$, and

$$E_n = 8 - 6\epsilon[1 - \gamma_2(\vec{k}_\parallel)](n = 2, 3, ...) \tag{11.289}$$

$$E_1 = 8 - 5\epsilon[1 - \gamma_2(\vec{k}_\parallel)] \tag{11.290}$$

$$\gamma_1(\vec{k}_\parallel) = \cos(\frac{k_x a}{2}) \cos(\frac{k_y a}{2}) \tag{11.291}$$

$$\gamma_2(\vec{k}_\parallel) = \frac{1}{2}[\cos(k_x a) + \cos(k_y a)] \tag{11.292}$$

To study bulk modes, we replace $U_{n\pm 1} = U_n \exp(\pm i k_z a/2)$ in Eq. (11.286) to obtain the energy of the bulk mode of wave vector $(\vec{k}_\parallel, k_z)$:

$$E = E_n - \left[8\gamma_1(\vec{k}_\parallel)\cos(k_z a/2) + 2\epsilon\cos(k_z a)\right].$$

For surface modes, we replace $U_{n\pm1} = U_1\phi^n$ in Eqs. (11.286)-(11.288) to obtain surface-mode energy E and the damping factor ϕ.

Problem 2. Critical next-nearest-neighbor interaction:

    Solution: If J_1 and J_2 are both ferromagnetic (> 0), the ferromagnetic state is stable. However, if J_2 becomes negative, the ferromagnetic state becomes unstable beyond a critical value of $\epsilon = |J_2|/J_1$. For an infinite crystal, the critical value is obtained by setting the energy E of the lowest magnon mode equal to zero. This corresponds to the instability due to a soft mode. The stable state is no more ferromagnetic for $\epsilon > \epsilon_c$ with $\epsilon_c = \frac{4}{3}$, namely $J_2 < -\frac{4}{3}J_1$

Problem 3. Uniform magnetization approximation:

    Solution: If we replace all $< S_n^z >$ by a unique value M, we see that all elements of the matrix $\mathbf{M}$ in section 9.5 are proportional to M. The eigenvalues E_i obtained by solving $\det\mathbf{M} = 0$ is therefore proportional to M.

Problem 4. Multilayers: critical magnetic field

    Solution: Without applied field, the spin configuration is A(up spin)$-B$(down spin)$-C$(up spin). In the very strong field applied along z, the B spins all turn up. We calculate the critical field beyond which the spin configuration is that state.

    The in-plane exchange energy for a B spin does not change whether the B film is in up or down state. We consider a column of spins in the z direction. The energy of the spins of B when they are antiparallel to $\vec{H}$ is

$$E_{AF} = -J_1(N_1 - 1) + J_s - HN_1 - J_2(N_2 - 2) + J_s$$
$$+HN_2 - J_3(N_3 - 1) - HN_3.$$

The energy of B spins when they are parallel to $\vec{H}$ is

$$E_F = -J_1(N_1 - 1) - J_s - HN_1 - J_2(N_2 - 2) - J_s$$

$$-HN_2 - J_3(N_3 - 1) - HN_3$$

(J_s is negative). We see that $E_F < E_{AF}$ when $H > H_c = \frac{2J_s}{N_2}$ (we have taken $J_1 = J_2 = J_3$).

Problem 5. Mean-field theory for thin films:

Solution: We assume the Hamiltonian $\mathcal{H} = -J\sum_{i,j} S_i S_j$ where $S_i = \pm 1$ (Ising spin at the lattice site i). We suppose a simple cubic lattice with a (001) surface. Using the mean-field theory we have the average values of the spins in the three layers

$$< S_1 > = \tanh\left[\beta J(Z < S_1 > + < S_2 >)\right]$$
$$< S_2 > = \tanh\left[\beta J(Z < S_2 > + < S_1 > + < S_3 >)\right]$$
$$< S_3 > = \tanh\left[\beta J(Z < S_3 > + < S_2 >)\right]$$

where $\beta = 1/(k_B T)$ and $Z = 4$ the number of nearest neighbors in the xy plane. By symmetry $< S_1 > = < S_3 >$, therefore we have only two equations to solve. Numerically, it can be easily done in a self-consistent manner.

Problem 6. Holstein-Primakoff method:

Guide: We can modify the equations (5.53)-(5.55) for a semi-infinite crystal: we write the equation of motion for a spin in each layer. We obtain a set of difference equations. We use next the Fourier transform in the xy plane. We can use (5.53)-(5.55) to make some comments: the lower the energy $\epsilon_{\vec{k}}$ is, the larger $< n_{\vec{k}} >$ becomes [see (5.53)]. As a consequence, M becomes smaller [see (5.55)]. Thus, the low-energy surface modes lower the layer magnetization near the surface with a stronger effect for surface magnetization because of the lack of neighbors.

Problem 7. Frustrated surface: surface spin rearrangement

Solution: If there is only the surface, then the surface spins form a planar 120° configuration (see section 7.5). We suppose this structure lies in the xy plane. Now, when the beneath layer acts on the surface spins with a ferromagnetic interaction, the surface spins on a triangle turn into the z direction to satisfy partially the ferromagnetic interaction: Let α be the projection on the (xy) plane the angle between two neighboring surface spins and β be the angle between a surface spin and its neighbor in the second layer (see Fig. 11.24). We have

$$\alpha_{1,2} = 0, \ \alpha_{2,3} = \frac{2\pi}{3}, \ \alpha_{3,1} = \frac{4\pi}{3} \qquad (11.293)$$

The energy of a cell formed by a surface triangle and the beneath triangle is, for spins 1/2,

$$H_p = -\frac{9J}{2} - \frac{3J}{2}\cos\beta - \frac{9J_s}{2}\cos^2\beta + \frac{9J_s}{4}\sin^2\beta. \qquad (11.294)$$

The minimization of this energy gives

$$\frac{\partial H_p}{\partial \beta} = \frac{27J_s}{2}\cos\beta\sin\beta + \frac{3J}{2}\sin\beta = 0 \qquad (11.295)$$

hence

$$\cos\beta = -\frac{J}{9J_s}. \qquad (11.296)$$

This solution is possible for $J_s < -J/9$. For $J_s > -J/9$, the stable state is ferromagnetic.

Note that we can use the steepest-descent method described in Problem 4 of chapter 8 to determine numerically the ground state of classical spin systems.

Problem 8. Ferrimagnetic film:

Solution: Let us take into account the difference between the average layer magnetizations, namely $< S_m^z >=< S_1^z >$ if m belongs to the first layer, $< S_l^z >=< S_2^z >$ if l belongs to the second layer, etc. Modifying (9.5)-(9.6) for the first two layers, we have

$$(E - E_1)U_1 = -< S_1^z > \left[4\gamma_1(\vec{k}_\parallel)U_2 + \epsilon U_3 \right]$$

$$(E + E_2)U_2 = -< S_2^z > \left[4\gamma_1(\vec{k}_\parallel)(U_1 + U_3) + \epsilon U_4 \right]$$

We can write, in the same manner, the two equations for layers 3 and 4 as

$$(E - E_3)U_3 = -< S_3^z > \left[4\gamma_1(\vec{k}_\parallel)(U_2 + U_4) + \epsilon U_1 \right]$$

$$(E + E_4)U_4 = -< S_4^z > \left[4\gamma_1(\vec{k}_\parallel)U_3 + \epsilon U_2 \right]$$

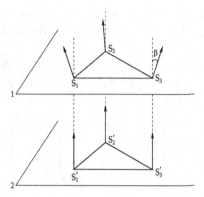

Fig. 11.24 Spin configuration of the ground state. The projection on the (xy) plane of the angle between two surface spins is $\alpha = 120°$. The angle between a surface spin and the beneath spin is β.

where the notations are defined as

$$E_1 = 4 < S_2^z > -4\epsilon < S_1^z > \left[1 - \gamma_2(\vec{k}_\parallel)\right] - \epsilon < S_3^z >$$

$$E_2 = -4(< S_1^z > + < S_3^z >) - 4\epsilon < S_2^z > \left[1 - \gamma_2(\vec{k}_\parallel)\right]$$
$$-\epsilon < S_4^z > \tag{11.297}$$

$$E_3 = 4(< S_2^z > + < S_4^z >) - 4\epsilon < S_3^z > \left[1 - \gamma_2(\vec{k}_\parallel)\right]$$
$$-\epsilon < S_1^z > \tag{11.298}$$

$$E_4 = -4 < S_3^z > -4\epsilon < S_4^z > \left[1 - \gamma_2(\vec{k}_\parallel)\right] - \epsilon < S_2^z >$$

Note that $< S_n^z > (n$: layer $n)$ is positive for $n = 2, 4$ (A sublattice) and negative for $n = 1, 3$ (B sublattice). Other notations in the above equations were defined after (9.6). We can put the above equations under a matrix form

$$\begin{pmatrix} x_{11} & x_{12} & x_{13} & 0 \\ x_{21} & x_{22} & x_{23} & x_{24} \\ x_{31} & x_{32} & x_{33} & x_{34} \\ 0 & x_{42} & x_{43} & x_{44} \end{pmatrix} \begin{pmatrix} U_1 \\ U_2 \\ U_3 \\ U_4 \end{pmatrix} = 0$$

where x represents a nonzero elements. Non trivial solutions impose that the determinant is zero. Solving numerically det $|...| = 0$ we obtain the energy eigenvalues, and by replacing them into the above matrix equation we get the spin-wave amplitudes. If $U_1 = U_2 = U_3 = U_4$, then the corresponding energy is a bulk mode. It is a surface mode otherwise. Note that the two surfaces of the 4-layer film are not symmetric. We should have four distinct modes.

For $\epsilon = 0$ and $k_x = k_y = 0$, we have $\gamma_1(\vec{k}_\parallel) = \cos(\frac{k_x a}{2})\cos(\frac{k_y a}{2}) = 1$ and $\gamma_2(\vec{k}_\parallel) = \frac{1}{2}[\cos(k_x a) + \cos(k_y a)] = 1$. We write for this case the matrix as follows:

$$\begin{pmatrix} x_{11} & x_{12} & 0 & 0 \\ x_{21} & x_{22} & x_{23} & 0 \\ 0 & x_{32} & x_{33} & x_{34} \\ 0 & 0 & x_{43} & x_{44} \end{pmatrix} \begin{pmatrix} U_1 \\ U_2 \\ U_3 \\ U_4 \end{pmatrix} = 0$$

where non zero elements are

$x_{11} = E - 4 < S_1^z >, \; x_{12} = < S_1^z >$

$x_{21} = 4 < S_2^z >, \; x_{22} = E - 4(< S_1^z > + < S_3^z >), \; x_{23} = 4 < S_2^z >$

$x_{32} = 4 < S_3^z >, \; x_{33} = E - 4(< S_2^z > + < S_4^z >), \; x_{34} = 4 < S_3^z >$

$x_{43} = < S_4^z >, \; x_{44} = E - 4 < S_3^z >$

This matrix can be diagonalized without difficulty.

For the case $k_x = k_y = \pi/a$, we proceed in the same manner.

11.10 Solutions of problems of chapter 10

Problem 1. Effect of magnetic field: demonstration of Eq. (10.24)

   Solution: When we apply a magnetic field $\vec{B}$, the force becomes $\vec{F} = -e[\vec{\varepsilon} + \vec{v} \wedge \vec{B}]$. We cannot therefore replace f by f_0 in $\frac{\vec{F}}{\hbar} \cdot \vec{\nabla}_{\vec{k}} f$ because the field $\vec{B}$ will not appear in the final equation. To see the effect of $\vec{B}$, we have to go to the second order: we have to replace f by $f_0 + \varphi$, not by f_0:

$$\begin{aligned} \frac{\vec{F}}{\hbar} \cdot \vec{\nabla}_{\vec{k}} f &= \frac{\vec{F}}{\hbar} \cdot \vec{\nabla}_{\vec{k}}(f_0 + \varphi) \\ &= \frac{\vec{F}}{\hbar} \cdot \left[\frac{\partial f_0}{\partial E} \vec{\nabla}_{\vec{k}} E + \vec{\nabla}_{\vec{k}} \varphi \right] \\ &= -e\vec{\varepsilon} \cdot \left[\frac{\partial f_0}{\partial E} \hbar \vec{v} + \frac{1}{\hbar} \vec{\nabla}_{\vec{k}} \varphi \right] - e(\vec{v} \wedge \vec{B}) \cdot \frac{\partial f_0}{\partial E} \vec{v} \\ &\quad - \frac{e}{\hbar}(\vec{v} \wedge \vec{B}) \cdot \vec{\nabla}_{\vec{k}} \varphi \end{aligned}$$

$$(11.299)$$

We see here that for the effect of the electric field we can neglect the second term in [...], namely $\vec{\nabla}_{\vec{k}}\varphi$. However, for the magnetic field the first term $(\vec{v} \wedge \vec{B}) \cdot \frac{\partial f_0}{\partial E}\vec{v}$ is zero because of the mixed product. We have to retain the last term of the above equation which depends on $\vec{B}$. We rewrite this term as

$$-\frac{e}{\hbar}(\vec{v} \wedge \vec{B}) \cdot \vec{\nabla}_{\vec{k}}\varphi = -\frac{e}{\hbar}\vec{B} \cdot [\vec{\nabla}_{\vec{k}}\varphi \wedge \vec{v}]$$

$$= \frac{e}{\hbar}\vec{B} \cdot [\vec{\nabla}_{\vec{k}}\phi\frac{\partial f_0}{\partial E} \wedge \frac{\vec{\nabla}_{\vec{k}}E}{\hbar}]$$

$$= \frac{e}{\hbar^2}\vec{B} \cdot [\vec{\nabla}_{\vec{k}}\phi \wedge \vec{\nabla}_{\vec{k}}E]\frac{\partial f_0}{\partial E} \quad (11.300)$$

The solution (10.23) becomes

$$\phi = \tau\vec{A} \cdot \vec{v} + \tau\frac{e}{\hbar^2}\vec{B} \cdot [\vec{\nabla}_{\vec{k}}\phi \wedge \vec{\nabla}_{\vec{k}}E] \quad (11.301)$$

This is a differential equation of ϕ because of the presence of $\vec{B}$.

Problem 2. Ohm's law: demonstrate Eq. (10.29)
Solution:
We suppose now that the electric field is weak, τ does not depend on the energy E of the electron. We can replace τ by an average value $< \tau >$. Using the expression (A.41) for $\rho(E)$, and integrating by parts, we obtain

$$j_z = -\frac{e^2\varepsilon_z < \tau >}{3\pi^2 m^*}(\frac{2m^*}{\hbar^2})^{3/2}\frac{3}{2}\left[f_0 E^{3/2}|_0^\infty - \int_0^\infty f_0 E^{1/2}dE\right]$$

$$= \frac{e^2\varepsilon_z < \tau >}{m^*}\frac{1}{2\pi^2}(\frac{2m^*}{\hbar^2})^{3/2}\int_0^\infty f_0 E^{1/2}dE$$

$$= \frac{e^2\varepsilon_z < \tau >}{m^*}\int_0^\infty f_0\rho(E)dE \quad (11.302)$$

where $f_0 E^{3/2}|_0^\infty = 0$ (f_0 tends to 0 faster than $E^{3/2}$ when $E \to \infty$). The last integral is the total number of electrons n per volume unit [cf. (2.1)]. We have finally

$$j_z = \frac{e^2\varepsilon_z < \tau >}{m^*}n = \sigma\varepsilon_z \quad (11.303)$$

where

$$\sigma \equiv \frac{e^2 <\tau>}{m^*} \tag{11.304}$$

σ is called "electric conductivity". Equation (11.303) is the "Ohm's law".

Problem 3. Hall effect — Magnetoresistance:

Solution:

$\vec{\epsilon}$ is supposed to be parallel to $\vec{Ox}$ and $\vec{B}$ parallel to $\vec{Oz}$. Since $\vec{\epsilon} \cdot \vec{B} = 0$, we have the geometry of the Hall effect.

a) Weak fields - linear approximation:

The Lorentz force is written as

$$\vec{F} = -e(\vec{\epsilon} + \frac{\vec{v} \wedge \vec{B}}{c}) \tag{11.305}$$

The equation of motion $\vec{F} = \frac{d\vec{k}}{dt}$ gives

$$m\dot{v}_x = -e(\epsilon_x + \frac{B}{c}v_y) \tag{11.306}$$

$$m\dot{v}_y = -e(\epsilon_y - \frac{B}{c}v_x) \tag{11.307}$$

$$m\dot{v}_z = -e\epsilon_z \tag{11.308}$$

where $v_y = 0$ (no current in the y direction). The first equation gives $v_x = -e\epsilon_x \tau/m$ when integrating from 0 to τ. The second equation with $\dot{v}_y = 0$ gives

$$\epsilon_y = \frac{m\omega_c v_x}{e} = -\frac{m\omega_c}{e}\frac{j_x}{ne} = -\omega_c \tau \epsilon_x.$$

ϵ_y is induced by the Lorentz force due to $\vec{B}$. We have

$$R_e = \frac{\epsilon_y}{\frac{j_x B}{c}} = -\frac{1}{ne}.$$

For holes, we replace n by p and $-e$ by e. We have $R_h = \frac{1}{pe}$. The mobilities are $\mu_e = \sigma_e/ne = -\sigma_e R_e$ and $\mu_h = \sigma_h R_h$. The Hall angle is defined as $\tan\theta_e = \epsilon_y/\epsilon_x = -\mu_e B/c = -\omega_c \tau$.

When there are two types of carriers, ϵ_x and ϵ_y remain unchanged but we decompose the current j_y into two currents j_{ye} and j_{yh}:

$$j_{ye} = \sigma_e \epsilon_y = -\sigma_e \frac{m\omega_c}{e}\frac{j_x}{ne} = -\mu_e \frac{B}{c}j_{xe}.$$

and similarly,

$$j_{yh} = \mu_h \frac{B}{c} j_{xh}.$$

Note that $\sigma = \sigma_e + \sigma_h = ne\mu_e + pe\mu_h$. We have $j_x = \sigma\epsilon_x$.

$$j_y = -\mu_e \frac{B}{c} j_{xe} + \mu_h \frac{B}{c} j_{xh}$$

$$= -\mu_e \frac{B}{c} \sigma_e \epsilon_x + \mu_h \frac{B}{c} \sigma_h \epsilon_x$$

$$= -ne\mu_e^2 \frac{B}{c} \frac{j_x}{\sigma} + pe\mu_h^2 \frac{B}{c} \frac{j_x}{\sigma}$$

$$\epsilon_y = \frac{j_y}{\sigma} = -(ne\mu_e^2 \frac{B}{c} \frac{j_x}{\sigma^2} - pe\mu_h^2 \frac{B}{c} \frac{j_x}{\sigma^2})$$

$$R = \frac{\epsilon_y}{\frac{Bj_x}{c}} = -(\frac{ne\mu_e^2}{\sigma^2} - \frac{pe\mu_h^2}{\sigma^2})$$

$$= -\frac{e}{\sigma^2}(n\mu_e^2 - p\mu_h^2) = -\frac{e(n\mu_e^2 - p\mu_h^2)}{e^2(n\mu_e + p\mu_h)^2}$$

$$= -\frac{1}{e} \frac{(n\mu_e^2 - p\mu_h^2)}{(n\mu_e + p\mu_h)^2} \tag{11.309}$$

Hence, the Hall mobility is

$$\mu_H = \sigma R = \frac{(p\mu_h^2 - n\mu_e^2)}{(n\mu_e + p\mu_h)};$$

We see that R and μ_H are zero if $p\mu_h^2 = n\mu_e^2$.

b) Moderate fields:

i) The equations of motion of an electron in the x and y directions are given by (11.306) and (11.307). For an electron, we rewrite

$$\frac{dv_x}{dt} = -\frac{e}{m_e}\epsilon_x - \omega_c v_y) \tag{11.310}$$

$$\frac{dv_y}{dt} = -\frac{e}{m_e}\epsilon_y + \omega_c v_x) \tag{11.311}$$

Putting $Z = v_x + iv_y$, we have, using the above equations,

$$\frac{dZ}{dt} = -\frac{e}{m_e}(\epsilon_x + i\epsilon_y) + i\omega_c Z.$$

The solution without the right-hand side is $Z = Ae^{i\omega_c t}$, and the stationary solution is

$$dZ/dt = 0 \rightarrow Z = \frac{e}{m_e} \frac{(\epsilon_x + i\epsilon_y)}{i\omega_c}.$$

The general solution is thus

$$Z = Z_0 + [\frac{1}{i\omega_c}\frac{e}{m_e}(\epsilon_x + i\epsilon_y)](1 - e^{i\omega_c t})$$

$$= Z_0 + C(1 - e^{i\omega_c t})$$

where C is the time-independent quantity in the square brackets and $Z_0 = v_x^0 + iv_y^0 = 0$ for simplicity.

ii) The time-averaged Z is given by

$$\bar{Z} = \frac{1}{\tau}\int_0^\infty Z e^{-t/\tau}\,dt$$

$$= \frac{C}{\tau}\int_0^\infty (1 - e^{i\omega_c t})e^{-t/\tau}$$

$$= C[1 + \frac{1}{i\omega_c\tau - 1}] = C\frac{i\omega_c\tau}{i\omega_c\tau - 1}$$

$$= \frac{e}{m_e}(\epsilon_x + i\epsilon_y)\frac{\tau}{i\omega_c\tau - 1}$$

$$= \frac{e}{m_e}(\epsilon_x + i\epsilon_y)\frac{\tau(i\omega_c\tau + 1)}{\omega_c^2\tau^2 + 1} \tag{11.312}$$

Identifying the real part with v_x and the imaginary part with v_y, we get

$$j_x = -nev_x = \frac{ne^2}{m_e}\left[\frac{\tau}{1 + \omega_c^2\tau^2}\epsilon_x - \frac{\omega_c\tau^2}{1 + \omega_c^2\tau^2}\epsilon_y\right] \tag{11.313}$$

$$j_y = -nev_y = \frac{ne^2}{m_e}\left[\frac{\tau}{1 + \omega_c^2\tau^2}\epsilon_y + \frac{\omega_c\tau^2}{1 + \omega_c^2\tau^2}\epsilon_x\right] \tag{11.314}$$

iii) If there is one type of carriers, Eq. (11.314) when $j_y = 0$ yields $\epsilon_y/\epsilon_x = -\omega_c\tau = \tan\theta_e$ as before. We have the same Hall coefficient, namely $R = -1/ne$. When both electrons and holes participate in the conduction, the Hall coefficient is different as seen in the following. Writing (11.313) and (11.314) for each kind of carriers, we have

$$j_{xe} = C_{1e}\epsilon_x - C_{2e}\epsilon_y$$

$$j_{ye} = C_{2e}\epsilon_x + C_{1e}\epsilon_y$$

$$j_{xh} = C_{1h}\epsilon_x + C_{2h}\epsilon_y$$

$$j_{yh} = -C_{2h}\epsilon_x + C_{1h}\epsilon_y$$

where the coefficients C_{1e} etc. are defined from the coefficients in (11.313) and (11.314)

$$C_{1e} = \frac{ne^2}{m_e} \frac{\tau_e}{1 + \omega_{ce}^2 \tau_e^2} \tag{11.315}$$

$$C_{2e} = \frac{ne^2}{m_e} \frac{\omega_{ce} \tau_e^2}{1 + \omega_{ce}^2 \tau_e^2} \tag{11.316}$$

$$C_{1h} = \frac{pe^2}{m_h} \frac{\tau_h}{1 + \omega_{ch}^2 \tau_h^2} \tag{11.317}$$

$$C_{2h} = \frac{pe^2}{m_h} \frac{\omega_{ch} \tau_h^2}{1 + \omega_{ch}^2 \tau_h^2} \tag{11.318}$$

We have

$$j_x = j_{xe} + j_{xh} = (C_{1e} + C1h)\epsilon_x - (C_{2e} - C_{2h})\epsilon_y$$

$$j_y = j_{ye} + j_{yh} = (C_{2e} - C2h)\epsilon_x + (C_{1e} + C_{1h})\epsilon_y$$

Setting $j_y = 0$, we have

$$\epsilon_x = -\frac{(C_{1e} + C_{1h})}{(C_{2e} - C_{2h})} \epsilon_y,$$

hence

$$j_x = -\frac{(C_{1e} + C_{1h})^2 + (C_{2e} - C_{2h})^2}{(C_{2e} - C_{2h})} \epsilon_y.$$

We obtain

$$R = \frac{\epsilon_y}{\frac{j_x B}{c}} = -\frac{C_{2e} - C_{2h}}{(C_{1e} + C_{1h})^2 + (C_{2e} - C_{2h})^2} \frac{1}{\frac{B}{c}} \tag{11.319}$$

Replacing the coefficients, we obtain

$$R = \frac{\sigma_e^2 R_e + \sigma_h^2 R_h + \sigma_e^2 \sigma_h^2 R_e R_h (R_e + R_h) B^2}{(\sigma_e + \sigma_h)^2 + \sigma_e^2 \sigma_h^2 (R_e + R_h)^2 B^2} \tag{11.320}$$

where $R_e = -1/ne$, $R_h = 1/pe$, $\sigma_e = ne\mu_e$, $\sigma_h = pe\mu_h$.

Comments:

- For small B, neglecting B^2 terms, we find the result of weak fields.
- If there is one kind of carriers the B terms in (11.320) are zero: R does not depend on B.
- If the material is intrinsic, namely $p = n$ and $R_e = -R_h$, then R does not depend on B, neither.
- For strong fields, (11.320) gives

$$R = \frac{R_e R_h}{R_e + R_h} = \frac{1}{(p-n)e}.$$

iv) The above results show that the presence of a magnetic field has no effect on the resistance when the conduction is due to only one type of carriers (n or p) and when the relaxation time is constant and when the effective mass is isotropic (spherical iso-energy surfaces). If one of these three conditions is not fulfilled, there is a correction to the initial resistivity ρ_0. We suppose that both electrons and holes participate to the conduction, keeping isotropic effective mass and constant τ, using R of (11.319) and $\sigma = j_x/\epsilon_x$ with j_x given by above, we write

$$\sigma = -\frac{(C_{1e} + C_{1h})^2 + (C_{2e} - C_{2h})^2}{C_{1e} + C_{1h}}$$

$$\sigma R = -\frac{C_{2e} - C_{2h}}{C_{1e} + C_{1h}} \frac{1}{\frac{B}{c}}$$

Replacing the coefficients and using R of (11.320) we obtain

$$\sigma = \frac{(\sigma_e + \sigma_h)^2 + \sigma_e^2 \sigma_h^2 (B/c)^2 (R_e + R_h)^2}{(\sigma_e + \sigma_h) + \sigma_e \sigma_h (B/c)^2 (\sigma_e R_e^2 + \sigma_h R_h^2)} \tag{11.321}$$

Putting $\sigma_0 \equiv \sigma_e + \sigma_h$ (at $B = 0$), and expanding the denominator with respect to B, we have

$$\frac{\Delta\sigma}{\sigma_0} = \frac{\sigma - \sigma_0}{\sigma_0}$$

$$= -\frac{np\mu_e\mu_h(\mu_e + \mu_h)^2(B/c)^2}{(n\mu_e + p\mu_h)^2}$$

$$\equiv -\xi R_0^2 \sigma_0^2 (B/c)^2 \tag{11.322}$$

where the coefficient of the transverse magneto-resistance ξ is

$$\xi = \frac{np\mu_e\mu_h(\mu_e + \mu_h)^2}{(p\mu_h^2 - n\mu_e^2)^2}$$

and R_0 given by the weak-field approximation [Eq. (11.309)]. Note that

$$\frac{\Delta\rho}{\rho_0} = -\frac{\Delta\sigma}{\sigma_0}$$

$$= \xi R_0^2 \sigma_0^2 (B/c)^2 \tag{11.323}$$

We see that if there is one kind of carriers, this correction is zero (n or p is zero).

Numerical application: The maximum of $\frac{\Delta\rho}{\rho_0}$ is at $n\mu_e = p\mu_h$. In order to have $\frac{\Delta\rho}{\rho_0} \simeq 0.1$ in an applied field of 10^3 Gauss,

$$\frac{\Delta\rho}{\rho_0} = \frac{(\mu_e + \mu_h)^2}{4}(B/c)^2 \simeq 0.1 \quad \rightarrow \quad (\mu_e + \mu_h) \simeq 10^2 \times 630 \text{ cm}^2/\text{V-sec}.$$

c) Effects of collisions:

One supposes again that the effective mass m_e is isotropic and one type of carriers (n), but the relaxation time τ depends on the electron energy under the form $\tau = aE^{-s}$. This form represents several types of collision. The results for j_x and j_x in the weak-field approximation shown above are still valid provided that the coefficients $\frac{\tau}{1+\omega_c^2\tau^2}$ and $\frac{\omega_c\tau^2}{1+\omega_c^2\tau^2}$ are replaced by their values averaged over all energies.

i) Hall coefficient:

Using (11.319) for R but with averaged values for the coefficients of (11.315)-(11.318), we have

$$R = -\frac{K}{ne} \quad \text{where } K \text{ is given by}$$

$$K = \frac{<\frac{\tau^2}{1+\omega_c^2\tau^2}>}{<\frac{\tau}{1+\omega_c^2\tau^2}>^2 +\omega_c^2 <\frac{\tau^2}{1+\omega_c^2\tau^2}>^2}$$

$$\simeq \frac{<\tau^2>}{<\tau>^2} \tag{11.324}$$

where we supposed $\omega_c^2\tau^2 \ll 1$ (weak fields).

ii) We calculate $<\tau^2>$ and $<\tau>$:

$$
\begin{aligned}
<\tau> &= \frac{\int_0^{E_{max}} \tau(E)e^{-E/k_BT}E^{3/2}dE}{\int_0^{E_{max}} e^{-E/k_BT}E^{3/2}dE} \\
&= \frac{A\int_0^{E_{max}} E^{-s}e^{-E/k_BT}E^{3/2}dE}{\int_0^{E_{max}} e^{-E/k_BT}E^{3/2}dE} \\
&\simeq a(k_BT)^{-s}\frac{\int_0^{\infty} u^{-s}e^{-u}u^{3/2}du}{\int_0^{\infty} e^{-u}u^{3/2}du} \\
&= a(k_BT)^{-s}\frac{\Gamma(5/2 - s)}{\Gamma(5/2)} \tag{11.325}
\end{aligned}
$$

where we used the Maxwell-Boltzmann statistics and put $u = E/k_BT$. The coefficient a is the conversion coefficient including A. The same calculation for $<\tau^2>$ leads to a similar result with $<\tau^2>= a^2(k_BT)^{-2s}\frac{\Gamma(5/2-2s)}{\Gamma(5/2)}$. Using the Γ integrals for

- $s = 1/2$ (phonon scattering), we have $K = 3\pi/8 \simeq 1.18$,
- $s = -3/2$ (scattering by charged impurities), we have $K = 315\pi/512 \simeq 1.93$.

We see that K has a strong deviation from the value 1 of the no collision case.

iii) One supposes there is only one kind of carriers, in the same manner we can show that

$$\frac{\Delta\rho}{\rho_0} = \frac{e^2 B^2}{(m_e)^2}\left[\frac{<\tau^3><\tau> - <\tau^2>^2}{<\tau>^2}\right]$$

where $<\tau^3>$ can be calculated as above.

Problem 4. Boltzmann's equation in the case of a strong field:
Solution:
One writes f as a series of the Legendre polynomials P_n as follows

$$f(E, \cos\theta) = \sum_{n=0}^{\infty} g_n(E)P_n(\cos\theta)$$

where $P_0 = 1$, $P_1 = \cos\theta$, $P_2 = 3(\cos^2\theta - 1/3)/2$, ... If one retains only the first two terms, then $f(E, \cos\theta) = g_0(E) + g_1(E)\cos\theta$.

a) Since $\cos\theta = k_z/k$, we can have

$$f(E, \cos\theta) = g_0(E) + \frac{g_1(E)}{k}k_z$$

$$\frac{\partial f(E, \cos\theta)}{\partial k_z} = \frac{\partial g_0}{\partial k_z} + g + k_z\frac{\partial g}{\partial k_z} \quad (g \equiv g_1/k)$$

$$= \frac{\hbar^2 k_z}{m^*}\frac{dg_0}{dE} + g + \frac{\hbar^2 k_z^2}{m^*}\frac{dg}{dE}$$

$$= \frac{\hbar^2 k_z}{m^*}g_0' + g + \frac{\hbar^2 k_z^2}{m^*}g' \qquad (11.326)$$

Since

$$\frac{\hbar^2 k_z^2}{m^*} = \frac{\hbar^2 k^2\cos^2\theta}{m^*} = \frac{\hbar^2 k^2}{m^*}\frac{1}{3} = \frac{2}{3}E$$

where $\cos^2\theta$ was replaced by $1/3$ because we have set $P_2 = 0$, the field term in the Boltzmann's equation is thus

$$-\left(\frac{\partial f}{\partial t}\right)_F = \frac{q\varepsilon}{\hbar}\frac{\partial f(E, \cos\theta)}{\partial k_z}$$

$$= \frac{q\varepsilon}{\hbar}(g + \frac{2}{3}Eg' + \frac{\hbar^2 k_z}{m^*}g_0') \qquad (11.327)$$

where $g = g_1/k$ ($k = k_z \cos\theta$), g' and g'_0 are the first derivatives with respect to E.

b) The collision term of the Boltzmann's equation is written as [see the right-hand side of Eq. (10.9)]

$$-\left(\frac{\partial f}{\partial t}\right)_C = \frac{\Omega}{(2\pi)^3} \int \{w(\vec{k}, \vec{k}')f(\vec{k}') - w(\vec{k}', \vec{k})f(\vec{k})\}d^3k' \quad (11.328)$$

where we took $1 - f \simeq 1$ for high temperatures. Replacing f by (11.326), we have

$$\left(\frac{\partial f}{\partial t}\right)_C = \phi_0 + \phi_1$$

where

$$\phi_0 = \frac{\Omega}{(2\pi)^3} \int \{w(\vec{k}, \vec{k}')g_0(E') - w(\vec{k}', \vec{k})g_0(E)\}d^3k'$$

$$\phi_1 = \frac{\Omega}{(2\pi)^3} \int \{w(\vec{k}, \vec{k}')k'_z g(E') - w(\vec{k}', \vec{k})k_z g(E)\}d^3k'$$

$w(\vec{k}, \vec{k}')$ is the probability per time unit of the transition $\vec{k} \to \vec{k}'$.

c) Taking into account the probability of an electron in the initial state $\vec{k}$ and that of the empty final state $\vec{k}'$, the transition probability is written as

$$\bar{w} = w(\vec{k}, \vec{k}')f(E_k)[1 - f(E_{k'})]$$

with

$$w(\vec{k}, \vec{k}') = \frac{\pi}{NM} \frac{|I|^2}{w_{\vec{q}_j}} \left[s(\omega_{\vec{q}_j}) + \frac{1}{2} \pm \frac{1}{2}\right] \delta(E_{k'} - E_k \pm \hbar\omega_{\vec{q}_j})$$

where I is the electron-phonon coupling, N the total number of electrons and M the mass of ions. The signs $\pm$ correspond respectively to emission and absorption of an phonon of energy $\hbar\omega_{\vec{q}_j}$ of the phonon branch j, of wave vector $\vec{q}$. $s(\omega_{\vec{q}_j})$ is the occupation number of mode $\omega_{\vec{q}_j}$. One supposes that $f[1 - f] \simeq f$ (high temperatures). For acoustic phonons, one shall use $\omega_{(\vec{q})} = v_s q$ (v_s: sound velocity) and $I = \pm iCq$ (C: deformation potential). Without demonstration (see Ref. [58] for details), we admit the following result

$$\phi_0 = \frac{2m^*v_s^2}{\tau} \left[Eg''_0 + \left(\frac{E}{k_BT} + 2\right)g'_0 + \frac{2g_0}{k_BT}\right] \quad (11.329)$$

$$\phi_1 = -\frac{k_z g}{\tau} \quad (11.330)$$

where τ is the relaxation time calculated for weak fields given by

$$\tau = \frac{2\pi M}{\Omega} \frac{\hbar v_s^2}{C^2} (\frac{\hbar^2}{2m^*})^{3/2} \frac{1}{\sqrt{E}} \frac{1}{k_B T} \qquad (11.331)$$

With Eqs. (11.327), (11.329), (11.330) and (11.331), we have

$$2m^* v_s^2 [E g_0'' + (\frac{E}{k_B T} + 2) g_0' + \frac{2 g_0}{k_B T}] - k_z g$$

$$= \tau \frac{q\varepsilon}{\hbar} (g + \frac{2}{3} E g' + \frac{\hbar^2 k_z}{m^*} g_0')$$

$$= \frac{\ell}{v} \frac{q\varepsilon}{\hbar} (g + \frac{2}{3} E g' + \frac{\hbar^2 k_z}{m^*} g_0')$$

$$= \ell \sqrt{\frac{m^*}{2E}} \frac{q\varepsilon}{\hbar} (g + \frac{2}{3} E g' + \frac{\hbar^2 k_z}{m^*} g_0')$$

$$(11.332)$$

where we replaced τ by ℓ/v because $\ell = v\tau \propto \sqrt{2E/m^*}/\sqrt{E}$ is independent of E. To solve (11.332), we equalize the coefficients of k_z and the coefficients of no k_z on the two sides of the above equation:

$$g = -\frac{q\varepsilon\hbar\ell}{\sqrt{2m^* E} g_0'} \qquad (11.333)$$

$$g' = -\frac{q\varepsilon\hbar\ell}{\sqrt{2m^*}} (\frac{g_0''}{\sqrt{E}} - \frac{1}{2} \frac{g_0'}{E^{3/2}}) \qquad (11.334)$$

Replacing these expressions in (11.332), we obtain the following differential equation for g_0

$$(E + \lambda k_B T) g_0'' + (\frac{E}{k_B T} + 2 + \frac{\lambda k_B T}{E}) g_0' + \frac{2}{k_B T} g_0 = 0 \quad (11.335)$$

where

$$\lambda = \frac{q^2 \ell^2}{6 m^* v_s^2 k_B T} \varepsilon^2 \equiv \frac{\varepsilon^2}{\varepsilon_0^2} \qquad (11.336)$$

d) In the strong-field limit $\lambda = \varepsilon^2/\varepsilon_0^2 \gg E/k_B T$, Eq. (11.335) becomes

$$\lambda k_B T g_0'' + (\frac{E}{k_B T} + 2 + \frac{\lambda k_B T}{E}) g_0' + \frac{2}{k_B T} g_0 = 0 \qquad (11.337)$$

where 2 was neglected in front of $\frac{\lambda k_B T}{E}$. We can verify by substitution that the following expression is the solution of (11.337)

$$g_0 = Ae^{-E^2/2\lambda k_B^2 T^2}$$

where A is a constant. Taking the derivative of g_0, we have from (11.333)

$$g = \frac{q\varepsilon\hbar\ell}{\lambda k_B^2 T^2}\sqrt{\frac{E}{2m^*}}Ae^{-E^2/2\lambda k_B^2 T^2}$$

e) To calculate A, we can calculate the number of electrons by integrating $f = g_0 + k_z g$ as follows

$$
\begin{aligned}
N &= \frac{\Omega}{(2\pi)^3}\int (g_0 + k_z g)d^3\vec{k} \\
&= \frac{\Omega}{(2\pi)^3}\int g_0 d^3\vec{k} \quad \text{second term=0 because odd in } k_z \\
&= A\frac{\Omega}{2\pi^2}\left(\frac{2m^*}{\hbar^2}\right)^{3/2}\int_0^\infty \sqrt{E}e^{-E^2/2\lambda k_B^2 T^2}dE \\
&= A\frac{\Omega}{2\pi^2}\left(\frac{2m^*}{\hbar^2}\right)^{3/2}\frac{1}{2}(2\lambda k_B^2 T^2)^{3/4}\Gamma 3/4
\end{aligned}
$$

$$A = \frac{4\pi^2}{\Gamma(3/4)}\frac{N}{\Omega(2\lambda)^{3/4}}\left(\frac{\hbar^2}{2m^*k_B T}\right)^{3/2} \tag{11.338}$$

where we transformed the k integral into an E integral with the help of the density of states $\rho \propto \sqrt{E}$ [see (A.41)].

f) Only the term $k_z g$ of f contributes to the current j. We have

$$
\begin{aligned}
j &= q\int v_z\, k_z g\, d^3\vec{k} \\
&= \frac{q\hbar}{4\pi^3 m^*}\int g\, k_z^2\, d^3\vec{k} \\
&= \frac{q^2\varepsilon\ell\hbar^2 A}{2\pi^3\lambda k_B^2 T^2(2m^*)^{3/2}} \\
&\quad \times \int \sqrt{E}e^{-E^2/2\lambda k_B^2 T^2}k^2\cos^2\theta\, k^2 dk\sin\theta d\theta d\phi \tag{11.339}
\end{aligned}
$$

The angular part gives $4\pi/3$. For the radial part, putting $E^2 = (\hbar^2 k^2/2m^*) = x$, we have

$$dk = \frac{1}{4}\sqrt{\frac{2m^*}{\hbar^2}}\frac{dx}{x^{3/4}}$$

so that

$$
j = \frac{q^2 \varepsilon \ell \hbar^2 A}{2\pi^3 \lambda k_B^2 T^2 (2m^*)^{3/2}} \frac{4\pi}{3} \left(\frac{2m^*}{\hbar^2}\right)^{5/2}
$$

$$
\times \frac{1}{4} \int_0^\infty \sqrt{x} e^{-x/2\lambda k_B^2 T^2} \, dx
$$

$$
= \frac{q^2 \varepsilon \ell A m^*}{3\pi^2 \lambda k_B^2 T^2 \hbar^3} \frac{\sqrt{\pi}}{2} (2\lambda k_B^2 T^2)^{3/2} \tag{11.340}
$$

Replacing q by $-e$, A and λ by (11.338) and (11.336), we have

$$
j = \gamma \sqrt{\varepsilon}
$$

where

$$
\gamma = en \frac{\sqrt{e\ell v_s}}{(m^* k_B T)^{1/4}} \frac{\sqrt{2\pi}}{3^{3/4} \Gamma(3/4)} \tag{11.341}
$$

The expression of j is quite different from the Ohm's law: it is valid for electron-acoustic phonon collisions in strong fields ("hot" electrons).

PART 4
Appendices

Appendix A

Introduction to Statistical Physics

A.1 Introduction

Statistical physics and quantum physics constitute the foundation of the modern physics. They provide methods to study properties of matter.

Methods of statistical physics allow us to study macroscopic properties of large systems using microscopic mechanisms and structures proposed by quantum mechanics. Thanks to a combination of quantum mechanics and statistical mechanics, we have seen since the second half of the 20-th century spectacular discoveries and progress in modern physics, in particular in the field of condensed matter, which have radically changed our way of life.

This Appendix aims at recalling elements of statistical physics which are used throughout this book.

Statistical physics for systems at equilibrium is based on one single postulate called "the fundamental postulate" introduced in the case of an isolated system at equilibrium. The complete properties of an isolated system are deduced from this postulate. Other systems, not isolated but in some special conditions, can be studied from methods derived from the fundamental postulate. One class of such systems includes systems in contact with a very large heat reservoir: a system of this class has a constant temperature fixed by the heat reservoir. The method used to study this class of systems is called the "canonical description" A second class of systems includes systems in contact with a large reservoir of heat and particles. A system of this class has a constant temperature and a constant chemical potential given by the reservoir. The number of particles of the system is not constant. This number fluctuates around a mean value when the system is at equilibrium. It is an "open system". The method used to study this class of systems is called the "grand-canonical description".

We consider a system of particles. In statistical physics the most fundamental quantity is the statistical entropy defined by

$$S = -k_B \sum_l P_l \ln P_l \tag{A.1}$$

where P_l is the probability of the microscopic state l of the system and k_B the Boltzmann constant. We shall use the statistical entropy to express various physical quantities in the following.

A.2   Isolated systems: Microcanonical description

A.2.1   *Fundamental postulate*

A system is said "isolated" when it has no interaction with the remaining universe. It is obvious that such a definition is not rigorous: we should understand that interactions are so small that they are not observable and the parameters imposed on the system from the outside world such as energy E, volume V and number of particles N, are constant for all time.

The accessible microscopic states of an isolated system are the states which obey the external constraints. Let Ω be the total number of accessible microscopic states. The fundamental postulate of the statistical mechanics of systems at equilibrium states that

"All accessible microscopic states of an isolated system at equilibrium have the same probability".

According the the above postulate, we have

$$P_l = \frac{1}{\Omega} \tag{A.2}$$

for any accessible microscopic state l. The above probability is called "microcanonical probability". Microscopic states which verify this probability form a "microcanonical ensemble". The description of properties of a system using the above probability is called "microcanonical description".

Statistical entropy S of an isolated system is thus

$$S = -k_B \sum_l P_l \ln P_l$$

$$= k_B \sum_l \frac{1}{\Omega} \ln \Omega$$

$$= k_B \frac{1}{\Omega} \ln \Omega \sum_l 1$$

$$= k_B \ln \Omega \tag{A.3}$$

Equation (A.3) is called "microcanonical entropy".
The microcanonical temperature T is defined by

$$\frac{1}{T} = \frac{\partial S}{\partial E} \tag{A.4}$$

The microcanonical pressure p is defined by

$$\frac{p}{T} = -\frac{\partial S}{\partial V} \tag{A.5}$$

The microcanonical chemical potential μ is defined by

$$\frac{\mu}{T} = \frac{\partial S}{\partial N} \tag{A.6}$$

We can show that these definitions correspond to physical quantities of the same names in thermodynamics. We can also show that the spontaneous evolution toward equilibrium of an isolated system when an external constraint is removed is always accompanied by an increase of statistical entropy S. Equilibrium is reached when S is maximum.

A.2.2 *Applications*

A.2.2.1 *Two-level systems*

We consider an isolated system of N independent, discernible particles. We suppose that $N \gg 1$. Each particle has two energy levels ϵ_1 and ϵ_2. The total energy of the system is equal to E. Using the microcanonical description, we calculate the total number of accessible microscopic states $\Omega(E)$, the microcanonical entropy $S(E)$ and the microcanonical temperature T as follows:

- the total number of accessible microscopic states $\Omega(E)$:
  System energy: $E = N_1\epsilon_1 + N_2\epsilon_2$ = constant, N_1=number of particles on ϵ_1, N_2=number of particles on ϵ_2, we have $N = N_1 + N_2$ = constant, hence $E = N_1\epsilon_1 + (N - N_1)\epsilon_2$. As E is constant, N_1 is thus determined. The total number of accessible microscopic states $\Omega(E)$ is equal to the number of ways to choose N_1 particles among N for the level ϵ_1. Thus, $\Omega = \frac{N!}{N_1!N_2!} = \frac{N!}{N_1!(N-N_1)!}$.

- the microcanonical entropy $S(E)$:
  $S = k_B \ln \Omega = k_B[\ln N! - \ln N_1! - \ln(N - N_1)!]$. Using the Stirling formula for $N \gg 1$: $\ln N! \simeq N \ln N - N$, we have $S \simeq k_B[N \ln N - N - N_1 \ln N_1 + N_1 - (N - N_1)\ln(N - N_1) + (N - N_1)] = k_B[N \ln N - N_1 \ln \frac{N_1}{N-N_1} - N \ln(N - N_1)]$

- the microcanonical temperature T as a function of ϵ_1 and ϵ_2:
  $T^{-1} = \frac{\partial k_B \ln \Omega}{\partial E} = \frac{\partial k_B \ln \Omega}{\partial N_1}\frac{\partial N_1}{\partial E} = \frac{1}{\epsilon_1-\epsilon_2}k_B[-\ln N_1 - 1 + \ln(N - N_1) + 1] = \frac{k_B}{\epsilon_1-\epsilon_2}\ln\frac{N-N_1}{N_1}$ hence $\frac{\epsilon_1-\epsilon_2}{k_B T} = \ln\frac{N-N_1}{N_1}$ or $\frac{N_1}{N-N_1} = \exp(-\beta(\epsilon_1 - \epsilon_2))$.
  We obtain the relation of N_1 and N_2 as functions of T, ϵ_1 and ϵ_2:
  $N_1 = N\frac{\exp(-\beta(\epsilon_1-\epsilon_2))}{1+\exp(-\beta(\epsilon_1-\epsilon_2))} = N\frac{\exp(2\beta\epsilon)}{1+\exp(2\beta\epsilon)}$

- With $\epsilon(> 0) \equiv -\epsilon_1 = \epsilon_2$, we have
  At low T, $N_1 \to N$ hence $N_2 \to 0$. This result is obvious since particles occupy low-energy level at $T = 0$. At high T, $N_1 \to N/2$ hence $N_2 \to N/2$: particles are equally distributed on the two levels.

A.2.2.2   *Classical ideal gas*

We consider a classical ideal gas of N particles, of volume V. The gas is isolated with a total energy E. Using the microcanonical description, we calculate

- the microcanonical entropy:
  The classical phase space (see section A.6 below) has $6N$ dimensions ($3N$ for particle positions and $3N$ for particle momenta). The number of states of energy $\leq E$ is the number of states in the sphere of radius $p = \sqrt{2mE}$

$$\mathcal{N} = \int_V d\vec{r}_1 \int_V d\vec{r}_2 ... \int_{p_1^2+p_2^2+....+p_N^2=2mE} d\vec{p}_1 d\vec{p}_2 ...$$

$$= V^N \int_{p_1^2+p_2^2+....+p_N^2=2mE} d\vec{p}_1 d\vec{p}_2 ... \qquad (A.7)$$

where $\vec{p}_i$ is the momentum of the i-th particle. The integration on $\vec{p}_i$

gives the volume of the sphere in $3N$ dimensions which is proportional to the radius to the power of $3N$, namely $V^N(2mE)^{3N/2}$ where V^N is the integration over N positions. The number of states of energy equal to E, namely $\Omega(E)$, is the number of states lying on the surface of the sphere of radius $V^N(2mE)^{3N/2}$, we have $\Omega(E) \propto V^N E^{3N/2-1} \simeq V^N E^{3N/2}$ (because $N \gg 1$). The microcanonical entropy is thus $S = A \ln V^N E^{3N/2}$ where A is a constant,

- the microcanonical temperature T:
  We have $T^{-1} = \frac{\partial k_B \ln \Omega}{\partial E} = k_B \frac{3N}{2E}$, hence $E = \frac{3}{2}Nk_BT$, namely the result of classical thermodynamics.

- the microcanonical pressure p:
  We have $\frac{p}{T} = \frac{\partial k_B \ln \Omega}{\partial V} = k_B N/V$ hence $pV = Nk_BT$. This is the equation of state found in thermodynamics using the kinetic theory of gas.

A.3 Systems at constant temperature: Canonical description

When a system is in contact with a heat reservoir much larger than the system, the reservoir imposes on the system its temperature T. The system energy is no more constant, it fluctuates by heat exchange with the reservoir. The equilibrium is reached when the system temperature is equal to that of the reservoir. If we know T, we can calculate main properties of the system as seen in the following.

The probability of the microscopic state l in the canonical situation is given by

$$P_l = \frac{e^{-\beta E_l}}{Z(T,V,N)} \tag{A.8}$$

where $\beta = \frac{1}{k_BT}$, and

$$Z = \sum_l e^{-\frac{E_l}{k_BT}} \tag{A.9}$$

We call Z the "partition function". This function depends on external variables imposed on the system such as T, V (system volume) and N (number of particles).

The probability (A.8) is called "canonical probability". The ensemble of microscopic states obeying this probability is called "canonical ensemble".

The description of properties of the system using (A.8) is called "canonical description". We see below that we can express various physical quantities in terms of Z:

- *Average energy and heat capacity:*
  The average energy $\overline{E}$ of the system is

$$\overline{E} = \sum_l E_l P_l = \sum_l \frac{E_l e^{-\frac{E_l}{k_B T}}}{Z}$$

$$= -\frac{1}{Z} \frac{\partial}{\partial \beta} \sum_l e^{-\beta E_l} = -\frac{1}{Z} \frac{\partial Z}{\partial \beta}$$

$$= -\frac{\partial \ln Z}{\partial \beta} \tag{A.10}$$

The heat capacity is

$$C_V = \frac{d\overline{E}}{dT} = \frac{d}{dT}[-\frac{\partial \ln Z}{\partial \beta}]$$

$$= -\frac{\partial^2 \ln Z}{\partial \beta^2}(\frac{d\beta}{dT})$$

$$= \frac{1}{k_B T^2} \frac{\partial^2 \ln Z}{\partial \beta^2} \tag{A.11}$$

- *Canonical entropy:*
  Replacing P_l by (A.8) in (A.1), we have

$$S = -k_B \sum_l \frac{e^{-\beta E_l}}{Z}(-\beta E_l - \ln Z)$$

$$= k_B[\beta \overline{E} + \ln Z \sum_l \frac{e^{-\beta E_l}}{Z}]$$

$$= \frac{\overline{E}}{T} + k_B \ln Z \tag{A.12}$$

- *Free energy:*
  The free energy F is defined by

$$F = -k_B T \ln Z \tag{A.13}$$

As Z, F is a function of T, V and N. This definition allows us to write

$$S = \frac{\overline{E}}{T} - \frac{F}{T} \tag{A.14}$$

or more often,

$$F = \overline{E} - TS \qquad (A.15)$$

- *Canonical pressure p:*
  p is defined by

$$p = -\frac{\partial F}{\partial V} \qquad (A.16)$$

- *Canonical chemical potential μ:*
  μ is defined by

$$\mu = \frac{\partial F}{\partial N} \qquad (A.17)$$

We can show that a system in a canonical situation tends to equilibrium, when an external constraint is removed, in the sense of decreasing F during the spontaneous evolution. The system reaches equilibrium when F is minimum.

A.3.1 Applications

A.3.1.1 Two-level systems

We consider the two-level system in A.2.2.1 using the canonical description at T. We calculate

- the partition function Z:
  We have $Z = z^N$ where z is the partition function of one particle,
  $z = \sum_i \exp(-\beta\epsilon_i) = \exp(-\beta\epsilon_1) + \exp(-\beta\epsilon_2) = 2\cosh(\beta\epsilon)$.
- the average energy $\overline{E}$ and the heat capacity C_V:
  $\overline{E} = -\frac{\partial \ln Z}{\partial \beta} = -N\epsilon\frac{\sinh(\beta\epsilon)}{\cosh(\beta\epsilon)} = -N\epsilon\tanh(\beta\epsilon)$.
  $C_V = \frac{d\overline{E}}{dT} = \frac{N}{k_B T^2}\frac{\epsilon^2}{\cosh^2(\beta\epsilon)}$
- the number of particles in each level:
  We have $\overline{E} = N_1\epsilon_1 + N_2\epsilon_2 = \epsilon(-N_1 + N_2) = \epsilon(-N_1 + N - N_1) = \epsilon(-2N_1 + N)$. Using $\overline{E}$ given above, we have $-N\epsilon\tanh(\beta\epsilon) = \epsilon(N - 2N_1)$, hence $N_1 = \frac{N}{2}[1 + \tanh(\beta\epsilon)]$
- At low T, $N_1 \to N$ (because $\tanh(\beta\epsilon) \to 1$) hence $N_2 \to 0$. At high T, $N_1 \to N/2$ (because $\tanh(\beta\epsilon) \to 0$) hence $N_2 \to N/2$. We have the same results as by the microcanonical description.

A.3.1.2   *Classical ideal gas*

We calculate with the canonical description the partition function, the average energy and the pressure as follows:

- the partition function $Z = z^N/N!$ (undiscernible particles):

$$z = \frac{1}{h^3} \int_V d\vec{r} \int d\vec{p}\,\exp(-\beta p^2/2m)$$

$$= \frac{V}{h^3} \int_{-\infty}^{\infty} dp_x \exp(-\beta p_x^2/2m) \int_{-\infty}^{\infty} dp_y \exp(-\beta p_y^2/2m)$$

$$\times \int_{-\infty}^{\infty} dp_z \exp(-\beta p_z x^2/2m)$$

$$= \frac{V}{h^3} (\sqrt{2\pi m/\beta})^3 \tag{A.18}$$

using $\int_{-\infty}^{\infty} du \exp(-au^2) = \sqrt{\pi/a}$.
- $\overline{E} = -\frac{\partial \ln Z}{\partial \beta} = \frac{3}{2\beta}N$ hence $\overline{E} = \frac{3Nk_BT}{2}$.
- $p = k_B T \frac{\partial \ln Z}{\partial V} = Nk_BT/V$.

Again we found here the same result as by the microcanonical description.

A.4   Open systems at constant temperature: Grand-canonical description

When a system is in contact with a reservoir of heat and particles much larger than the system, the reservoir imposes on the system its temperature T and its chemical potential. The system is in the "grand-canonical situation". The energy and the number of particles of the system fluctuate by exchange of heat and particles with the reservoir. The system reaches equilibrium when its temperature is equal to T.

The grand-canonical probability of the microscopic state l is given by

$$P_l = \frac{e^{-\beta(E_l-\mu N_l)}}{\mathcal{Z}} \tag{A.19}$$

where

$$\mathcal{Z} = \sum_l e^{-\beta(E_l-\mu N_l)} \tag{A.20}$$

$\mathcal{Z}$ is called "grand-partition function". The ensemble of microscopic states obeying the probability (A.19) is called "grand-canonical ensemble".

$\mathcal{Z}$ plays an important role in the calculation of principal properties of the system. As $\ln \mathcal{Z}$ appears often in the calculation, we define a new function J, called "grand potential", by

$$J = -k_B T \ln \mathcal{Z} \qquad (A.21)$$

We can express the following quantities as functions of J:

- *Average number of particles:*

$$\begin{aligned}
\overline{N} &= \sum_l N_l P_l \\
&= \frac{1}{\mathcal{Z}} \frac{1}{\beta} \frac{\partial}{\partial \mu} \sum_l e^{-\beta(E_l - \mu N_l)} \\
&= k_B T \frac{\partial \ln \mathcal{Z}}{\partial \mu} \\
&= -\frac{\partial J}{\partial \mu} \qquad (A.22)
\end{aligned}$$

- *Average energy:*

$$\begin{aligned}
\overline{E} - \mu \overline{N} &= \sum_l (E_l - \mu N_l) P_l \\
&= -\frac{1}{\mathcal{Z}} \frac{\partial}{\partial \beta} \sum_l e^{-\beta(E_l - \mu N_l)} \\
&= -\frac{\partial \ln \mathcal{Z}}{\partial \beta} \\
&= \frac{\partial}{\partial \beta}(\beta J) \\
&= J + \beta \frac{\partial J}{\partial \beta} \qquad (A.23)
\end{aligned}$$

We deduce

$$\begin{aligned}
\overline{E} &= J + \beta \frac{\partial J}{\partial \beta} + -\mu \frac{\partial J}{\partial \mu} \\
&= J + (\mu \frac{\partial}{\partial \mu} + \beta \frac{\partial}{\partial \beta})J \qquad (A.24)
\end{aligned}$$

- *Grand-canonical pressure p:*
  p is defined by

$$p = -\frac{\partial J}{\partial V} \tag{A.25}$$

- *Grand-canonical entropy:*
  Using P_l of (A.19), we have

$$
\begin{aligned}
S &= -k_B \sum_l P_l \ln P_l \\
&= -k_B \frac{1}{\mathcal{Z}} \sum_l e^{-\beta(E_l - \mu N_l)} [-\beta(E_l - \mu N_l - \ln \mathcal{Z}] \\
&= k_B \beta(\overline{E} - \mu \overline{N}) + k_B \ln \mathcal{Z}) \\
&= \frac{1}{T}(\overline{E} - \mu \overline{N}) + k_B \ln \mathcal{Z}) \\
&= \frac{1}{T}(J + \beta \frac{\partial J}{\partial \beta}) - \frac{J}{T} \\
&= \frac{1}{k_B T^2} \frac{\partial J}{\partial \beta} \\
&= -\frac{\partial J}{\partial T} \tag{A.26}
\end{aligned}
$$

We can show that in the spontaneous evolution when an external constraint is removed, the system tends to equilibrium in the sense of decreasing J. The new equilibrium is attained when J is minimum.

A.4.1   *Applications*

We consider again a classical ideal gas studied above by the microcanonical and canonical descriptions. We show below that the grand-canonical description gives the same results. We calculate the grand-partition function, the average number of particles of the system, the energy and the pressure as follows:

- the grand-partition function:
  $\mathcal{Z} = \sum_N e^{N\beta\mu} Z(T, N, V)$ where Z is the partition function. We have $Z = z^N/N! =$, therefore
  $\mathcal{Z} = \sum_{N=0}^{\infty} e^{N\beta\mu} z^N/N! = \sum_{N=0}^{\infty} (\lambda z)^N/N! = \exp(\lambda z)$ where $\lambda = e^{\beta\mu}$
  (fugacity).
- the average number of particles:

$$\bar{N} = \frac{1}{\beta}\frac{\partial \ln \mathcal{Z}}{\partial \mu} = \frac{1}{\beta}\frac{\partial \lambda z}{\partial \mu}$$

$$= \frac{1}{\beta}z\frac{\partial e^{\beta\mu}}{\partial \mu} = ze^{\beta\mu} = \lambda z.$$

We have

$$\bar{E} - \mu\bar{N} = -\frac{\partial \ln \mathcal{Z}}{\partial \beta} = -\frac{\partial \lambda z}{\partial \beta}$$

$$= -\frac{\partial(e^{\beta\mu}z)}{\partial \beta} = -\mu\lambda z - \lambda\frac{\partial z}{\partial \beta}$$

$$= -\mu\bar{N} + \frac{3\bar{N}k_BT}{2} \text{ hence } \bar{E} = \frac{3\bar{N}k_BT}{2}.$$

- $p = -\frac{\partial J}{\partial V} = k_BT\frac{\partial J}{\partial V} = k_BT\frac{\partial \ln \mathcal{Z}}{\partial V}$

$$= k_BT\lambda\frac{\partial z}{\partial V} = k_BT\lambda z/V \text{ [see } z \text{ in (A.18)] hence } pV = \bar{N}k_BT.$$

We obtain thus the same equation of state for the gas as before.

A.5 Fermi-Dirac and Bose-Einstein statistics

We can use the grand-canonical description to demonstrate the Fermi-Dirac and Bose-Einstein distributions.

We consider a system of identical, independent and indiscernible particles. In such a hypothesis, each particle has the same "list" of individual states. We can define each individual state by the number of particles present in that state. Let k be an individual state of energy ϵ_k and n_k its number of particles. The total energy and the total number of particles in the microscopic state l of the system are

$$E_l = \sum_k n_k\epsilon_k \tag{A.27}$$

$$N_l = \sum_k n_k \tag{A.28}$$

The grand-partition function reads

$$\mathcal{Z} = \sum_l e^{-\beta(E_l - \mu N_l)}$$

$$= \sum_{\{n_k\}} e^{-\beta\sum_k n_k(\epsilon_k - \mu)} \tag{A.29}$$

where, for a given distribution of particles $\{n_k\}$ on individual states, we make the sum in the argument of the exponential, then we repeat for another distribution $\{n'_k\}$ until all distributions have been considered. This procedure is equivalent to the sum on the states l. In doing so, we can express $\mathcal{Z}$ as

$$\mathcal{Z} = \prod_k z_k \tag{A.30}$$

where

$$z_k = \sum_{n_k} e^{-\beta n_k(\epsilon_k - \mu)} \tag{A.31}$$

It is noted that the sum in z_k is performed on all possible values of n_k for the level ϵ_k. We distinguish two cases:

- Bosons (particles of spin 0 or integer):
  In this case, $n_k = 0, 1, 2, ...$ (no limit). We have

$$
\begin{aligned}
z_k^{BE} &= \sum_{n_k=0}^{\infty} e^{-\beta n_k(\epsilon_k - \mu)} \\
&= \frac{1}{1 - e^{-\beta(\epsilon_k - \mu)}}
\end{aligned} \tag{A.32}
$$

- Fermions (particles of spin half-integer):
  In this case, $n_k = 0, 1$. We have

$$
\begin{aligned}
z_k^{FD} &= \sum_{n_k=0}^{1} e^{-\beta n_k(\epsilon_k - \mu)} \\
&= 1 + e^{-\beta(\epsilon_k - \mu)}
\end{aligned} \tag{A.33}
$$

To calculate the average number of particles $\overline{n}_k$ of the state of energy ϵ_k, we use (A.22), (A.28) and (A.30):

$$
\begin{aligned}
\overline{N} = \sum_k \overline{n}_k &= -\frac{\partial J}{\partial \mu} \\
&= k_B T \sum_k \frac{\partial \ln z_k}{\partial \mu}
\end{aligned} \tag{A.34}
$$

We obtain

$$\overline{n}_k = k_B T \frac{\partial \ln z_k}{\partial \mu} \tag{A.35}$$

Using (A.32) and (A.33), we have, for the boson case,

$$f(\epsilon)^{BE} \equiv \overline{n}_k^{BE} = \frac{1}{e^{\beta(\epsilon_k - \mu)} - 1} \tag{A.36}$$

and for the fermion case,

$$f(\epsilon)^{FD} \equiv \overline{n}_k^{FD} = \frac{1}{e^{\beta(\epsilon_k - \mu)} + 1} \tag{A.37}$$

The distributions $f(\epsilon)^{BE}$ and $f(\epsilon)^{FD}$ are called "Bose-Einstein and Fermi-Dirac distributions", respectively.

A.6 Phase space — Density of states

The phase space is defined by the number of degrees of freedom which characterize the microscopic states of the system. In a quantum case, each state is defined by some quantum numbers. For example, each of the microscopic states of a free particle in a box is defined by a wave vector $\vec{k}$ which is quantified by the boundary conditions, the state of an electron in an atomic orbital is given by four quantum numbers (n, l, m_l, m_s). In the case of a system of classical particles each of the microscopic states of the system is defined by the momentum and the position of each particle, $\vec{p}_i$ and $\vec{r}_i$. These variables constitute the phase space. The sum on the microscopic states is taken over all of these variables.

We consider a system of N classical particles in three dimensions. The number of degrees of freedom is $6N$ because each particle is defined by 6 variables: three of $\vec{p}_i$ and three of $\vec{r}_i$. We attribute a "volume" of $6N$ dimensions for each microscopic state in the phase space. The elementary volume occupied by a particle is chosen as small as allowed by the uncertainty principle of quantum mechanics. The smallest elementary volume occupied by a particle is equal to $(\sqrt{h})^6 = h^3$ where h is the Planck constant. This choice is made to discretize the classical "continuous" phase space in order to count the number of states: it suffices to divide a chosen volume in the phase space by the elementary volume to find the number of states contained in that volume. To find results for classical particles, we let $h \to 0$ at the end of the calculation.

The sum on the microscopic states is written as an integral in the classical phase space as follows:

$$\sum_k ... = \frac{1}{h^{3N}} \int d\vec{p}_1 \int d\vec{p}_2 ... \int d\vec{p}_N \int d\vec{r}_1 \int d\vec{r}_2 ... \int d\vec{r}_N ... \quad (A.38)$$

We consider now a quantum system. When the size of the system is large (thermodynamic limit) we can consider the energy as a continuous variable. We can replace the sum on discrete microscopic states $\vec{k}$ by an integral on the energy ϵ but we have to take into account the degeneracy of each energy. For a continuous energy, the degeneracy is the density of states . We write

$$\overline{E} = \sum_k \overline{n}_k \epsilon_k = \int_{\epsilon_0}^{\infty} d\epsilon \rho(\epsilon) \overline{n}_k \epsilon \quad (A.39)$$

where ϵ_0 is the lowest energy and $\rho(\epsilon)$ the density of states. According to the studied case, we replace in this integral $\overline{n}_k$ by $\overline{n}_k^{BE}$ or $\overline{n}_k^{FD}$.

For a free particle in three dimensions, we can show that (see Problem 3 of chapter 1)

$$\rho(\epsilon) = \frac{\Omega}{4\pi^2} \left(\frac{2m}{\hbar^2}\right)^{3/2} \epsilon^{1/2} \tag{A.40}$$

where Ω the system volume, and m the particle mass. If the particle has a spin s then the density of states is

$$\rho(\epsilon) = (2s+1)\frac{\Omega}{4\pi^2} \left(\frac{2m}{\hbar^2}\right)^{3/2} \epsilon^{1/2} \tag{A.41}$$

$(2s+1)$ is the spin degeneracy. For electrons, $s = 1/2$.

Appendix B

A Simple Monte Carlo Program for Ising Model

We give here a simple program for Monte Carlo simulation of the Ising model on the square lattice. This program calculates only the energy at a given temperature. The reader should complete the program for various cases and spin models (see exercises of chapter 8 and their solutions).

The program is written in Fortran. The reader can translate it into the language of his or her choice.

```
CCCCCC ISING SPINS ON SQUARE LATTICE 2D
CCCCCC
       PROGRAM IRC

       PARAMETER(N = 36, NN = N * N)
       DIMENSIONS(N, N)
CCCCCC PARAMETERS
       AJ1 = 1
       ITIME1 = 1
       NE = 10000
       ITIME = 1
       INTV = 10000
       T = 1.
CCCCCC
CCCCCC *****INITIAL CONDITION *****
CCCCCC FERROMAGNETIC STATE
       DO I = 1, N
       DO J = 1, N
       S(I, J) = 1
       ENDDO
       ENDDO
CCCCCC MONTE CARLO SIMULATION
CCCCCC EQUILIBRATING
       DO 1903 IE = 1, ITIME1
       CALL MC1(AJ1, NE, T, S)
       WRITE(6, 481)IE
   481 FORMAT('ITERATION', I9)
  1903 CONTINUE
CCCCCC AVERAGING
       ISTEP = ITIME1 * NE
       ENERGY = 0.
       IN = 0
```

```
CCCCCC
        DO 480 IK = 1, ITIME
        CALL MC2(AJ1, INTV, T, E, E1, S)
        IN = IN + INTV
        ISTEP = ISTEP + INTV
        WRITE(6, 478)IK
   478  FORMAT('ITERATION', I9)

        OPEN(UNIT = 3, file =' irc.fig')
        REWIND (3)
        ENERGY = ENERGY + E1
        U = ENERGY/IN
        WRITE(6, 1901)ISTEP, T, U
  1901  FORMAT(1X, I7,' T', E15.7,' U =', E15.7)
        WRITE(3, 1902)ISTEP, T, U
  1902  FORMAT(I9, 1X, 2(E13.7, 1X))
        CLOSE(3)
   480  CONTINUE
        STOP
        END
CCCCCC
CCCCCC  SUBROUTINE FOR EQUILIBRATING
        SUBROUTINE MC1(AJ1, NE, T, S)
        PARAMETER (N = 36, NN = N * N)
        DIMENSION S(N, N)
        E = 0.
        DO 460 I = 1, NE
        CALL IMC(AJ1, E, T, S)
   460  CONTINUE
        RETURN
        END
```

```
CCCCCC
CCCCCC     SUBROUTINE FOR AVERAGING
           SUBROUTINE MC2(AJ1, INTV, T, E, E1, S)
           PARAMETER (N = 36, NN = N * N)
           DIMENSION S(N, N)
           E1 = 0.
           DO 490 IJ = 1, INTV
           CALL IMC(AJ1, E, T, S)
           E1 = E1 + E/(2 * NN)
   490     CONTINUE
           RETURN
           END
CCCCCC     MONTE - CARLO UPDATING
           SUBROUTINE IMC(AJ1, E, T, S)
           PARAMETER (N = 36, NN = N * N)
           DIMENSION S(N, N)
           E = 0.
           DO 9200 J1 = 1, N
           JM = J1 - 1 + (1/J1) * N
           JP = J1 + 1 - (J1/N) * N
           DO 5200 I1 = 1, N
           IM = I1 - 1 + (1/I1) * N
           IP = I1 + 1 - (I1/N) * N
           ENGY = -S(I1, J1) * AJ1 * (S(IP, J1) + S(IM, J1)
      &    +S(I1, JP) + S(I1, JM))
           SI = S(I1, J1)
           EN = -ENGY
           IF(EN.LE.ENGY)THEN
           S(I1, J1) = -SI
           ER = EN
           ELSE
```

```
CCCCCC
        IF(RAND().LT.EXP(-(EN - ENGY)/T))THEN
        S(I1, J1) = -SI
        ER = EN
        ELSE
        ER = ENGY
        ENDIF
        ENDIF
        E = E + ER
5200    CONTINUE
9200    CONTINUE
        RETURN
        END
```

Bibliography

[1] K. Akabli, H. T. Diep and S. Reynal, J. Phys.: Condens. Matter **19**, 356204 (2007).

[2] K. Akabli and H. T. Diep, J. Appl. Phys. **103**, 07F307 (2008).

[3] K. Akabli and H. T. Diep, Phys. Rev. B **77**, 165433 (2008).

[4] K. Akabli, Y. Magnin, M. Oko, I. Harada and H. T. Diep, Phys. Rev. B **84**, 024428 (2011).

[5] J. W. Allen, G. Locovsky, and J. C. Mikkelsen Jr., Solid State Commun. **24**, 367 (1977).

[6] S. Alexander, J. S. Helman, and I. Balberg, Phys. Rev. B **13**, 304 (1975).

[7] D. J. Amit, *Field Theory, Renormalization Group and Critical Phenomena*, World Scientific, Singapore (1984).

[8] S. A. Antonenko and A.I. Sokolov, Phys. Rev. B **49**, 15901 (1994).

[9] S. S. Aplesnin, L. I. Ryabinkina, O. B. Romanova, D. A. Balaev, O. F. Demidenko, K. I. Yanushkevich and N. S. Miroshnichenko, Phys. Solid State **49**, Number 11, 2080-2085 (2007).

[10] N. W. Ashcroft et N. D. Mermin, *Solid State Physics*, Saunders College, Philadelphia (1976).

[11] P. Azaria, H. T. Diep and H. Giacomini, Phys. Rev. Lett. **59**, 1629 (1987).

[12] M. N. Baibich, J. M. Broto, A. Fert, F. Nguyen Van Dau, F. Petroff, P. Etienne, G. Creuzet, A. Friederich, and J. Chazelas, Phys. Rev. Lett. **61**, 2472 (1988).

[13] M. N. Barber, *Finite-size Scaling*, in *Phase Transitions and Critical Phenomena*, Vol. 8, Ed. C. Domb and J. L. Lebowitz, Academic Press, London (1983).

[14] A. Barthélémy *et al.*, J. Magn. Magn. Mater. **242-245**, 68 (2002).

[15] See review on Oxide Spintronics by Manuel Bibes and Agnès Barthélémy, in a Special Issue of IEEE Transactions on Electron Devices on Spintronics, IEEE Trans. Electron. Devices **54**, 1003 (2007).

[16] R. J. Baxter, *Exactly Solved Models in Statistical Mechanics*, Academic, New York (1982).

[17] K. Binder and D. W. Heermann, *Monte Carlo Simulation in Statistical Physics*, Springer-Verlag, Berlin (1992).

[18] K. Binder, *Critical Hehaviour at Surfaces*, in *Phase Transitions and Critical Phenomena*, Vol. 8, Ed. C. Domb and J. L. Lebowitz, Academic Press, London (1983).

[19] J. A. C. Bland and B. Heinrich (Editors), *Ultrathin Magnetic Structures*, vol. I and II, Springer-Verlag (1994).

[20] N. N. Bogolyubov and S. V. Tyablikov, Doklady Akad. Nauk S.S.S.R. **126**, 53 (1959) [translation: Soviet Phys.-Doklady **4**, 604 (1959)].

[21] M. Born and K. Huang, *Dynamical Theory of Crystal Lattices*, Clarendon Press, Oxford (1954).

[22] G. Brown and T.C. Schulhess, J. Appl. Phys. **97**, 10E303 (2005).

[23] A. Bunker, B. D. Gaulin, and C. Kallin, Phys. Rev. B **48**, 15861 (1993).

[24] J. Cardy, *Scaling and Renormalization in Statistical Physics*, Cambridge Lecture Notes in Physics, Cambridge University Press (2000).

[25] C. Cercignani, *The Boltzmann Equation and Its Applications*, Springer-Verlag, New York (1988).

[26] P. M. Chaikin and T. C. Lubensky, *Principles of Condensed Matter Physics*, Cambridge University Press (1995).

[27] S. Chandra, L. K. Malhotra, S. Dhara and A. C. Rastogi, Phys. Rev. B **54**, 13694 (1996).

[28] P. P. Craig *et al.*, Phys. Rev. Lett. 19, 1334 (1967).

[29] M. Debauche, H. T. Diep, H. Giacomini and P. Azaria, Phys. Rev. B **44**, 2369 (1991).

[30] M. Debauche and H. T. Diep, Phys. Rev. B **46**, 8214 (1992).

[31] P.-G. de Gennes and J. Friedel, J. Phys. Chem. Solids **4**, 71 (1958).

[32] H. W. Diehl, *Field-theoretic Approach to Critical Behaviour at Surfaces*, in *Phase Transitions and Critical Phenomena*, Vol. 10, Ed. C. Domb and J. L. Lebowitz, Academic Press, London (1986).

[33] H. W. Diehl, Int. J. Mod. Phys. B **11**, 3503 (1997).

[34] See *Review on Semiconductor Spintronics* by T. Dietl, in Lectures Notes, vol. 712, Springer-Verlag, Berlin, pp. 1-46 (2007).

[35] Diep-The-Hung, J.C. S. Levy and O. Nagai, Phys. Stat. Solidi (b) **93**, 351 (1979).

[36] Diep-The-Hung, Phys. Status Solidi (b) **103**, 809 (1981).

[37] H. T. Diep, Phys. Rev. B **43**, 8509 (1991).

[38] H. T. Diep, M. Debauche and H. Giacomini, Phys. Rev. B (rapid communication) **43**, 8759 (1991).

[39] H. T. Diep (Editor), *Magnetic Systems with Competing Interactions* , World Scientific, Singapore (1994).

[40] H. T. Diep (Editor), *Frustrated Spin Systems*, 2nd edition, World Scientific, Singapore (2013).

[41] H. T. Diep and H. Giacomini, *Frustration - Exactly Solved Models*, p. 1-58, in *Frustrated Spin Systems*, Ed. H. T. Diep, World Scientific, Singapore (2013).

[42] See the series *Phase Transitions and Critical Phenomena*, Ed. C. Domb and J. L. Lebowitz, Academic Press, London.

[43] J. Du, D. Li, Y. B. Li, N. K. Sun, J. Li and Z. D. Zhang, Phys. Rev. B **76**, 094401 (2007).

[44] J. B. C. Efrem D'Sa, P. A. Bhobe, K. R. Priolkar, A. Das, S. K. Paranjpe, R. B. Prabhu, P. R. Sarode, J. Mag. Mag. Mater. **285** , 267 (2005).

[45] Fabian H. L. Essler, Holger Frahm, Frank Göhmann, Andreas Klümper and Vladimir E. Korepin, *The One-Dimensional Hubbard Model*, Cambridge University Press, London (2005).

[46] A. M. Ferrenberg and R. H. Swendsen, Phys. Rev. Let. **61**, 2635 (1988).

[47] A. M. Ferrenberg and R. H. Swendsen, Phys. Rev. Let. **63**, 1195 (1989).

[48] A. M. Ferrenberg and D. P. Landau, Phys. Rev. B **44**, 5081 (1991).

[49] A. Fert and I. A. Campbell, Phys. Rev. Lett. **21**, 1190 (1968); I. A. Campbell, Phys. Rev. Lett. **24**, 269 (1970).

[50] A. L. Fetter and J. D. Walecka, *Quantum Theory of Many-Particle Systems*, McGraw-Hill, New York (1971).

[51] M. E. Fisher and J. S. Langer, Phys. Rev. Lett. **20** , 665 (1968).

[52] M. E. Fisher and R. J. Burford, Phys. Rev. **156**, 583 (1967).

[53] A. Gaff and J. Hijmann, Physica A **80**, 149 (1975).

[54] A. Georges, G. Kotliar, W. Krauth and M. J. Rozenberg, Rev. Mod. Phys, **68**, 13 (1996).

[55] M. C. Gutzwiller, Phys. Rev. Lett. **10**, 159 (1963); M. C. Gutzwiller, Phys. Rev. **137**, A1726 (1965).

[56] P. Grunberg, R. Schreiber, Y. Pang, M. B. Brodsky, and H. Sowers, Phys. Rev. Lett. **57**, 2442 (1986); G. Binasch, P. Grunberg, F. Saurenbach, and W. Zinn, Phys. Rev. B **39**, 4828 (1989).

[57] C. Haas, Phys. Rev. **168**, 531 (1968).

[58] A. Haug, *Theoretical Solid State Physics*, vol. II, p. 258, Pergamon Press (1972).

[59] X. He, Y.Q. Zhang and Z.D. Zhang, J. Mater. Sci. Technol. **27**, 64 (2011).

[60] C. Henley, Phys. Rev. Lett. **62**, 2056 (1989).

[61] B. Hennion, W. Szuszkiewicz, E. Dynowska, E. Janik, T. Wojtowicz, Phys. Rev. B **66**, 224426 (2002).

[62] A. Herpin, *Théorie du Magnétisme* (in French), Presse Universitaire de France (1968).

[63] Danh-Tai Hoang, Y. Magnin and H. T. Diep, Mod. Phys. Lett. B **25**, 937 (2011).

[64] Danh-Tai Hoang and H. T. Diep, Phys. Rev. E **85**, 041107 (2012).

[65] P. C. Hohenberg and B. I. Halperin, Rev. Mod. Phys. **49**, 435 (1977).

[66] J. Hoshen and R. Kopelman, Phys. Rev. B **14**, 3438 (1974).

[67] J. Hubbard, Proc. Roy. Soc. London A **276**, 238 (1963); J. Hubbard, Proc. Roy. Soc. London A **277**, 237 (1964).

[68] M. Itakura, J. Phys. Soc. Jpn **72**, 74 (2003).

[69] L. P. Kadanoff *et al.*, Reviews of Modern Physics **39**, 395 (1967).

[70] H. Kadowaki, K. Ubukoshi, K. Hirakawa, J. L. Martinez and G. Shirane, J. Phys. Soc. Jpn. **56**, 4027 (1987).

[71] J. Kanamori, Prog. of Theor. Phys. (Kyoto) **30**, 275 (1963).

[72] Kazuki Kanki, Damien Loison and Klaus-Dieter Schotte, J. Phys. Soc. Jpn. **75**, 015001 (2006).

[73] K. Kano and S. Naya, Prog. Theor. Phys. **10**, 158 (1953).

[74] T. Kasuya, Prog. Theor. Phys. **16**, 58 (1956).

[75] M. Kataoka, Phys. Rev. B **63**, 134435 (2001).

[76] M. Kaufman, Phys. Rev. **36**, 3697 (1987).

[77] M. P. Kawatra, J. A. Mydosh, and J. I. Budnick, Phys. Rev. B **2**, 665 (1970).

[78] C. Kittel, *Introduction to Solid State Physics*, 8-th edition, John Wiley & Sons, New York (2008).

[79] C. Kittel, *Quantum Theory of Solids*, John Wiley & Sons, New York (1987).

[80] J. M. Kosterlitz and D. J. Thouless, J. Phys. C **6**, 1181 (1973); J. M. Kosterlitz, J. Phys. C **7**, 1046 (1974).

[81] D. P. Landau and K. Binder, *A Guide to Monte Carlo Simulations in Statistical Physics*, Cambridge University Press, Cambridge (2000).

[82] Review by Ph. Lecheminant, in *Frustrated Spin Systems*, ed. H. T. Diep, 2nd edition, World Scientific (2013).

[83] S. Legvold, F. H. Spedding, F. Barson and J. F. Elliott, Rev. Mod. Phys. **25**, 129 (1953).

[84] Y. B. Li, Y. Q. Zhang, N. K. Sun, Q. Zhang, D. Li, J. Li and Z. D. Zhang, Phys. Rev. B **72**, 193308 (2005).

[85] E. H. Lieb and F. Y. Wu, Phys. Rev. Lett. **20**, 1445 (1967).

[86] R. Liebmann, *Statistical Mechanics of Periodic Frustrated Ising Systems*, Lecture Notes in Physics, vol. 251 Springer-Verlag, Berlin (1986).

[87] C. L. Lu, X. Chen, S. Dong, K. F. Wang, H. L. Cai, J.-M. Liu, D. Li and Z. D. Zhang, Phys. Rev. B **79**, 245105 (2009).

[88] Y. Magnin, K. Akabli, H. T. Diep and I. Harada, Computational Materials Science **49**, S204-S209 (2010).

[89] Y. Magnin, K. Akabli and H. T. Diep, Phys. Rev. B **83**, 144406 (2011).

[90] Y. Magnin, Danh-Tai Hoang and H. T. Diep, Mod. Phys. Lett. B **25**, 1029 (2011).

[91] N. Majlis, *The Quantum Theory of Magnetism*, World Scientific, Singapore (2000).

[92] A. Malakis, S. S. Martinos, I. A. Hadjiagapiou, N. G. Fytas, and P. Kalozoumis, Phys. Rev. E **72**, 066120 (2005).

[93] F. Matsukura, H. Ohno, A. Shen and Y. Sugawara, Phys. Rev. B **57** , R2037 (1998).

[94] D. C. Mattis, *The Theory of Magnetism I: Statics and Dynamics*, 2nd ed., Springer-Verlag, Berlin (1988).

[95] N. D. Mermin and H. Wagner, Phys. Rev. Letters **17**, 1133 (1966).

[96] N. Metropolis, A. W. Rosenbluth, M. N. Rosenbluth, A. H. Teller, E. Teller, J. Chem. Phys. **21**, 1087 (1953).

[97] A. Montorsi, *The Hubbard Model*, World Scientific, Singapore (1992),

[98] G. Misguich and C. Lhuillier, *Two-dimensional Quantum Antiferromagnets*, in *Frustrated Spin Systems*, Ed. H. T. Diep, World Scientific, Singapore (2013).

[99] S. R. Mobasser and T. R. Hart, Proceed. Society of Photo-Optical Instrumentation Engineers (SPIE), Conference Series **524**, 137 (1985).

[100] M.E.J. Newman and G. T. Barkema, *Monte Carlo Methods in Statistical Physics*, Clarendon Press, Oxford (2002).

[101] V. T. Ngo and H. T. Diep, J. Appl. Phys. **103**, 07C712 (2008).

[102] V. T. Ngo and H. T. Diep, Phys. Rev. E **78**, 031119 (2008).

[103] V. T. Ngo and H. T. Diep, Phys. Rev. **B75**, 035412 (2007).

[104] V. Thanh Ngo, D. Tien Hoang and H. T. Diep, J. Phys.: Cond. Matt. **23**, 226002 (2011).

[105] V. Thanh Ngo, D. Tien Hoang and H. T. Diep, Phys. Rev. E **82**, 041123 (2010).

[106] V. Thanh Ngo, D. Tien Hoang and H. T. Diep, Modern Phys. Letters B **25**, 929-936 (2011).

[107] T. Oguchi, H. Nishimori andY. Taguchi, J . Phys. Jpn. **54**, 4494 (1985).

[108] P. Peczak and D. P. Landau, J. Appl. Phys. **67**, 5427 (1990).

[109] A. Peles, B. W. Southern, B. Delamotte, D. Mouhanna, and M. Tissier, Phys. Rev. B **69**, 220408 (2004).

[110] A. E. Petrova, E. D. Bauer, V. Krasnorussky and S. M. Stishov, Phys. Rev. B **74**, 092401 (2006).

[111] C. Pinettes and H. T. Diep, J. Appl. Phys. **83**, 6317 (1998).

[112] H. Puszkarski, Acta Physica Polonica A **38**, 217 (1970); *ibid.* A **38**, 899 (1970).

[113] See, for example, R. Quartu and H.T. Diep, Phys. Rev. B **55**, 2975 (1997).

[114] M. Rasetti (Ed.), *The Hubbard Model, Recent Results*, World Scientific, Singapore (1991).

[115] J. E. Sacco and F. Y. Wu, J. Phys. A **8**, 1780 (1975).

[116] S. Sachdev and K. Park, Ann. Phys. **298**, 58 (2002).

[117] Subir Sachdev, *Quantum Phase Transitions*, Handbook of Magnetism and Advanced Magnetic Materials, John Wiley & Sons, Ltd (2007).

[118] C. Santamaria, R. Quartu and H. T. Diep, J. Appl. Physics **84**, 1953 (1998).

[119] T. S. Santos, S. J. May, J. L. Robertson and A. Bhattacharya, Phys. Rev. B **80**, 155114 (2009).

[120] B. J. Schulz, K. Binder, M. Müller, and D. P. Landau, Phys. Rev. E **67**, 067102 (2003).

[121] F. C. Schwerer and L. J. Cuddy, Phys. Rev. **2**, 1575 (1970).

[122] K. Seeger, *Semiconductor Physics*, Springer-Verlag, Berlin (1982).

[123] J. Stephenson, J. Math. Phys. **11**, 420 (1970); Can. J. Phys. **48**, 2118 (1970); Phys. Rev. B **1**, 4405, (1970).

[124] S. M. Stishov, A.E. Petrova, S. Khasanov, G. Kh. Panova, A.A.Shikov, J. C. Lashley, D. Wu, and T. A. Lograsso, Phys. Rev. B **76**, 052405 (2007).

[125] R. H. Swendsen and J.-S. Wang, Phys. Rev. Lett. **58**, 86 (1987).

[126] S. M. Sze and Kwok K. Ng, *Physics of Semiconductor Devices*, Third Edition, John Wiley & Sons (2006).

[127] W. Szuszkiewicz, E. Dynowska, B. Witkowska and B. Hennion, Phys. Rev. B **73**, 104403 (2006).

[128] R. A. Tahir-Kheli and D. Ter Haar, Phys. Rev. **127**, 88 (1962).

[129] M. Tissier, B. Delamotte and D. Mouhana, Phys. Rev. Lett. **84**, 5208 (2000).

[130] G. Toulouse, Commun. Phys. **2**, 115 (1977).

[131] See review by E. Y. Tsymbal and D. G. Pettifor, *Solid State Physics* (Academic Press, San Diego), Vol. 56, pp. 113-237 (2001).

[132] V. Vaks , A. Larkin and Y. Ovchinnikov, Sov. Phys. JEPT **22**, 820 (1966).

[133] J. Villain, J. Phys. C**10**, 1717 (1977).

[134] J. Villain, Phys. Chem. Solids **11**, 303 (1959).

[135] J. Villain, R. Bidaux, J.P. Carton, and R. Conte, J. Physique **41**, 1263 (1980).

[136] F. Wang and D. P. Landau, Phys. Rev. Lett. **86**, 2050 (2001); Phys. Rev. E **64**, 056101 (2001).

[137] X. F. Wang, T. Wu, G. Wu, H. Chen, Y. L. Xie, J. J. Ying, Y. J. Yan, R. H. Liu and X. H. Chen, Phys. Rev. Lett. **102**, 117005 (2009).

[138] G. H. Wannier, Phys. Rev. **79**, 357 (1950); Phys. Rev. B **7**, 5017 (E) (1973).

[139] K. G. Wilson, Phys. Rev. B **4**, 3174 (1971).

[140] K. G. Wilson and M. E. Fisher, Phys. Rev. Lett. **28**, 240 (1972).

[141] K. G. Wilson, Rev. Mod. Phys. **47**, 773 (1975).

[142] U. Wolff, Phys. Rev. Lett. **62**, 361 (1989); Phys. Lett. B **228**, 379 (1989).

[143] J. Wosnitza, R. Deutschmann, H. von Löhneysen and R. K. Kremer, J. Cond. Matter. Phys. **6**, 8045 (1994).

[144] A. L. Wysocki, K. D. Belashchenko, J. P. Velev, and M. van Schilfgaarde, J. Appl. Phys. **101**, 09G506 (2007).

[145] J. Xia, W. Siemons, G. Koster, M. R. Beasley and A. Kapitulnik, Phys. Rev. B **79**, R140407 (2009).

[146] A. Yoshimori, J. Phys. Soc. Jpn. **14**, 807 (1959).

[147] A. Zangwill, *Physics at Surfaces*, Cambridge University Press (1988).

[148] G. Zarand, C. P. Moca and B. Janko, Phys. Rev. Lett. **94**, 247202 (2005).

[149] Y. Q. Zhang, Z. D. Zhang and J. Aarts, Phys. Rev. B **79**, 224422 (2009).

[150] J. Zinn-Justin, *Quantum Field Theory and Critical Phenomena*, Oxford Unversity Press (2002).

[151] D. N. Zubarev, Usp. Fiz. Nauk **187**, 71 (1960)[translation: Soviet Phys.-Uspekhi **3**, 320 (1960)].

Index